Nutritional Value and Uses of Microalgae in Aquaculture

Nutritional Value and Uses of Microalgae in Aquaculture

Editor:

Jaspreet Banga, Ph.D.

Delve Publishing
www.delvepublishing.com

Nutritional Value and Uses of Microalgae in Aquaculture

Editor: Jaspreet Banga, Ph.D.

708 3rd Avenue, 6th Floor
New York, NY 10017
United States of America
www.delvepublishing.com

ISBN: 978-1-68095-847-8

Library of Congress Control Number: 2016951600

About the Editor

Jaspreet Banga, Ph.D.

Jaspreet Banga has completed PhD from Central Drug Research Institute, India and Master of Technology from Indian Institute of Technology. Jaspreet's interests range from Drug development, Purification, Fermentation technology, Plant Biotechnology and Oncology.

Contents

List of Contributors

A. Catarina Guedes
CIMAR/CIIMAR – Centro Interdisciplinar de Investigação Marinha e Ambiental, Porto

F. Xavier Malcata
CIMAR/CIIMAR – Centro Interdisciplinar de Investigação Marinha e Ambiental, Porto
ISMAI – Instituto Superior da Maia, Castelo da Maia, Avioso S. Pedro
Instituto de Tecnologia Química e Biológica, Universidade Nova de Lisboa, Avenida da República, Oeiras Portugal

Silvio Matassa
Center of Microbial Ecology and Technology (CMET), Ghent University, Coupure Links 653, 9000 Gent, Belgium.
Avecom NV, Industrieweg 122P, 9032 Wondelgem, Belgium.

Nico Boon
Center of Microbial Ecology and Technology (CMET), Ghent University, Coupure Links 653, 9000 Gent, Belgium.

Ilje Pikaar
The School of Civil Engineering, The University of Queensland, St. Lucia, QLD 4072, Australia.

Willy Verstraete
Center of Microbial Ecology and Technology (CMET), Ghent University, Coupure Links 653, 9000 Gent, Belgium.
Avecom NV, Industrieweg 122P, 9032 Wondelgem, Belgium.
KWR Watercycle Research Institute, Post Box 1072, 3430, BB Nieuwegein, The Netherlands

Zahira Yaakob
Department of Chemical & Process Engineering Faculty of Engineering & Built Environment, University Kebangsaan Malaysia, 43600 Bangi, Malaysia.

Ehsan Ali
Centre for Energy Systems, National University of Sciences and Technology, Sector H-12, Islamabad, Pakistan.

Afifi Zainal
Department of Chemical & Process Engineering Faculty of Engineering & Built Environment, University Kebangsaan Malaysia, 43600 Bangi, Malaysia.

Masita Mohamad
Department of Chemical & Process Engineering Faculty of Engineering & Built Environment, University Kebangsaan Malaysia, 43600 Bangi, Malaysia.

Mohd Sobri Takriff[1]
Department of Chemical & Process Engineering Faculty of Engineering & Built Environment, University Kebangsaan Malaysia, 43600 Bangi, Malaysia.

Patricia Martínez Cruz
Departamento de Sistemas Biológicos, Universidad Autónoma Metropolitana-Xochimilco, Calzada del Hueso 1100, 04960 Mexico City, Mexico
Universidad Autónoma Metropolitana-Xochimilco, Calzada del Hueso 1100, 04960 Mexico City, Mexico

Ana L. Ibáñez
Departamento de Hidrobiología, Universidad Autónoma Metropolitana-Iztapalapa, Av. San Rafael Atlixco 186, 09340 Mexico City, Mexico

Oscar A. Monroy Hermosillo
Departamento de Biotecnología, Universidad Autónoma Metropolitana-Iztapalapa, Av. San Rafael Atlixco 186, 09340 Mexico City, Mexico

Hugo C. Ramírez Saad
Departamento de Sistemas Biológicos, Universidad Autónoma Metropolitana-Xochimilco, Calzada del Hueso 1100, 04960 Mexico City, Mexico

Viswanath Kiron
Faculty of Biosciences and Aquaculture, Nord University, Bodø, Norway

Mette Sørensen[1]
Faculty of Biosciences and Aquaculture, Nord University, Bodø, Norway

Mark Huntley
Department of Biological and Environmental Engineering, Cornell University, Ithaca, NY, USA
Marine Laboratory, Nicholas School of the Environment, Duke University, Beaufort, NC, USA

Ghana K. Vasanth
Faculty of Biosciences and Aquaculture, Nord University, Bodø, Norway

Yangyang Gong
Faculty of Biosciences and Aquaculture, Nord University, Bodø, Norway
Key Laboratory of East China Sea and Oceanic Fishery Resources Exploitation, East China Sea Fisheries Research Institute, Shanghai, China

Dalia Dahle
Faculty of Biosciences and Aquaculture, Nord University, Bodø, Norway

Anjana M. Palihawadana
Faculty of Biosciences and Aquaculture, Nord University, Bodø, Norway

Antonín Kouba
Research Institute of Fish Culture and Hydrobiology, South Bohemian Research Center of Aquaculture and Biodiversity of Hydrocenoses, Faculty of Fisheries and Protection of Waters, University of South Bohemia in České Budějovice, Zátiší 728/II, 389 25 Vodňany, Czech Republic

Josef Velíšek
Research Institute of Fish Culture and Hydrobiology, South Bohemian Research Center of Aquaculture and Biodiversity of Hydrocenoses, Faculty of Fisheries and Protection of Waters, University of South Bohemia in České Budějovice, Zátiší 728/II, 389 25 Vodňany, Czech Republic

Alžběta Stará
Research Institute of Fish Culture and Hydrobiology, South Bohemian Research Center of Aquaculture and Biodiversity of Hydrocenoses, Faculty of Fisheries and Protection of Waters, University of South Bohemia in České Budějovice, Zátiší 728/II, 389 25 Vodňany, Czech Republic

Jiří Masojídek
Department of Phototrophic Microorganisms, AlgaTech, Institute of Microbiology, Academy of Sciences, Opatovický mlýn, 379 81 Třeboň, Czech Republic

Pavel Kozák
Research Institute of Fish Culture and Hydrobiology, South Bohemian Research Center of Aquaculture and Biodiversity of Hydrocenoses, Faculty of Fisheries and Protection of Waters, University of South Bohemia in České Budějovice, Zátiší 728/II, 389 25 Vodňany, Czech Republic

M.E.B. Seiffert
Departamento de Aquicultura, Centro de Ciências Agrárias, and

V.R. Cerqueira
Departamento de Aquicultura, Centro de Ciências Agrárias

L.A.S. Madureira
Central de Análises, Departamento de Química, Centro de Ciências Físicas e Matemáticas, Universidade Federal de Santa Catarina, Florianópolis, SC, Brasil

Harry W. Palm*
Department of Aquaculture and Sea-Ranching, University of Rostock, Germany

Hendrik Sörensen
Department of Aquaculture and Sea-Ranching, University of Rostock, Germany

Ulrich Knaus
Department of Aquaculture and Sea-Ranching, University of Rostock, Germany

Osamah A Ahmad
Department of Marine Biology, Faculty of Marine Science, King Abdulaziz University, Jeddah, Saudi Arabia

Adnan J Salama
Department of Marine Biology, Faculty of Marine Science, King Abdulaziz University, Jeddah, Saudi Arabia

Sambhu Chithambaran*
Department of Marine Biology, Faculty of Marine Science, King Abdulaziz University, Jeddah, Saudi Arabia

Stefano Marras
IAMC-CNR, Institute for the Coastal Marine Environment, National Research Council, Localitá Sa Mardini, Torregrande, OR 09170, Italy

Andrea Cucco
IAMC-CNR, Institute for the Coastal Marine Environment, National Research Council, Localitá Sa Mardini, Torregrande, OR 09170, Italy

Fabio Antognarelli
IAMC-CNR, Institute for the Coastal Marine Environment, National Research Council, Localitá Sa Mardini, Torregrande, OR 09170, Italy

Ernesto Azzurro
Institute for Environmental Protection and Research (ISPRA), Piazzale dei Marmi 2, Livorno 57123, Italy

Marco Milazzo
Dipartimento di Scienze Della Terra e del Mare, University of Palermo, via Archirafi 28, Palermo 90123, Italy

Michel Bariche
Department of Biology, FAS, American University of Beirut, Riad El-Solh, Beirut 1107 2020, Lebanon

Momme Butenschön5
Plymouth Marine Laboratory, Prospect Place, The Hoe, Plymouth PL1 3DH, UK

Susan Kay
Plymouth Marine Laboratory, Prospect Place, The Hoe, Plymouth PL1 3DH, UK

Massimiliano Di Bitetto
Central Management for Programming and Infrastructures, National Research Council, Piazzale Aldo Moro 7, Roma 00185, Italy

Giovanni Quattrocchi[1]
IAMC-CNR, Institute for the Coastal Marine Environment, National Research Council, Localitá Sa Mardini, Torregrande, OR 09170, Italy

Matteo Sinerchia[1]
IAMC-CNR, Institute for the Coastal Marine Environment, National Research Council, Localitá Sa Mardini, Torregrande, OR 09170, Italy

Paolo Domenici
IAMC-CNR, Institute for the Coastal Marine Environment, National Research Council, Localitá Sa Mardini, Torregrande, OR 09170, Italy

McDowall S
Department of Business and Primary Industries, Australia

Awal S
Department of Business and Primary Industries, Australia

Christie A
Department of Business and Primary Industries, Australia

Duur K Aanen
Laboratory of Genetics, Wageningen University

Zaleha Kassim
Department of Fisheries, Faculty of Fisheries and Aqua-Industry, University Malaysia Terengganu, Terengganu, Malaysia

Akbar John
Kuliyyah of Science, International Islamic University Malaysia, Kuantan Campus, Malaysia

Lim Keng Chin
Institute of Tropical Aquaculture University Malaysia Terengganu, Terengganu, Malaysia

Nur Farahiyah Zakaria
Unit of Farmer's Organisation of South Johor Bahru, Johor, Malaysia

Nur Hidayah Asgnari
Department of Fisheries, Faculty of Fisheries and Aqua-Industry, University Malaysia Terengganu, Terengganu, Malaysia

Rajrupa Ghosh
Department of Bioscience, Jogesh Chandra Chaudhury College, Kolkata, India

Abhijit Mitra
Department of Oceanography, Techno India University, Salt lake Campus, Kolkata, India

Preface

Microalgae, are enormously bio-diverse microscopic algae that are primary producers of biomass and oxygen by photosynthesis, ie from carbon dioxide in the presence of sunlight, in marine or freshwater systems. It is un-disputed that they are the basic sustenance units of earth's atmosphere, as they produce approximately half of the atmospheric oxygen and use simultaneously the greenhouse gas carbon dioxide to grow photoautotrophically. In addition to biological diversity, the chemical composition of microalgae can be diversely varied by cultivation conditions making it ideal for biomass production as a nutrient source or for bulk production of bio-chemicals. The microalgae are an untapped resource of nutrition and protein for not only human, animal feel but also for the aquaculture. This book details the use of microalgae as food source in the rearing of all stages of marine clams, oysters, scallops or the the larval stages of some marine gastropods, fish species and shrimp.

Editor
Jaspreet Banga, Ph.D.

INTRODUCTION

Microalgae are utilized in aquaculture as live feeds for all growth stages of bivalve molluscs (eg. oysters, scallops, clams and mussels), for the larval/early juvenile stages of abalone, crustaceans and some fish species, and for zooplankton used in aquaculture food chains. Over the last four decades, several hundred microalgae species have been tested as food, but probably less than twenty have gained widespread use in aquaculture. Microalgae must possess a number of key attributes to be useful aquaculture species. They must be of an appropriate size for ingestion, e.g. from 1 to 15 μm for filter feeders; 10 to 100 μm for grazers and readily digested. They must have rapid growth rates, be amenable to mass culture, and also be stable in culture to any fluctuations in temperature, light and nutrients as may occur in hatchery systems. Finally, they must have a good nutrient composition, including an absence of toxins that might be transferred up the food chain.

Table 1. Microalgae commonly used in aquaculture, either as individual diets or components of mixed diets. (++ denotes more popular than +).

	Bivalve molluscs	Crustacean larvae	Juvenile abalone	Zooplankton (used for crustacean, fish larvae)
Isochrysis sp. (T.ISO)	++	+		++
Pavlova lutheri	++	+		++
Chaetoceros calcitrans	++	++		+
C. muelleri or C. gracilis	+	++		+
Thalassiosira pseudonana	+	+		
Skeletonema spp.	+	++		
Tetraselmis suecica	+	+		++
Rhodomonas spp.	+			
Pyramimonas spp.	+			
Navicula spp.	+	+	++	
Nitzschia spp.		+	++	
Cocconeis spp.			+	
Amphora spp			+	
Nannochloropsis spp.				++

Isochrysis sp. (T.ISO), Pavlova lutheri and Chaetoceros calcitrans are the most common species used to feed the larval, early juvenile and broodstock (during hatchery conditioning) stages of bivalve molluscs; these are usually fed together as a mixed diet (O'Connor & Heasman, 1997; Richard Pugh, Shellfish Culture Ltd., pers. comm.). Many of the strains successfully used for bivalves are also used as direct feed for crustaceans (especially shrimp) during the early larval stages, especially diatoms such as Skeletonema spp. and Chaetoceros spp. Benthic diatoms such as Navicula spp. and Nitzschia are commonly mass-cultured and then settled onto plates as a diet for grazing juvenile abalone. Isochrysis sp. (T.ISO), Pavlova lutheri, T. suecica or Nannochloropsis spp. are commonly fed to Artemia or rotifers, which are then fed on to later larval stages of crustacean and fish larvae.

Nutritional Properties of Microalgae

Microalgal species can vary significantly in their nutritional value, and this may also change under different culture conditions. Nevertheless, a carefully selected mixture of microalgae can offer an excellent nutritional package for larval animals, either directly or indirectly (through enrichment of zooplankton). Microalgae that have been found to have good nutritional properties - either as monospecies or within a mixed diet - include C. calcitrans, C. muelleri, P. lutheri, Isochrysis sp. (T.ISO), T. suecica, S. costatum and Thalassiosira pseudonana shape, digestibility (related to cell wall structure and composition), biochemical composition (eg. nutrients, enzymes, toxins if present) and the requirements of the animal feeding on the alga. Since the early reports that demonstrated biochemical differences in gross composition between microalgae (Parsons, Stephens & Strickland, 1963) and fatty acids (Webb & Chu, 1983), many studies have attempted to correlate the nutritional value of microalgae with their biochemical profile. However, results from feeding experiments that have tested microalgae differing in a specific nutrient are often difficult to interpret because of the confounding effects of other microalgal nutrients. Nevertheless, from examining all the literature data, including experiments where algal diets have been supplemented with compounded diets or emulsions, some general conclusions can be reached. Microalgae grown to late-logarithmic growth phase typically contain 30 to 40% protein, 10 to 20% lipid and 5 to 15% carbohydrate. When cultured through to stationary phase, the proximate composition of microalgae can change significantly; for example when nitrate is limiting, carbohydrate levels can double at the expense of protein. There does not appear to be a strong correlation between the proximate composition of microalgae and nutritional value, though algal diets with high levels of carbohydrate are reported to produce the best growth for juvenile oysters and larval scallops provided polyunsaturated fatty acids (PUFAs) are also present in adequate proportions.

Production Systems

Typical systems used indoors for microalgal mass culture include carboys (10 to 20 L), polythene bags (100 to 500 L) and tubs (1000 to 5000 L). These are usually operated in batch or continuous mode. For larger volumes, out-door tanks or ponds are used, operated semi-continuously. Depending on their scale, hatcheries may produce between several hundred to tens of thousands of liters of algae daily. Cells densities range from 105 to 107 cells mL^{-1} with these standard systems, and production costs can range from US $50 to 200 kg^{-1} dry weight, or 20 to 50% of hatcheries operating costs. There are clear economies of scale with algal production, so that production costs become especially significant for small hatcheries. Consequently, there has been much effort directed at examining alternatives to production of fresh algae, and also more cost-efficient production systems. Large-scale photo bioreactors, either for indoor or outdoor production, have been assessed. Essentially these can be considered as variations of the standard culture systems, but with a much higher surface area to volume ratio (SA:VOL). Consequently, light is less likely to become limiting and systems are characterized by high productivity and cell biomass at harvest – potentially eliciting a low production cost. However, these systems do have some disadvantages. Oxygen concentrations (resulting from photosynthesis) can build up because of the high biomass, and therefore give rise to photo inhibition - thus restricting productivity. Because of the high SA:VOL overheating of cultures can be a problem in outdoor systems. Also, because of the high biomass the systems need turbulent flow to ensure nutrient exchange and to avoid light-limitation, thereby making them unsuitable for fragile species. In fact, most aquaculture strains have not been effectively cultured in such systems. Fermentation technology is well established for low-cost production of bacteria and yeast, and there are some microalgae capable of heterotrophic growth. The advantages include a high-density and biomass production, and elimination of light – a major cost for phototrophic production. Because lower volumes are required for producing the same biomass (compared to conventional algal systems) this provides a greater degree of control and potentially, better economy of production. Production costs of between US $2 to 25 kg^{-1} dry weight microalgae have been projected for this technology. Unfortunately, few aquaculture species have been identified that can grow heterotrophically. Tetraselmis spp. are exceptions, though these are generally recognized as having moderate food value, unless forming part of a mixed diet. There is also a high capital cost associated with fermentation; 2 L units can cost US $5 to 10 K, whereas 10,000 L units may exceed US $1M.

Alternatives to Fresh Algae

Alternatives to microalgae have been sought that may be more cost-effective. Those tested have included compounded microcapsules, lipid emulsions, yeasts or yeast-based diets and bacteria. More recently, several products based on thraustochytrids (microorganisms whose taxonomy may be related to certain algal classes) from the genus Schizochytrium have been marketed through Aquafauna Biomarine Inc. and Sanders Brine Shrimp Co. These products have high concentrations of DHA, and so are being applied as alternatives to commercial oil enrichments for zooplankton fed to larvae (discussed more in the next section). Generally, as direct feeds most of the above products have had lower nutritional value than mixtures of microalgae commonly used in aquaculture, though some performed well as components of a mixed diet with live microalgae.

Use of Algae to Enrich Zooplankton

Microalgae have an important role in aquaculture as a means of enriching zooplankton for on-feeding to fish and other larvae. In addition to providing protein (essential amino acids) and energy, they provide other key nutrients such as vitamins, essential PUFAs, pigments and sterols, which are transferred through the food chain. For example, rotifers fed microalgae become rapidly enriched with ascorbic acid (AsA). After 24 h, rotifers fed on Isochrysis sp. (T.ISO) and Nannochloropsis oculata contained 2.5 and 1.7 mg g-1 DW, respectively, whereas rotifers fed on baker's yeast (itself deficient in AsA) contained only 0.6 mg g-1 DW.

PUFA-rich microalgae, such as Pavlova spp. and Isochrysis sp. (T.ISO) can be fed to zooplankton to enrich them in DHA. However, often these do not provide the level of enrichment often sought for zooplankton, and commercial oilemulsions are often used. Recently, "algal-like" products such as AlgaMac 2000 and Docosa Gold (dried preparations of the thraustochytrid Schizochytrium sp.) – which contain 5-15% of their DW as DHA – have been utilized. These have produced similar levels of enrichment of DHA within the zooplankton compared to the commercial oils, and also produce DHA to EPA ratios of between 1 and 2, which are considered favorable for fish larval nutrition. Research is also in progress assessing alternate live and dried thraustochytrids as dietary constituents and for enrichment demonstrated that microalgal pigments transferred through to zooplankton may contribute to nutritional value. They found the dominant pigments in the copepod Temora sp. were lutein and astaxanthin, whereas in Artemia it was canthaxanthin. When these prey items were fed to halibut larvae, adequate amounts of vitamin A were found in halibut fed on copepods, but not with halibut fed on Artemia. The authors ascribed this to the ability of the larvae to convert lutein and/or astaxanthin, but not canthaxanthin, into vitamin A. They recommended that Artemia should routinely be enriched with astaxanthin and lutein to improve their nutritional value.

Avenues for Future Research

The high production cost of microalgae remains a constraint to many hatcheries. Despite efforts over several decades to develop cost-effective artificial diets to replace microalgae as hatchery feeds, on-site microalgal production remains a critical element for most marine hatcheries. Improvements in alternative diets may continue, but production costs of microalgae may also decrease due to the uptake of new technology by hatcheries (eg. continuous bag system, from Sea salter Shellfish). Therefore it is unlikely that microalgae will be totally replaced, at least in the medium term. A good selection of microalgal species is available to support the aquaculture industry. However for some particular applications or industry sectors, new species with improved nutritional quality or growth characteristics could improve hatchery efficiency. For example, copepods are recognized as excellent feeds for fish larvae, but they have proven difficult to produce in intensive systems. The utilization of alternate microalgal species could improve their production rates. Apart from improvements in the cost-efficiencies of on-site algal production, an alternative is the centralization of algal production at specialized mass-culture facilities, using heterotrophic methods or photo bioreactors to produce cheaper algal biomass. These technologies could be married with post-harvest processing such as spray-drying, or algal concentration (centrifugation or flocculation) to develop off-the-shelf algal biomass for distribution to hatcheries.

The use of microalgae either as a full or partial (i.e. in conjunction with products like Selco and AlgaMac 2000) enrichment should be considered for improving the nutritional quality of zooplankton. Microalgae contain an array of essential nutrients that may be transferred through food chains, especially PUFAs. Microalgae and P. lutheri) can provide a moderate enrichment of DHA, though not as effective as commercial oil emulsions like DHA Selco. The new "algal-like" thraustochytrid products are extremely high in DHA and provide an effective means of enriching zooplankton to produce good DHA: EPA ratios. New thraustochytrids are being investigated with other nutritional characteristics, eg. high concentrations of AA. Some work has been documented on the transfer of AsA from microalgae through to zooplankton and fish larvae, but much less is known about other vitamins. Though microalgae have generally been proposed here as good sources of vitamins, they can vary significantly in their composition. Therefore zooplankton could be deficient in one or more vitamins when enriched using certain dietary regimens. Future research should focus on this issue, and the transfer of other essential nutrients (pigments, sterols) to zooplankton fed different diets and grown under different culture conditions.

Use of micro-algae in Aquaculture

Micro-algae are an essential food source in the rearing of all stages of marine bivalve molluscs (clams, oysters, and scallops), the larval stages of some marine gastropods (abalone, conch), larvae of several marine fish species and penaeid shrimp, and zooplankton.

Bivalve molluscs

Intensive rearing of bivalves has so far relied on the production of live algae, which comprises on average 30% of the operating costs in a bivalve hatchery. The relative algal requirements of the various stages of the bivalve culture process depend on whether the operation aims at the mass-production of larvae for remote setting or growing millions of seed till planting size. In either case, the juveniles, representing the largest biomass in the hatchery and demanding the highest weight-specific rations, consume the largest volumes of algal culture (Fig. 2.16.). The algal species that were reported in an international survey among hatchery operators in 1991 are listed in Table 2.14. Eight algal species (Isochrysis sp., clone T-Iso; C. gracilis; C. calcitrans; T. suecica; T. pseudonana, clone 3H; P. lutheri; I. galbana; S. costatum) were widely used and represented over 90% of the volume of algal culture produced in 23 facilities.

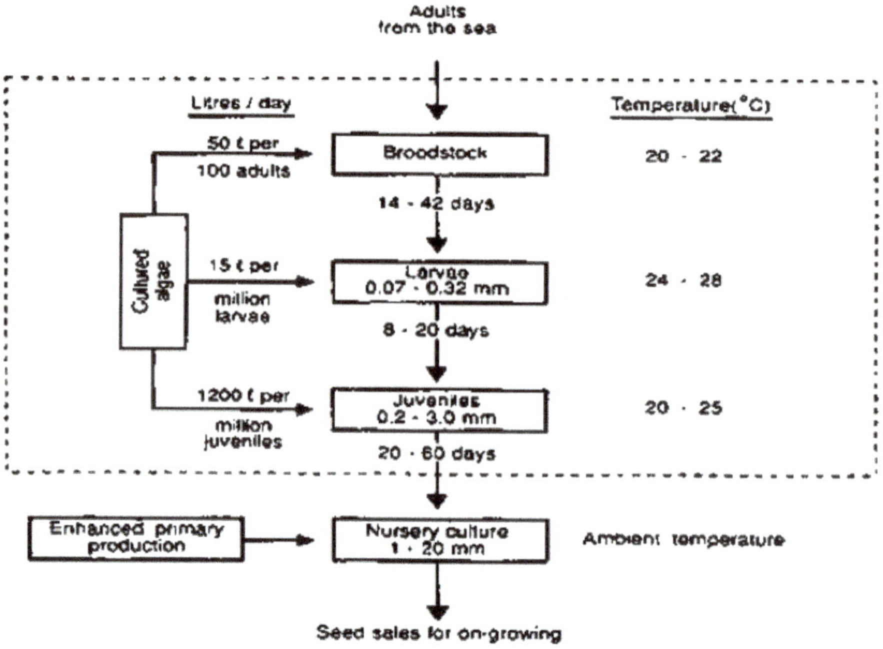

Figure: Requirements for cultured algae in hatchery and nursery culture of bivalve molluscs (Utting and Spencer, 1991).

Table 2. Algal species used in hatchery and nursery rearing of bivalve molluscs as reported in an international questionnaire. Species are ranked according to decreasing frequency of use.

Algal species	frequency of use†	total daily production n‡	volume (m^3)
Isochrysis sp., clone T-Iso	31	18	23.8
Chaetoceros gracilis	23	11	14.1
Chaetoceros calcitrans	16	10	6.0
Tetraselmis suecica	15	10	39.1
Thalassiosira pseudonana, clone 3H	14	9	112.0
Pavlova lutheri	11	7	11.7
Isochrysis galbana	8	5	9.1
Skeletonema costatum	6	3	58.8
Chroomonas salina	5	3	0.76
Dunaliella tertiolecta	4	2	2.2
Chaetoceros simplex	3	3	1.76
Chaetoceros muelleri	3	2	5.0
Nannochloropsis sp.	3	2	0.20
Cyclotella sp.	2	1	0.36
Phaeodactylum tricornutum	2	1	2.0
Tetraselmis chui	2	0	-
Pavlova salina	1	1	3.18
Dicruteria sp.	1	1	4.07
Tetraselmis levis	1	0	-
Dunaliella perva	1	1	0.012
Thalassiosira weissfloggii	1	1	0.12
Chlamydomonas sp.	1	1	0.52
Chlorella sp.	1	1	0.36
TOTAL	43	23	295

The larvae of most bivalve species have similar food preferences; suitable algal species including C. calcitrans, T. pseudonana (3H), I. galbana, and T. suecica (for larvae > 120 µm in length). Combinations of flagellates and diatoms provide a well-balanced diet which will generally accelerate the rate of larval development to metamorphosis in comparison with unialgal diets. The quantity fed depends upon the larval density, but suitable cell concentrations (expressed as cells.µl-1) are given by each of the following combinations:

- I. galbana; 50
- C. calcitrans; 250
- I. galbana/C. calcitrans; 25/125
- I. galbana/C. calcitrans/T. suecica; 33/83/3.3 (larvae > 120µm)

Table 3. Typical algal feeding regimes (cells.ml-1) for penaeid larvae (N: nauplius, P: protozoea, M: mysis, PL: postlarva stage) (modified from Smith et al.

Substage	*Chaetoceros neogracile (C. gracilis)*	*Tetraselmis chuii*
N_5 or N_6	60,000	0-15,000
P_1	100,000-120,000	30,000
P_2	120,000	35,000
P_3	120,000	35,000
M_1	100,000	30,000
M_2	75,000	20,000
M_3	50,000-75,000	20,000
PL_1 to PL_5	20,000-75,000	5,000-20,000

Marine fish

Apart from the requirement for micro-algae for culturing and/or enriching live prey organisms such as Artemia and rotifers (see Chapters 3. and 4.3.), algae are often used directly in the tanks for rearing marine fish larvae. This "green water technique" is part of the commonly applied techniques for rearing larvae of gilthead seabream Sparus aurata (50,000 cells ml^{-1} of Isochrysis sp. + 400,000 cells.ml^{-1} of Chlorella sp. per day), milkfish Chanos chanos (between 500 and 3,500 Chlorella cells.ml^{-1} are added from hatching till day 21), Mahimahi Coryphaena hippurus (200,000 cells.ml^{-1} of either Chaetoceros gracilis, Tetraselmis chui, or Chlorella sp.), halibut Hippoglossus hippoglossus (Tetraselmis sp.), and turbot Scophthalmus maximus (60,000 cells.ml^{-1} of Tetraselmis sp. or 130,000 cells.ml^{-1} of I. galbana).

The effects of the presence of micro-algae in the larval rearing tank are still not fully understood and include:

- stabilizing the water quality in static rearing systems (remove metabolic by-products, produce oxygen);
- a direct food source through active uptake by the larvae with the polysaccharides present in the algal cell walls possibly stimulating the non-specific immune system in the larvae;
- an indirect source of nutrients for fish larvae through the live feed (i.e. by maintaining the nutritional value of the live prey organisms in the tank);
- increasing feeding incidence by enhancing visual contrast and light dispersion, and
- microbial control by algal exudates in tank water and/or larval gut.

Nutritional value of micro-algae

The nutritional value of any algal species for a particular organism depends on its cell size, digestibility, production of toxic compounds, and biochemical composition. The gross composition of 16 species of micro-algae is compared in Table. Although there are marked differences in the compositions of the micro-algal classes and species, protein is always the major organic constituent, followed usually by lipid and then by carbohydrate. Expressed as percentage of dry weight, the range for the level of protein, lipid, and carbohydrate are 12-35%, 7.2-23%, and 4.6-23%, respectively. The content of highly unsaturated fatty acids (HUFA), in particular eicosapentaenoic acid (20:5n-3, EPA), arachidonic acid (20:4n-6, ARA), and docosahexaenoic acid (22:6n-3, DHA), is of major importance in the evaluation of the nutritional composition of an algal species to be used as food for marine organisms. The fatty acid composition of 10 species of micro-algae grown under defined conditions and harvested during the log phase is presented in Fig. Significant concentrations of EPA are present in the diatom species (Chaetoceros calcitrans, C. gracilis, S. costatum, T. pseudonana) and the prymnesiophyte Platymonas lutheri, whereas high concentrations of DHA are found in the prymnesiophytes (P. lutheri, Isochrysis sp.) and Chroomonas salina.

The nutritional value of micro-algae can vary considerably according to the culture conditions. For example the effect of the composition of the culture medium on the proximate composition of various species of micro-algae is demonstrated in Table 3. The protein content per cell, which is considered as one of the most important factors determining the nutritional value of micro-algae as feed in aquaculture, was found to be more susceptible to medium-induced variation than the other cellular constituents.

Table 4. Concentrations of chlorophyl a, protein, carbohydrate and lipid in 16 species of micro-algae commonly used in aquaculture.

Algal class Species	Dry weight ($pg.cell^{-1}$)	Chl a	Protein	Carbohydrate	Lipid
	Weight of constituent ($pg.cell^{-1}$)				
Bacillariophyceae					
Chaetoceros calcitrans	11.3	0.34	3.8	0.68	1.8
Chaetoceros gracilis	74.8	0.78	9.0	2.0	5.2
Nitzchia closterium	-	-	-	-	-
Phaeodactylum tricornutum	76.7	0.41	23.0	6.4	10.7
Skeletonema costatum	52.2	0.63	13.1	2.4	5.0
Thalassiosira pseudonana	28.4	0.27	9.7	2.5	5.5
Chlorophyceae					
Dunaliella tertiolecta	99.9	1.73	20.0	12.2	15.0
Nannochloris atomus	21.4	0.080	6.4	5.0	4.5
Cryptophyceae					
Chroomonas salina	122.5	0.98	35.5	11.0	14.5
Eustigmatophyceae					
Nannochloropsis oculata	6.1	0.054	2.1	0.48	1.1
Prasinophyceae					
Tetraselmis chui	269.0	3.83	83.4	32.5	45.7
Tetraselmis suecica	168.2	1.63	52.1	20.2	16.8
Prymnesiophyceae					
Isochrysis galbana	30.5	0.30	8.8	3.9	7.0
Isochrysis aff. Galbana (T-iso)	29.7	0.29	6.8	1.8	5.9
Pavlova lutheri	102.3	0.86	29.7	9.1	12.3
Pavlova salina	93.1	0.34	24.2	6.9	11.2
	Percentage of dry weight				
Bacillariophyceae					
Chaetoceros calcitrans	11.3	3.01	34	6.0	16
Chaetoceros gracilis	74.8	1.04	12	4.7	7.2
Nitzchia closterium	-	-	26	9.8	13
Phaeodactylumtricornutum	76.7	0.53	30	8.4	14
Skeletonema costatum	52.2	1.21	25	4.6	10
Thalassiosira pseudonana	28.4	0.95	34	8.8	19
Chlorophyceae					
Dunaliella tertiolecta	99.9	1.73	20	12.2	15
Nannochloris atomus	21.4	0.37	30	23.0	21
Cryptophyceae					
Chroomonas salina	122.5	0.80	29	9.1	12
Eustigmatophyceae					
Nannochloropsis oculata	6.1	0.89	35	7.8	18
Prasinophyceae					
Tetraselmis chui	269.0	1.42	31	12.1	17
Tetraselmis suecica	168.2	0.97	31	12.0	10
Prymnesiophyceae					
Isochrysis galbana	30.5	0.98	29	12.9	23
Isochrysis aff. Galbana (T-iso)	29.7	0.98	23	6.0	20
Pavlova lutheri	102.3	0.84	29	9.0	12
Pavlova salina	93.1	0.98	26	7.4	12

Moreoever, the growth of animals fed a mixture of several algal species is often superior to that obtained when feeding only one algal species. A particular alga may lack a nutrient, while another alga may contain that nutrient and lack a different one. In this way, a mixture of both algal species supplies the animals with an adequate amount of both nutrients. An extensive review of the nutritional aspects of micro-algae used in mariculture of bivalve molluscs, crustaceans, and fish is presented in Brown *et al.*

Table 5. Cellular density (10^6 cells.ml^{-1}) and proximate composition (pg.cell^{-1}) of four marine micro-algae grown in different culture media (Algal^{-1} is a commercial nutrient)

	Cellular density	Protein	Carbohydrates	Lipids
T. suecica				
Walne	2.29	13.31	6.20	7.04
ES	2.58	16.98	6.93	7.22
F/2	2.38	21.75	8.37	7.92
Algal-1	4.11	32.22	8.83	8.65
D. tertiolecta				
Walne	4.04	13.37	13.22	22.28
ES	4.24	14.88	15.73	23.94
F/2	4.97	13.26	17.91	23.67
Algal-1	8.45	18.82	11.08	18.18
I. galbana				
Walne	10.11	5.17	4.28	25.95
ES	12.09	7.23	5.21	28.38
F/2	10.81	8.13	5.59	26.82
Algal-1	16.15	9.57	4.28	20.68
P. tricornutum				
Walne	19.01	2.65	6.42	6.51
ES	16.23	5.21	9.20	6.45
F/2	24.65	3.34	6.90	5.52
Algal-1	39.04	4.20	5.98	5.79

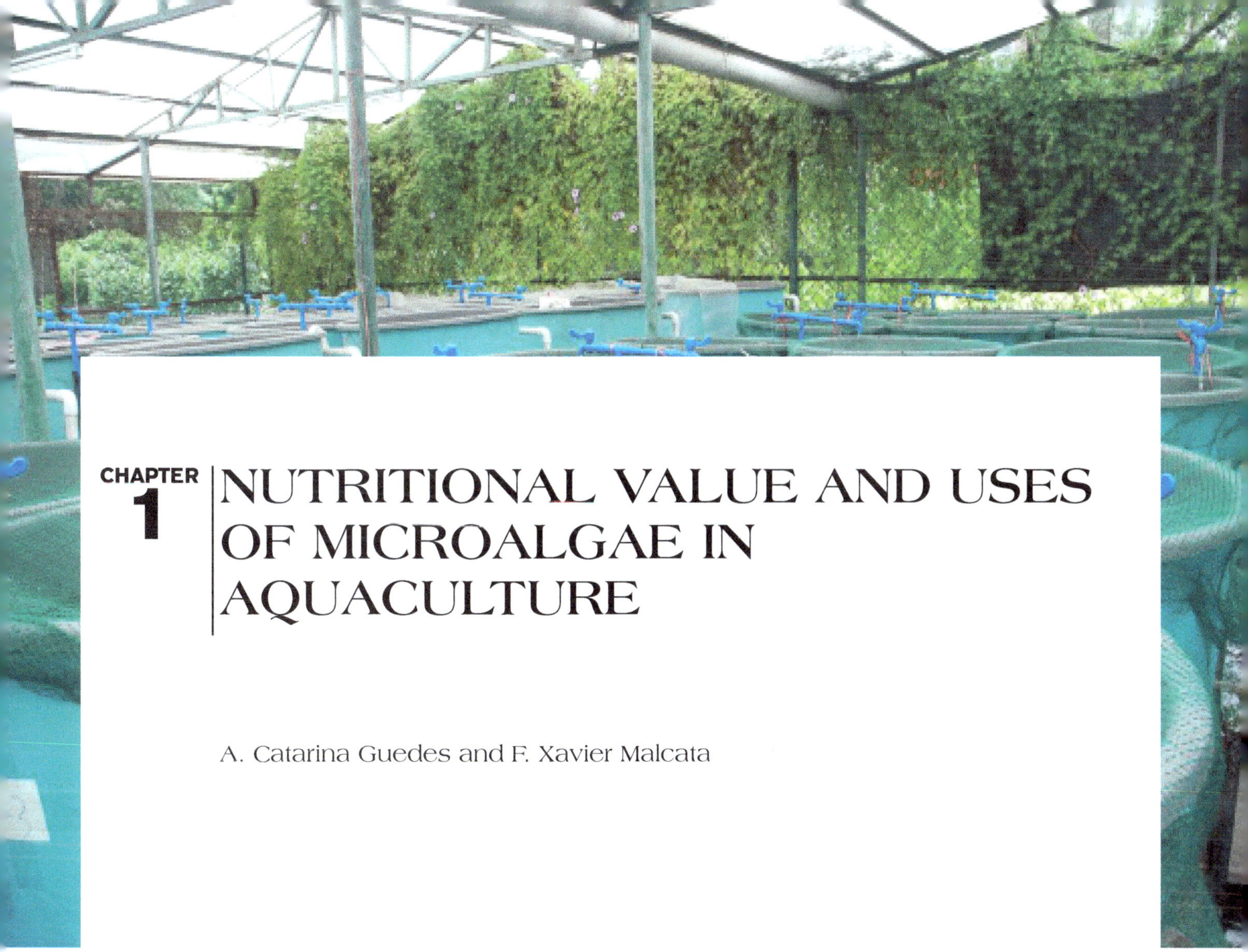

CHAPTER 1

NUTRITIONAL VALUE AND USES OF MICROALGAE IN AQUACULTURE

A. Catarina Guedes and F. Xavier Malcata

INTRODUCTION

Microalgae (i.e. single-celled algae or phytoplankton) represent the largest, yet one of the most poorly understood groups of microorganisms on Earth. As happens with plants relative to terrestrial animals, microalgae represent the natural nutritional base and primary source of bulk nutrients in the aquatic food chain. Microalgae play indeed a crucial nutritional role with regard to marine animals in the open sea, and consequently in aquaculture. Most marine invertebrates depend on microalgae for their whole life cycle, so commercial and experimental mollusc or fish hatcheries have included a microalga production system in parallel to their animal production itself. Microalgae are utilized as live feed for all growth stages of bivalve molluscs (e.g. oysters, scallops, clams and mussels), for larval/early juvenile stages of abalone, crustaceans and some fish species, and for zooplankton used in aquaculture food webs at large. It should be emphasized that the productivity of any hatchery is directly related to the quantity and quality of the food source used therein. On the other hand, the concept of aquaculture as a set of engineered systems in terms of wastewater treatment and recycling has received an impetus over the past few years. They are designed to meet specific treatment and wastewater specifications, and may simultaneously solve environmental and sanitary problems along with economic feasibility [1,2]. A renewed interest has also been experienced by high rate microalgal ponds for treatment of wastewater – where photosynthetic microalgae supply oxygen to heterotrophic bacteria, and where wastewater-borne nutrients are converted into biomass protein [2,3]. Therefore, microalga culturing is likely to play an increasingly important role in aquatic food production modules, specifically to produce (or be used as) feed for fish, convert CO_2 to O_2 and remediate water quality.

GENERAL ATTRIBUTES OF MICROALGAL SPECIES IN AQUACULTURE

Unlike air-breathing animals, those living in aquatic media and used for large scale human consumption as food are seldom herbivorous at the adult stage; most farmed animals are indeed carnivorous from their post-larval stage on, or omnivorous at best. The associated food web is accordingly longer, so only filtering molluscs and a few other animals truly depend on plankton throughout their lifetime. However, microalgae are required for larva nutrition during a brief period – either for direct consumption in the case of molluscs and penaeid shrimps, or indirectly for the live prey fed to small-larval fish. In these cases, the post-larvae specimens are hatched, bred and raised by specialized establishments (hatcheries) – which are

Citation: A. Catarina Guedes and F. Xavier Malcata, Nutritional Value and Uses of Microalgae in Aquaculture, DOI: 10.5772/30576

particularly complex to operate because they involve intensive production of microalgae and, in the case of small-larval fish, production of such small live prey as rotifers. Aquacultured animals for which rearing does not exhibit these constraints are seldom found; this is the case of salmonids, whose eggs have sufficient reserves to hatch big larvae capable of feeding directly on dry particles [4]. Over the last decades, several hundred microalga species have been tested as feed, but probably less than twenty have experienced a widespread application in aquaculture. In fact, microalgal species vary significantly in their nutritional value – which is also dependent on culturing conditions [5,6]. To provide a better balanced nutrition package and more effectively improve animal growth, a carefully selected mixture of microalgae should be fed to fish, directly or indirectly (through enrichment of zooplankton) – as this leads to better results than a diet composed of a single microalga [7,8]. Microalga production for use as feed is divided into intensive monoculture – for larval stages of bivalves, shrimp and certain fish species, and extensive culture – for growth of bivalves, carp and shrimp. Favored genera for the former include Chaetoceros, Thalassiosira, Tetraselmis, Isochrysis, Nannochloropsis, Pavlova and Skeletonema [6,9,10]. These organisms are fed directly or indirectly to the cultured larval organism; indirect means of providing them are usually through artemia, rotifers and Daphnia – which are, in turn, fed to the target larval organisms. It is widely accepted that microalgae are actively taken up by shrimp larvae, and play an important role in nutrition at that life stage; however, it is uncertain whether juveniles and adults do actively feed on microalgae as well. Some reports suggest that microalgae are found in their gut because shrimp accidentally ingest them together with debris [11,12]. The nutritional value of a microalgal diet is critically related to its ability to supply essential macro- and micronutrients to the target animal consumer. As emphasized above, a mixed microalgal diet – as routinely used in the hatchery and nursery phases of oyster cultivation [13], is likely to outperform monoalgal diets [14]. However, the nutritional requirements of bivalves are poorly defined; feeding experiments with microalgae of partially defined compositions have shown that carbohydrate and polyunsaturated fatty acid (PUFA) levels are major factors for growth of oysters [5]. Supply of additional dietary carbohydrate was found to increase oyster growth rate, provided that adequate protein and essential fatty acids were concomitantly supplied [9]. Supplementation of juvenile mussel diets with protein microcapsules led to a positive growth response, and indicated that a protein content below 40 %(w/w) significantly contrains mussel growth rates [15]

Diatoms and haptophytes (prymnesiophytes) are nutritious microalgae that are frequently used as feed for oysters [13]. The prymnesiophytes Isochrysis sp. and P. lutheri are rich sources of docosahexaenoic acid (DHA, 22:6n-3) – comprising 8-10% total fatty acids [16], while diatoms are a rich source of eicosapentaenoic acid (EPA, 20:5n-3) [17]. Mixed microalgal diets of prymnesiophytes and diatoms are common in bivalve hatcheries, and considered as highly nutritious in terms of requirements for essential PUFAs [18]. Microalgae should, in general, possess a number of key attributes to be useful for aquacultured species: they should be of an appropriate size and shape for ingestion and ready digestion (i.e. they should have a digestible cell wall to make nutrients easily available); they should undergo fast growth rates, and be amenable to mass culture; they should be stable to fluctuations in temperature, light and nutrient profile, as often occur in hatchery systems; and they should exhibit appropriate nutritional qualities, including absence of toxins (that might otherwise accumulate through the food chain). A major challenge faced by algologists is thus to reduce production costs, while maintaining reliability of microalgal feed. Microalgae provide food for zooplankton, but they can also help stabilize (and even improve) the quality of the culture medium. For numerous freshwater and seawater animal species, introduction of phytoplankton to rearing ponds (the so-called green-water technique) produces much better results in terms of survival, growth and transformation index than the classical clear-water technique [19-21]. The rationale behind this observation is not entirely known, yet it may include water quality improvement by oxygen production and pH stabilization, and action of some excreted biochemical compounds, along with induction of behavioral processes such as initial prey catching and regulation of bacterial population [4,22], probiotic effects [23], and stimulation of immunity [24].

NUTRITIONAL FEATURES OF MICROALGAE

Microalgal species can vary significantly in nutritional value, as a function of the prevailing culture conditions. Only a reduced number of species have been used, primarily for historical reasons and ease of cultivation – rather than supported by scientific evidence of any superior performance as nutritional or therapeutical supplements. Hence, formulations more carefully selected of microalgal origin may offer the opportunity for development of improved nutritional packages aimed at larval animals. Several factors contribute to the nutritional value of a microalga – including its size and shape, and digestibility as related to cell wall structure and composition (as mentioned above), as well as biochemical composition (e.g. accumulation compounds, enzymes and toxins) and specific requirements of the target animal. For this reason, several studies have attempted to correlate the nutritional value of microalgae to their chemical profile. However, results from feeding experiments are often difficult to interpret because of the confounding effects of other formulation additives. An examination of literature data – including those pertaining to microalga-based, compounded diet emulsions, have meanwhile allowed a few general conclusions to be reached [25]. As primary producers in the aquatic food chain, microalgae provide many phytonutrients, including in particular PUFAs – e.g. EPA, arachidonic acid (AA) and DHA, which are known to be essential for various marine animals [25], as well as for growth and metamorphosis of many larvae [8,26]. However, the ratios of DHA, EPA and AA may actually be more important than their absolute levels [24,27]. Most microalgal species exhibit moderate to high percents of EPA (7 to 34%); and prymnesiophytes (e.g. Pavlova spp. and Isochrysis sp.) and cryptomonads are relatively rich in DHA (0.2 to 11%), whereas eustigmatophytes (e.g. Nannochloropsis spp.) and diatoms have the highest percentages of AA (up to 4%). Chlorophytes (Dunaliella spp. and Chlorella spp.) are deficient in both C20 and C22 PUFAs, although some species have small amounts of EPA (up to 3.2%); because of such a PUFA deficiency, chlorophytes are in general ascribed a poor nutritional value, so they are not suitable for use as single species-diet [6]. Prasinophyte species contain significant proportions of C20 (Tetraselmis spp.) or

C22 (Micromonas spp.), but rarely of both. Therefore, the fatty acid contents of microalgae exhibit systematic differences according to taxonomic group – although there are examples of significant differences between microalgae, even within the same class. The contents of antioxidants are also not uniform among microalgae; e.g. the concentrations of vitamins and carotenoids convey significant variations among species. Note that any mixed-algal diet should provide adequate concentrations of vitamins and carotenoids to be effective in aquaculture; unfortunately, the nutritional requirements of larval or juvenile animals that feed directly on microalgae are still poorly understood at present. In fact, artificial diets often lack natural pigments that allow such organisms as salmon or trout acquire their characteristic red color (muscle), which, in nature, is a result of eating microalgae containing red pigments; without such a color, a lower market value results. One way to alleviate this shortcoming is by adding astaxanthin to fish feed, with a consequently growing market for microalga-based sources, e.g. Haematococcus pluvialis [24,28]. On the other hand, the amino acid composition of microalgal proteins is rather similar between species [29], and relatively unaffected by their intrinsic growth phase and extrinsic light conditions [30,31]. Furthermore, the content in essential amino acids of microalgae is similar to that of oyster larvae. Overall, this indicates that protein quality is unlikely a factor that contributes to differences in nutritional value among microalgae. Finally, sterols [32], minerals [33] and pigments of microalgae also contribute to their nutritional performance in aquaculture. Several studies have indicated that, in the late-logarithmic growth phase, microalgae contain typically 30-40 %(w/w) protein, 10-20 %(w/w) lipids and 5-15 %(w/w) carbohydrates [6,34]. When cultured through the stationary phase, the proximate composition of microalgae may significantly change; e.g. nitrate limitation leads carbohydrate levels to double at the expense of protein [31,35]. Hence, a strong correlation exists between composition of microalgae and their measurable nutritional value – even though diets containing high levels of carbohydrates have been reported to produce the best growth of juvenile oysters [9] and larval scallops [36], as long as PUFAs are also present to adequate proportions. Conversely, high dietary protein provides maximum growth for juvenile mussels [15] and oysters [18]. Another relevant issue is that marine environments are typically filled with bacteria and viruses that can attack fish and shellfish, and thus potentially devastate aquaculture farms. Bacteria and viruses can also attack single-celled microalgae, so these microorganisms have developed biochemical mechanisms for self-defense; such mechanisms involve secretion of compounds that inhibit bacterial growth or viral attachment. For instance, compounds synthesized by Scenedesmus costatum, and partially purified from its organic extract exhibited activity against aquacultured bacteria because of their fatty acids longer than 10 carbon atoms in chain length – which apparently induce lysis of bacterial protoplasts. The ability of fatty acids at large to interfere with bacterial growth and survival has been known for some time, and recent structure-function relationship studies have proven that said ability depends on both their chain length and degree of unsaturation. Cholesterol and other compounds can antagonize antimicrobial features [37], so both composition and concentration of free lipids should be taken into account [38]. The activity of extracts of Phaeodactylum tricornutum against Vibrio spp. was attributed to EPA – a compound synthesized de novo by diatoms [39]; this PUFA is found chiefly as a polar lipid species in structural cell components (e.g. membranes) and is toxic to grazers [40], as well as a precursor of aldehydes with deleterious effects upon such consumers as copepods [41]. Similarly, unsaturated and saturated long chain fatty acids isolated from S. costatum [42] and organic extracts from Euglena viridis [43] display activity against that bacterial genus.

MICROALGAL BIOMASS PRODUCTION SYSTEMS

Commercial culture of microalgae targeted at their metabolites has been taking place for over 40 years, and the main microalgal species grown are Chlorella and Spirulina for healthy foods, Dunaliella salina for β-carotene, H. pluvialis for astaxanthin and several species for aquaculture [44]. There are several reactor configurations that met with success in mass cultivation of microalgae – chosen according to such factors as physiology of the microalga, cost of layout land, intensity of labour, cost of energy, availability of water, cost of nutrients, suitability of climate (if the culture is implemented outdoors) and specification of final product(s). Largescale culture systems should be compared on the basis of such indicators as efficiency of light utilization, controllability of temperature, hydrodynamic stress allowable, ability to maintain unimicroalgal and/or axenic cultures and feasibility of scale-up. A major decision to be made is whether to use closed photobioreactors (PBRs) or open ponds to cultivate a given microalga. The latter may entertain a large area, and are relatively cheap to build and easy to operate – but contamination is hard to control, stable environmental conditions (particularly temperature) are difficult to maintain, and the attainable cell density is relatively low because of mutual shading effects. On the other hand, extensive areas of land will be needed for commercial exploitation, besides substantial costs of harvesting afterwards [45,46]. The final choice of system is always a compromise between these parameters, aimed at achieving an economically acceptable outcome [44]. A common feature of most microalgal species produced commercially (i.e. Chlorella, Spirulina and Dunaliella) is that they grow in highly selective environments – which means that they can be grown in open air cultures and still remain relatively free of contamination by other microalgae and protozoa [47-49]. Species of microalgae that do not possess this selective advantage must be grown in closed systems; this includes most marine algae grown as aquaculture feeds (e.g. Skeletonema, Chaetoceros, Thalassiosira, Tetraselmis and Isochrysis), as well as the dinoflagellate Crypthecodinium cohnii [44].

Typical systems used indoors for microalgal mass culture include carboys (10 to 20 L), polythene bags (100 to 500 L) and tubs (1000 to 5000 L); these are usually operated batch- or continuouswise [44]. For larger volumes, outdoor tanks or ponds are preferred, which are operated semicontinuously; depending on their scale, hatcheries may produce between several hundreds to tens of thousands of liters of microalgal biomass per day. However, the culture systems employed at present are still fairly unsophisticated: e.g. D. salina is cultured in large (up to ca. 250 ha) shallow open-air ponds with no artificial mixing; and Chlorella and Spirulina are also grown outdoors, in either paddle-wheel mixed ponds or circular ponds (up to 1 ha each) with a rotating mixing arm. The production of microalgae for aquaculture occurs generally on a much lower scale. Other commercial large-scale systems include tanks used in aquaculture, the cascade system developed in the Czech Republic [50] and heterotrophic fermenter devices used for culture of Chlorella in Japan and Taiwan [47,51], and for culture of C. cohnii in USA [52,53]. The choice of which configuration is preferable depends obviously on the objective function; e.g. wastewater treatment might preclude open systems, owing to the unacceptably high costs arising from the large volumes to be processed and the low added value of the resulting products [54]. There has been a major effort directed at examining alternatives for the production of fresh microalgae, and also at more costefficient production systems.

Open cultivation systems

Microalgae cultivation in open ponds has been in current used since the 1950s [44]; these systems have been categorized as natural waters (lakes, lagoons and ponds), and artificial ponds or containers – with raceway ponds being the most frequently used artificial system [55]. The four major types of open-air systems currently in use (i.e. shallow big ponds, tanks, circular ponds and raceway ponds) have all advantages and disadvantages. This type of system usually consists of either circular ponds with a rotating arm to mix the culture, or long channels in a single or multiple loop configuration stirred by paddle wheels [56] – although simpler configurations have also met with sucess [54]. Raceway ponds are usually built in concrete, but compacted earth-lined ponds with (white) plastic have also been proposed. In a continuous production cycle, broth and nutrients required by microalgal growth are introduced in front of the paddlewheel, and circulated through the loop to the harvest extraction point; said paddlewheel undergoes a continuous motion to prevent sedimentation. The CO_2 requirement is usually satisfied using the open atmosphere as source – yet submerged aerators may be installed to enhance CO_2 supply, and thus absorption yield [57]. Compared to closed photobioreactors, open ponds represent a less expensive investment for large-scale production of microalgal biomass. On the other hand, open pond production does not necessarily compete with agricultural crops for land, since it can be implemented in areas with marginal crop production potential [58]. Open ponds also have low energy input requirements [59], and regular maintenance and cleaning are easier [60]. Open ponds and raceways were the first large-scale designs implemented, and are still the most widely applied in industrial processing. The main constraints related to their operation are the difficulty to control contamination and to keep the culture environment steady, and the cost associated with harvesting. Furthermore, the open character of the system makes it possible for naturally occurring microalgae or their predators to infiltrate, and thus compete with microalgae intended for cultivation. Therefore, a monoculture can only be maintained under extreme conditions of pH, salinity or temperature that guarantee dominance by the desired strain (e.g. D. salina dominance requires highly salted media, whereas Spirulina platensis demands high pH values). Unfortunately, high pH, temperature and salt concentration are not compatible with most microalgal species of interest. Regarding biomass productivity, however, open pond systems are less efficient than closed photobioreactors [61]. This can be attributed to such parameters as evaporation losses, temperature variation, CO_2 deficiency, inefficient mixing and light limitation. Although evaporation losses make a net contribution to cooling, they may also cause significant changes in the ionic composition of the medium – with detrimental effects upon microalga growth [62]. Although this type of reactor is extensively used in industrial microalgal production – e.g. to produce Spirulina and Dunaliella spp. up to worldwide totals of 5000 and 1200 ton/yr, respectively [24], open systems have apparently reached their upper limit – with little room for further technological improvement.

Closed cultivation systems

Despite the success of open systems, future advances in microalgal mass culture will require improved closed systems, as the most interesting microalgal species cannot grow in highly selective environments [44]. Hence, photobioreactor technology is on the rise, which is designed to overcome the major constraints associated with open pond production systems [63]; recall that both pollution and contamination risks preclude use of open ponds to prepare high-value products for eventual use as active ingredients in aquaculture feed formulation [60]. Closed systems include tubular, flat plate and fermenter types, among other possibilities. The former two are specifically designed for efficient recovery of sunlight, whereas the latter may require artificial illumination. Owing to the higher cell mass productivities attained, harvesting costs can be significantly reduced. Closed photobioreactors also provide reproducible cultivation conditions, good heat transfer, better biomass yield, higher product quality and opportunity for flexible technical design [44,60]. Note, however, that the costs of closed systems are substantially higher than their open pond counterparts [54]. A variety of closed photobioreactors have been tested (or at least proposed) for industrial microalgal biomass production [64,65], but engineering and economic analyses of such reactors still lag behind the open ponds [66-70]. A typical photobioreactor is essentially a four-phase system, consisting of solid microalgal cells, a liquid growth medium, a gaseous phase and a light radiation field [71]. Its productivity is limited by various design features – but, most importantly, the reactor is to be operated under favorable illumination conditions, with optimized surface-to-volume ratio and light/dark cycle, coupled with adequate mass transfer features [72]. The current consensus is that commercial (photoautotrophic) production of metabolites with interest for aquaculture by microalgae should resort to outdoor enclosed photobioreactors [56,62,65,73,74]. Tredici [65] reviewed the development of those type of reactors over the last decade; while many types of experimental PBRs have been considered, built and tested, very few have actually succeeded on a commercial level. Commercial application of photobioreactor technology remains indeed restricted to the production of two Chlorophyte microalgae: Chlorella and Haematococcus [62,75]. Scale up of photobioreactors from bench to commercial scale is not trivial – since it needs changes in illumination, gas transfer and temperature to be taken into account, all of which are severely affected by turbulence in the reactor, and consequently require a tight control. Therefore, scale up appears to be much more of an engineering problem than a biological one; and general recommendations as to possible maximum scales have accordingly been produced [75,76].

ALTERNATIVES TO FRESH MICROALGAE

Marine microalgae have been the traditional food component in finfish and shellfish aquaculture, e.g. for larval and juvenile animals [77]; they are indeed essential in hatchery and nursery of bivalves, shrimp and some finfish cultures. Microalgae are also used to produce zooplankton – typically rotifers, which are in turn fed to freshly hatched carnivorous fish [78]. As aquaculture industry expands [79] – and since microalgal biomass cultivation on-site may represent up to 30% of the operating costs [13], there is a demand for marine microalgae that cannot be met by the conventional methods used in hatcheries – thus forcing one to resort to substitutes with mediocre results that bring about several problems [44,80]. Despite the obvious advantages of alive microalgae in aquaculture, the current trend is to avoid using them because of their high cost and difficulty in producing, concentrating and storing them [8,81]. Alternatives that are potentially more cost-effective have been investigated – including nonliving food, viz. microalga pastes, dried microalgae, microencapsulates, cryopreservation, flocculation, bacteria or yeasts; they have been tested in vitro and in actual hatcheries, but met with variable degrees of success [82,83]. For instance, in Japan, where Nannochloropsis oculata is the most important cultured feed for the rotifer Brachionus plicatilis, concentrated suspensions and frozen biomass of this microalga are commercially available [84]; and partial replacement of alive microalgae by microencapsulated and yeast-based diets is indeed a routine practice in hatcheries for penaeid shrimp [24,85]. However, most these approaches have proven unsuitable as major dietary components, because of their lower nutritional value than mixtures of alive microalgae. Several criteria should be addressed in attempts to find substitutes for alive microalgae as diet in aquaculture. From a nutrition standpoint, alive microalgae possess higher nutritional value and better digestibility than most substitutes; note that the nutritional quality depends critically on such biochemical constituents as PUFAs, vitamins, sterols and carbohydrates [86]. Useful bacteria can provide only a part of the metabolic requirements in aquaculture – by supplying a few organic molecules and vitamins. Under conditions close to those found in rearing facilities, the bacterial input should not represent more than 15% of the microalgal contribution for mollusk larvae and juveniles of many species [87,88]. Yeasts were as well investigated as an alternative food source – but poor results were observed [83,89]. Therefore, these two alternatives are not suitable to fully replace alive microalgae.

An alternative diet with an apparently better potential is microalgal pastes or concentrates [90-92]; these are prepared by centrifugation (up to 1:500 concentration) or flocculation (up to 1:100 concentration). Concentrates prepared from distinct microalgae vary in their suitability – with diatoms being the most promising; and they have a shelf life of between 2 and 8 weeks, when stored below 4°C. Commercially, microalgal concentrates can be prepared under two different scenarios: (a) by hatcheries on-site, which prepare concentrates as back-up or as a means to store overproduction; or (b) by remote production, centralized at a large facility – with a greater economy of scale, with the resulting concentrates dispatched to hatcheries upon request. The advantage of such concentrates is that they can be used "off-the-shelf", thus contributing favorably to the cost-efficiency in hatcheries. On the other hand, the lower nutritional value of most dried microalgae compared to alive feed, and the limited availability of commercial dried products appear as main shortcomings. Globally speaking, concentrates have low levels (or even absence) of ω3-PUFA, and lead to a difficult digestion by bivalve larvae [93]. The genus Tetraselmis seems to be a good candidate for microalgal paste, but its nutritional quality deteriorates quite rapidly [94]; experiments have indicated that such substitutes should be used as supplement only when rations of live microalgae are insufficient. Furthermore, spray-dried microalgae and microalga paste may be useful to replace up to 50% of alive microalgae. Coutteau and Sorgeloos [13] reported that artificial or non-living diets are rarely applied in routine processing of bivalves, and are mostly considered as a backup food source only. Centrifuged concentrates of P. lutheri, in combination with Chaetoceros calcitrans or S. costatum, lead to 85-90% of the growth when a mixed diet of alive microalgae for oyster Saccostrea glomerata larvae is used [92]. Centrifugation has been successfully applied to prepare concentrates, but it has some limitations – i.e. the process involves exposing cells to high gravitational and shear forces that damage the cell structure. On the other hand, processing of large culture volumes is time-consuming and requires costly equipment, i.e. a specialized continuous centrifuge. Research on post-harvest preservation is required to extend shelf-life beyond 4 to 8 weeks, and also to prepare concentrates from flagellate species (e.g. Isochrysis sp. and P. lutheri). Alternative processes have meanwhile been developed that are potentially less damaging to cells – including foam fractionation [95], flocculation [96,97] and filtration [98]. Sandbank [99] fed microalgae, grown in waste-water and flocculated with aluminium sulfate, to common carp (Cyprinus carpia); a diet containing 25% of microalgal meal led to a growth comparable to that by the control diet, with no harmful effects detected upon long term health of the fish. Millamena et al. [96] successfully fed Penaeus monodon larvae with dried, flocculated C. calcitrans and Tetraselmis chuii cells. However, a common disadvantage encountered was that the harvested cells are difficult to disaggregate back to single cells, which is a requirement to feed them to filter-feeding species such as bivalves [100]. A novel technique was developed for flocculation of marine microalgae that appears useful in aquaculture: it entails adjustment of pH of the culture using NaOH, followed by addition of a non-ionic polymer, Magnafloc LT-25; the ensuing flocculate is then harvested and neutralized, thus leading to a final concentration of between 200- and 800-fold. This process was successfully applied to harvest cells of C. calcitrans and C. muelleri, Thalassiosira pseudonana, Attheya septentrionalis, Nitzschia closterium, Skeletonema sp., Tetraselmis suecica and Rhodomonas salina, with efficiencies above 80%; it proved rapid, simple and inexpensive, and relatively independent of processed volume (unlike concentration by centrifugation). The harvested material was readily disaggregated to single cell suspensions by dilution in seawater, coupled with mild agitation. Microscopic examination proved that the final cells are indistinguishable from the nonflocculated ones; and assay for chlorophyll of the concentrates prepared from cultures of up to 130 L showed marginal degradation by 2 weeks of storage [100]. Cryopreservation has been thoroughly adopted by culture collections to preserve strains, but may also find an application in aquaculture [80]. Viable cryostorage of biological specimens has followed various protocols of cooling/thawing rates and cryoprotectant addition, which have been developed and tuned more or less empirically [101]. Recall that temperatures used for cryostorage are well below freezing – down to even -196 °C in liquid helium, when biological specimens are to be stored without limit [102]. While cryostorage is generally thought to be innocuous to the cell, the events occurring upon freezing or thawing can lead to severe damage, or even cell death. Moreover, cryoprotectants that enhance the cell viability at cryogenic temperatures are usually toxic at physiological temperatures [103] – an obstacle that is overcome by reducing the exposure time or the temperature of incubation prior to cryopreservation [104]. Knowledge of cryoprotectant tolerance levels for microalgae is still limited [105], as well as for early larval stages and for zooplankton that are cultivated and rely on the availability of microalgae for growth. In general, cryopreservation possesses a high potential for culture collections, and may also offer a solution for reliable supply of microalgae in aquaculture. For instance, marine microalgae used in aquaculture were successfully cryopreserved under 4, -20 and -80 °C using common cryoprotectants (i.e. methanol, dimethylsulfoxide, propylene glycol and polyvinylpyrrolidone), with promising results at least for Chlorella minutissima, Chlorella stigmatophora, Isochrysis galbana and Dunaliella tertiolecta [80]. Several products based on thraustochytrids (i.e. microorganisms with a taxonomy related to certain microalgal classes), from the genus Schizochytrium, have been marketed through Aquafauna Biomarine and Sanders Brine Shrimp. These products have high concentrations of DHA [106], and have accordingly been applied as alternatives to commercial oil enrichment of zooplankton fed to larvae. As direct feeds, most such products have a lower nutritional value than mixtures of microalgae, yet some performed well as components of a mixed diet with alive microalgae [83,107]. In general, substitutes of alive microalgae should present an appropriate physical behavior – and this constitutes a significant challenge; in particular, they should not aggregate or easily break apart. Drying microalgae can cause, due to oxidation, a loss of PUFAs [108], which are essential components for larval growth [87]; the poor performance reported for dried microalgae was associated chiefly with the difficulty to keep cells in suspension without disintegrating them, so as to avoid said oxidation [13]. Moreover, when cell walls are broken, a high fraction of water-soluble components cannot be ingested by the organism, and may consequently interfere with the water quality of the aquaculture [109]. Therefore, pathogenic bacterial proliferation may occur, and cause costly production losses. Similar difficulties arise when using microalga paste, because the preparation procedures (i.e. centrifugation, flocculation or filtration) and/or preservation techniques (i.e. additives or freezing) must ensure that cell wall integrity is essentially preserved.

Products other than alive microalgae must obviously be free of bacterial contamination and devoid of toxicity. Consequently, the use of alive bacteria as a food source in hatcheries seems somehow inappropriate, since physical and chemical treatments are often used to limit bacterial contamination that would otherwise be responsible for drastic larval mortality [110]. Oyster larvae fed with alive microalga diets underwent improved growth via addition of some bacterial isolates [111,112], but this advantage may obviously not be possible in a treated microalgal product. However, in alive microalgae, the natural bacterial flora was proven to enhance the health of molluscs. Langdon and Bolton [88] showed that antibiotic suppression of the bacterial flora in artificial feed of juvenile oysters reduced their growth. In conclusion, mitigated or unsuccessful results when using nonliving microalgae have turned alive microalgae into the first choice in aquaculture feeding. Only partial replacement thereof has been possible in studies encompassing preserved non-living algae [113], microencapsulated diets [88] or spray-dried algae [114]; but no whole replacement can be recommended, despite intensive research efforts in that direction [107]. Consequently, novel solutions to totally replace microalgae in aquaculture diets cannot at present be widely adopted [4,24,81].

USE OF MICROALGAE TO ENRICH ZOOPLANKTON

Microalgae have an important role in aquaculture, also as a means to enrich zooplankton for feeding fish and other larvae afterwards. In addition to providing proteins (that contain essential amino acids) and energy, they carry such other key nutrients as vitamins, essential PUFAs, pigments and sterols – which are transferred up through the food chain. For instance, PUFA-rich microalgae, such as Pavlova sp. and Isochrysis sp., have been successfully fed to zooplankton to enrich them in DHA [115]. However, when the level of enrichment attained is not sufficient, commercial oil-emulsions are often used. Recently, such products as dried preparations of Schizochytrium sp. (which contain 5-15% of their DW as DHA) have been utilized, which produce levels of DHA enrichment in zooplankton comparable to use of commercial oils [116] – and also produce DHA to EPA ratios of 1–2, which are considered favorable for fish larval nutrition [117]. Brown, Skabo & Wilkinson [118] described that rotifers fed with microalgae (e.g. Isochrysis sp. and N. oculata) become rapidly enriched with ascorbic acid (AsA), whereas rotifers fed on baker's yeast (which itself is deficient in AsA) contained only residual amounts of AsA.; after 16 h of starving, rotifers lost ca. 10% of their AsA, while retaining ca. 50% of the total AsA ingested. Similarly, the concentration of AsA in Artemia sp. may be increased by feeding with microalgae [119]. However, little information is available on the transfer of other vitamins from microalgae to fish larvae. Rønnestad, Helland & Lie [120] demonstrated that microalgal pigments transferred to zooplankton may add to their nutritional value; recall that the dominant pigments in the copepod Temora sp. are lutein and astaxanthin, whereas in Artemia it is canthaxanthin. When these microalgae were fed to copepods and then to halibut larvae, adequate amounts of vitamin A were found, but not when halibut was fed on Artemia; this was attributed to the ability of larvae to convert lutein and/or astaxanthin, but not canthaxanthin to vitamin A. They accordingly recommended that Artemia should routinely be enriched with astaxanthin and lutein (the latter pigment is common in "green" microalgae, e.g. Tetraselmis sp.) to improve their nutritional value. A common procedure during culture of both larval fish and prawns is to add microalgae (i.e. "green water") to intensive culture systems, together with the zooplankton prey [121]. The most popular microalga species used for this purpose are N. oculata and T. suecica. Addition of microalgae to larval tanks can also improve the production of larvae, but their exact mechanism of action remains unclear. Light attenuation (i.e. shading effects) may have a beneficial effect on larvae; however, maintenance of nutritional quality of the zooplankton, excretion of vitamins or other growth-promoting substances by the microalgae, and probiotic effects of the microalgae have also been hypothesized. Maintenance of NH3- and O2-balances has also been proposed, but this assumption failed to be supported by experimental evidence [121]. More research is still needed on the application of other microalgae – especially those species rich in DHA, to green water systems. Green water may also be applied to extensive outdoor production facilities, by fertilizing ponds in attempts to stimulate microalgal growth, and consequently zooplankton production.

AVENUES FOR FUTURE RESEARCH ON MICROALGAE

The high production costs of microalgae remain a constraint to many hatcheries. Despite efforts developed over the latest decades toward cost-effective artificial diets to replace microalgae, on-site microalgal production still remains a critical element for operation of most marine hatcheries. Improvements in alternative diets will surely continue, but production costs of microalgae will also likely decrease – so it is not expected that microalgae will be replaced in full, at least on the medium run. A wide selection of microalgal species is already available to support aquaculture activities. However, specific applications in industrial subsectors demand novel species with improved nutritional quality or growth characteristics, which are compatible with attempts to improve hatchery efficiency and yield. Appart from improvements in cost-efficiency of on-site microalgal production, an alternative is centralizing microalga production in dedicated mass-culture facilities, using heterotrophic methods or nonconventional photobioreactors. These technologies may be coupled with post-harvest processing (e.g. spray-drying) or concentration (e.g. centrifugation or flocculation) to develop off-the-shelf microalgal biomass for ready distribution to hatcheries. On the other hand, antifouling activity of extracts from some microalgae has been observed in microalga culture tanks, which are better (and less toxic) than common biocides. Those natural compounds could therefore be considered as good substitutes of commercial biocides in antifouling paints. Furthermore, as paint coatings remain the predominant preventative technique of marine biofouling, coatings adapted to the needs of aquaculture apparatuses – containing an active product from microalgae and able to inhibit the major microorganisms causing trouble in cultivation, are a potentially good solution to fight fouling. The exact substances that exhibit antifouling activities in microalgae are not yet known, so this type of study is warranted – to purify and identify the active compounds involved.

The need to reduce water consumption in aquaculture has long been recognized, so a great deal of effort has been directed toward development of recirculating systems. Unfortunately, current research and development encompassing aquaculture water re-use is largely devoted to bacteria-based systems – and the possibility of using microalga-based water re-use has been essentially neglected. The bacterial component in a water re-use system dedicates itself in full to excessive nutrient removal; conversely, a microalga-based water re-use system produces microalgae that can be used to produce a second crop, such as bivalve seed or Artemia – which may thus be sold to generate extra income. The main difficulty faced in development of microalga-based water re-use systems is the inability to maintain the desired microalgal species in an open system. A breakthrough in marine diatom production technology may allow one to focus on development of water re-use systems where the 'effluent' becomes itself a valuable resource: an integrated shrimp/microalga/oyster production system reduces water consumption, and turns effluent 'waste' into a profitable item – while taking advantage of the antibacterial properties of the marine diatom to control diseases, and thus reduce susceptibility of the shrimp to viral infections.

ACKNOWLEDGEMENTS

A postdoctoral fellowship (ref. SFRH/BPD/72777/2010), supervised by author F.X.M., was granted to author A.C.G., under the auspices of ESF (III Quadro Comunitário de Apoio) and the Portuguese State.

REFERENCES

1. G. Oron, L. R. Wildschut, D. Porath, "Wastewater recycling by duckweed for protein production and effluent renovation," Water Science and Technology, vol. 17, pp. 803- 818, 1985.
2. O. Hammouda, A. Gaber, N. Abdel-Raouf, "Microalgae and wastewater treatment," Ecotoxicology and Environmental Safety, vol. 31, pp. 205-210, 1995.
3. W. J. Oswald, "Algal Pond Systems–Habitats," in Proceedings of a Seminar held at Murdoch University, Western Australia, 1991, pp. 61-69.
4. Muller-Feuga, "The role of microalgae in aquaculture: situation and trends," Journal of Applied Phycology, vol. 12, pp. 527-534, 2000.
5. T. Enright, G. F. Newkirk, J. S. Craigie, J. D. Castell, "Evaluation of phytoplankton as diets for juvenile Ostrea edulis L.," Journal of Experimental Marine Biology and Ecology, vol. 96, pp. 1-13, 1986.
6. M. R. Brown, S. W. Jeffrey, J. K. Volkman, G. A. Dunstan, "Nutritional properties of microalgae for mariculture," Aquaculture, vol. 151, pp. 315–331, 1997.
7. K. Yamaguchi, "Recent advances in microalgal bioscience in Japan, with special reference to utilization of biomass and metabolites: a review," Journal of Applied Phycology, vol. 8, pp. 487–502, 1997.
8. W. Becker, "Microalgae for aquaculture. The nutritional value of microalgae for aquaculture." In Richmond, A. (ed.), Handbook of Microalgal Culture. Blackwell, Oxford, pp. 380–391, 2004.
9. C. T. Enright, G. F. Newkirk, J. S. Craigie, J. D. Castell, "Growth of juvenile Ostrea edulis L. fed Chaetoceros gracilis Schütt of varied chemical composition," Journal of Experimental Marine Biology and Ecology, vol. 96, pp. 15-26, 1986.
10. P. A. Thompson, M. Guo, P. J. Harrison, "The influence of irradiance on the biochemical composition of three phytoplankton species and their nutritional value for larvae of the Pacific oyster (Crassostrea gigas)," Marine Biology, vol. 117, pp. 259-268, 1993.

11. L. R. Marínez-Córdova, E. Peña-Messina, "Biotic communities and feeding habits of Litopenaeus vannamei (Boone 1931) and Litopenaeus stylirostris (Stimpson 1974) in monoculture and polyculture semi-intensive ponds," Aquaculture Research, vol. 36, pp. 1075-1084, 2005.

12. M. Kent, C. L. Browdy, J. W. Leffler, "Consumption and digestion of suspended microbes by juvenile Pacific white shrimp Litopenaeus vannamei," Aquaculture, doi: 10.1016/j.aquaculture.2011.06.048, 2011.

13. P. Coutteau, P. Sorgeloos, "The use of algal substitutes and the requirement for live algae in the hatchery and nursery rearing of bivalve molluscs: an international survey," Journal of Shellfish Research, vol.11, pp. 467-476, 1992.

14. C. E. Epifanio, "Growth in bivalve molluscs: nutritional effects of two or more species of algae in diets fed to the American oyster Crassostrea virginica (Gmelin) and the hard clam Mercenaria mercenaria (L.)," Aquaculture, vol. 18, pp. 1-12, 1979.

15. D. A. Kreeger, C. J. Langdon, "Effect of dietary protein content on growth of juvenile mussels, Mytilus trossulus (Gould 1850)," Biological Bulletin, vol. 185, pp. 123-139, 1993.

16. J. K. Volkman, S. W. Jeffrey, P. D. Nichols, G. I. Rogers, C. D. Garland, "Fatty acid and lipid composition of 10 species of microalgae used in mariculture," Journal of Experimental Marine Biology and Ecology, vol. 128, pp. 219-240, 1989.

17. G. A. Dunstan, J. K. Volkman, S. M. Barrett, J.-M. Leroi, S. W. Jeffrey, "Essential polyunsaturated fatty acids from 14 species of diatom (Bacillariophyceae)," Phytochemistry, vol. 35, pp. 155-161, 1994.

18. R. M. Knuckey, M. R. Brown, S. M. Barrett, G. M. Hallegraeff, "Isolation of new nanoplanktonic diatom strains and their evaluation as diets for juvenile Pacific oysters (Crassostrea gigas)," Aquaculture, vol. 211, pp. 253-274, 2002.

19. D. Chuntapa, S. Powtongsook, P. Menasveta, "Water quality control using Spirulina platensis in shrimp culture tanks," Aquaculture, vol. 220, pp. 355-366, 2003.

20. G. D. Lio-Po, E. M. Leaño, M. M. D. Peñaranda, A. U. Villa- Franco, C. D. Sombito, N. G. Guanzon, "Antiluminous Vibrio factors associated with the 'green water' grow out culture of the tiger shrimp Penaeus monodon," Aquaculture, vol. 250, pp. 1-7, 2005.

21. L. Rodolfi, G. C. Zittelli, L. Barsanti, G. Rosati, M. R. Tredici, "Growth medium recycling in Nannochloropsis sp. mass cultivation," Biomolecular Engineering, vol. 20, pp. 243-248, 2003.

22. Muller-Feuga, "Microalgae for aquaculture. The current global situation and future trends." In Richmond, A. (ed.), Handbook of Microalgal Culture. Blackwell, Oxford, 2004, pp. 352-364.

23. Irianto, B. Austin, "Probiotics in aquaculture," Journal of Fish Diseases, vol. 25, pp. 633-642, 2002.

24. P. Spolaore, C. Joannis-Cassan, E. Duran, A. Isambert, "Review: commercial applications of microalgae," Journal of Bioscience and Bioengineering, vol. 101, pp. 87- 96, 2006.

25. D. S. Nichols, "Prokaryotes and the input of polyunsaturated fatty acids to the marine food web," FEMS Microbiology Letters, vol. 219, pp. 1-7, 2003.

26. Aragão, L. E. C. Conceição, M. T. Dinis, H.-J. Fyhn," Amino acid pools of rotifers and Artemia under different conditions: nutritional implications for fish larvae," Aquaculture, vol. 234, pp. 429-445, 2004.

27. K. E. Apt, P. W. Behrens, "Commercial developments in microalgal biotechnology," Journal of Phycology, vol. 35, pp. 215-226, 1999.

28. L. Waldenstedt, J. Inborr, I. Hansson, K. Elwinger, "Effects of astaxanthin-rich algal meal (Haematococcus pluvalis) on growth performance, faecal campylobacter and clostridial counts and tissue astaxanthin concentration of broiler chickens," Animal Feed Science and Technology, vol. 108, pp. 119-132, 2003.

29. M. R. Brown, "The amino acid and sugar composition of 16 species of microalgae used in mariculture," Journal of Experimental Marine Biology and Ecology, vol. 145, pp. 79- 99, 1991.

30. M. R. Brown, G. A. Dunstan, S. W. Jeffrey, J. K., Volkman, S. M. Barrett, J. M. Leroi, "The influence of irradiance on the biochemical composition of the prymnesiophyte Isochrysis sp. (clone T-ISO)," Journal of Phycology, vol. 29, pp. 601- 612, 1993.

31. M. R. Brown, C. D. Garland, S. W. Jeffrey, I. D. Jameson, J. M. Leroi, "The gross and amino acid compositions of batch and semi-continuous cultures of Isochrysis sp. (clone T.ISO), Pavlova lutheri and Nannochloropsis oculata," Journal of Applied Phycology, vol. 5, pp. 285-296, 1993.

32. J. Knauer, S. M. Barrett, J. K. Volkman, P. C. Southgate, "Assimilation of dietary phytosterols by Pacific oyster Crassostrea gigas spat," Aquaculture Nutrition, vol. 5, pp. 257-266, 1999.

33. J. Fabregas, C. Herrero, "Marine microalgae as a potential source of minerals in fish diets," Aquaculture, vol. 51, pp. 237-243, 1986.

34. S. M. Renaud, L. V. Thinh, D. L. Parry, "The gross composition and fatty acid composition of 18 species of tropical Australia microalgae for possible use in mariculture," Aquaculture, vol. 170, pp. 147-159, 1999.

35. P. J. Harrison, P. A. Thompson, G. S. Calderwood, "Effects of nutrient and light limitation on the biochemical composition of phytoplankton," Journal of Applied Phycology, vol. 2, pp. 45-56, 1990.

36. J. N. C. Whyte, N. Bourne, C. A. Hodgson, "Influence of algal diets on biochemical composition and energy reserves in Patinopecten yessoensis (Jay) larvae," Aquaculture, vol. 78, pp. 333-347, 1989.

37. J. A. Mendiola, C. F. Torres, P. J. Martín-Alvarez, S. Santoyo, A. Toré, B. O. Arredondo, F. J. Señoráns, A. Cifuentes, E. Ibáñez, "Use of supercritical CO2 to obtain extracts with antimicrobial activity from Chaetoceros muelleri microalga. A correlation with their lipidic content," European Food Research and Technology, vol. 224, pp. 505-510, 2007.

38. K. Benkendorff, A. R. Davis, C. N. Rogers, J. B. Bremner, "Free fatty acids and sterols in the benthic spawn of aquatic molluscs, and their associated antimicrobial properties," Journal of Experimental Marine Biology and Ecology, vol. 316, pp. 29-44, 2005.

39. V. J. Smith, A. P. Desbois, E. A. Dyrynda, "Conventional and unconventional antimicrobials from fish, marine invertebrates and micro-algae," Marine Drugs, vol. 8, pp. 1213-1262, 2010.

40. F. Jüttner, "Liberation of 5,8,11,14,17-eicosapentaenoic acid and other polyunsaturated fatty acids from lipids as a grazer defense reaction in epiphithic diatom biofilms," Journal of Phycology, vol. 37, pp. 744-755, 2001.

41. G. d'Ippolito, S. Tucci, A. Cutignano, G. Romano, G. Cimino, A. Miralto, A. Fontana, "The role of complex lipids in the synthesis of bioactive aldehydes of the marine diatom Skeletonema costatum," Biochimica et Biophysica Acta, vol. 1686, pp. 100-107, 2004.

42. M. Naviner, J.-P. Bergé, P. Durand, H. le Bris, "Antibacterial activity of the marine diatom Skeletonema costatum against aquacultural pathogens," Aquaculture, vol. 174, pp. 15-24, 1999.

43. K. Das, J. Pradhan, P. Pattnaik, B. R. Samantaray, S. K. Samal, "Production of antibacterials from the freshwater alga Euglena viridis (Ehren)," World Journal of Microbiology and Biotechnology, vol. 21, pp. 45-50, 2005.

44. M. A. Borowitzka, "Commercial production of microalgae: ponds, tanks, tubes and fermenters," Journal of Biotechnology, vol. 70, pp. 313–321, 1999.

45. H. M. Amaro, A. C. Guedes, F. X. Malcata, "Advances and perspectives in using microalgae to produce biodiesel," Applied Energy, vol. 88, pp. 3402-3410, 2011.

46. S. A. Scott, M. P. Davey, J. S. Dennis, I. Horst, C. J. Howe, D. J. Lea-Smith, "Biodiesel from algae: challenges and prospects," Current Opinion in Biotechnology, vol. 21, pp. 277-286, 2010.

47. P. Soong, "Production and development of Chlorella and Spirulina in Taiwan." In: Shelef, G., Soeder, C. J. (Eds.), Algae Biomass. Elsevier, Amsterdam, 1980, pp. 97-113.

48. L. J. Borowitzka, M. A. Borowitzka, "Industrial production: methods and economics." In: Cresswell, R. C., Rees, T. A. V., Shah, N. (Eds.), Algal and Cyanobacterial Biotechnology. Longman Scientific, London, 1989, pp. 294–316.

49. Belay, "Mass culture of Spirulina outdoors—the Earthrise Farms experience". In: Vonshak, A. (Ed.), Spirulina platensis (Arthrospira): Physiology, Cell-biology and Biotechnology. Taylor & Francis, London, 1997, pp. 131-158.

50. Setlík, S. Veladimir, I. Malek, "Dual purpose open circulation units for large scale culture of algae in temperate zones. I. Basic design considerations and scheme of pilot plant," Algology Studies (Trebon), vol. 1, pp. 11, 1970.

51. Kawaguchi, "Microalgae production systems in Asia." In: Shelef, G., Soeder, C. J. (Eds.), Algae Biomass Production and Use. Elsevier, Amsterdam, 1980, pp. 25-33.

52. Kyle, R. M. Gladue, "Eicosapentaenoic acids and methods for their production," World Patent, vol. 9, pp. 114,427, 1991.

53. J. Kyle, S. E. Reeb, V. J. Sicotte, "Dinoflagellate biomass, methods for its production, and compositions containing the same," USA Patent 5,711,983, 1998.

54. P. Carvalho, L. A. Meireles, F. X. Malcata, "Microalgal reactors: a review of enclosed system designs and performances," Biotechnology Progress, vol. 22, pp. 1490-1506, 2006.

55. Jiménez, B. R. Cossío, D. Labella, N. F. Xavier, "The feasibility of industrial production of Spirulina (Arthrospira) in southern Spain," Aquaculture, vol. 217, pp. 179-190, 2003.

56. Chaumont, "Biotechnology of algal biomass production: a review of systems for outdoor mass culture," Journal of Applied Phycology, vol. 5, pp. 593-604, 1993.

57. Terry, L. P. Raymond, "System design for the autotrophic production of microalgae," Enzyme and Microbial Technology, vol. 7, pp. 474-487, 1985.

58. Y. Chisti, "Biodiesel from microalgae beats bioethanol," Trends in Biotechnology, vol. 26, pp. 126-131, 2008.

59. Rodolfi, G. C. Zittelli, N. Bassi, G. Padovani, N. Biondi, G. Bonini, "Microalgae for oil: strain selection, induction of lipid synthesis and outdoor mass cultivation in a lowcost photobioreactor," Biotechnology and Bioengineering, vol. 102, pp. 100-112, 2008.

60. U. Ugwu, H. Aoyagi, H. Uchiyama, "Photobioreactors for mass cultivation of algae," Bioresource Technology, vol. 99, pp. 4021-4028, 2008.

61. Y. Chisti, "Biodiesel from microalgae," Biotechnology Advances, vol. 25, pp. 294-306, 2007.

62. Pulz, "Photobioreactors: production systems for phototrophic microorganisms," Applied Microbiology and Biotechnology, vol. 57, pp. 287-293, 2001.

63. Metting, "Biodiversity and application of microalgae," Journal of Industrial Microbiology, vol. 17, pp. 477-489, 1996.

64. Molina-Grima, "Microalgae, mass culture methods." In M. C. Flickinger, & S. W. Drew (Eds.), Encyclopedia of Bioprocess Technology: Fermentation, Biocatalysis and Bioseparation. New York: Wiley, 1999, vol. 3, pp. 1753-1769.

65. R. Tredici, "Bioreactors, photo." In Flickinger, M. C. & Drew, S. W. (Eds.), Encyclopedia of Bioprocess Technology: Fermentation, Biocatalysis and Bioseparation, vol. 1. New York, Wiley, 1999, pp. 395-419.

66. G. Acién-Fernández, F. García-Camacho, J. A. Sánchez-Pérez, J. M. Fernández-Sevilla, E. Molina-Grima, "Modelling of biomass productivity in tubular photobioreactors for microalgal cultures: effects of dilution rate, tube diameter and solar irradiance," Biotechnology and Bioengineering, vol. 58, pp. 605-616, 1998.

67. S. Aiba, "Growth kinetics of photosynthetic microorganisms," Advances in Biochemical Engineering, vol. 23, pp. 85-156, 1982.

68. L. E. Erickson, H. Y. Lee, "Process analysis and design of algal growth system." In: Barclay, W. R. and McIntosh, R. P. (Eds), Algal Biomass Technologies: an Interdisciplinary Perspective. J. Cramer, Berlin-Stuttgart, 1986, pp. 197–206.

69. S. L. Pirt, Y. K. Lee, M. R. Walach, M. W. Pirt, H. H. Balyuzi, M. J. Bazin, "A tubular bioreactor for photosynthetic production of biomass from carbon dioxide: design and performance," Journal of Chemical Technology and Biotechnology, vol. 33B, pp. 35- 58, 1983.

70. L. Rorrer, R. K. Mullikin, "Modeling and simulation of a tubular recycle photobioreactor for macroalgal cell suspension cultures," Chemical Engineering Science, vol. 54, pp. 3153-3162, 1999.

71. C. Posten, "Design principles of photo-bioreactors for cultivation of microalgae," Engineering in Life Sciences, vol. 9, pp. 165-177, 2009.

72. A. Tsygankov, "Laboratory scale photobioreactors," Applied Biochemistry and Microbiology, vol. 37, pp. 333-341, 2001.

73. Borowitzka, "Microalgae as sources of pharmaceuticals and other biologically active compounds," Journal of Applied Phycology, vol. 7, pp. 3-15, 1995.

74. Vonshak, "Tubular photobioreactors for algal mass production; prospects and achievements," Israeli Journal of Aquaculture (Bamidgeh), vol. 44, pp. 151, 1992.

75. M. Olaizola, "Commercial development of microalgal biotechnology: from the test tube to the marketplace," Biomolecular Engineering, vol. 20, pp. 459-466, 2003.

76. Molina-Grima, F. G. Acién-Fernández, F. García-Camacho, F. Camacho-Rubio, Y. Chisti, "Scale-up of tubular photobioreactors," Journal of Applied Phycology, vol. 12, pp. 355-368, 2000.

77. de Pauw, G. Persoone, "Micro-algae for aquaculture." In: Borowitzka, M. A., Borowitzka, L. J. (Eds.), Micro-Algal Biotechnology. Cambridge University Press, Cambridge, 1988, pp. 197-221.

78. R. Benemann, W. J. Oswald, "Systems and economic analysis of microalgae ponds for conversion of CO2 to biomass," Final Report, Subcontract XK 4-04136- 06, Pittsburgh Energy Technology Center Grant No. DE-FG22-93PC93204, p. 260, 1996.

79. FAO, 2000. Aquaculture Production 1992–1998. FAO, Rome.

80. Tzovenis, G. Triantaphyllidis, X. Naihong, E. Chatzinikolaou, K. Papadopoulou, G. Xouri, T. Tafas, "Cryopreservation of marine microalgae and potential toxicity of cryoprotectants to the primary steps of the aquacultural food chain," Aquaculture, vol. 230, pp. 457-473, 2004.

81. M. A. Borowitzka, "Microalgae for aquaculture: opportunities and constraints," Journal of Applied Phycology, vol. 9, pp. 393-401, 1997.

82. Knauer, P. C. Southgate, "A review of the nutritional requirements of bivalves and the development of alternative and artificial diets for bivalve aquaculture," Reviews in Fisheries Science, vol. 7, pp. 241–280, 1999.

83. R. Robert, P. Trintignac, "Substitutes for live microalgae in mariculture: a review," Aquatic Living Resources, vol. 10, pp. 315–327, 1997.

84. Chini-Zittelli, F. Lavista, A. Bastianini, L. Rodolfi, M. Vincenzini, M. R. Tredici, "Production of eicosapentaenoic acid by Nannochloropsis sp. cultures in outdoor tubular photobioreactors," Journal of Biotechnology, vol. 70, pp. 299-312, 1999.

85. Spolaore-Robinson, T. M. Samocha, J. M. Fox, R. L. Gandy, D. A. McKee, "The use of inert artificial commercial food sources as replacements of traditional live food items in the culture of larval shrimp, Farfantepenaeus aztecus," Aquaculture, vol. 245, pp. 135-147, 2005.

86. Seguineau, A. Laschi-Loquerie, J. Moral, J.-F. Samain, "Vitamin requirements in great scallop larvae," Aquaculture International, vol. 4, pp. 315-324, 1996.

87. R. Brown, S. M. Barret, J. K. Volkman, S. P. Nearhos, J. Nell, G. L. Allan, "Biochemical composition of new yeasts and bacteria evaluated as food for bivalve aquaculture," Aquaculture, vol. 143, pp. 341-360, 1996.

88. Langdon, E. T. Bolton, "A microparticulate diet for suspension-feeding bivalve mollusc, Crassostrea virginica (Gmelin)," Journal of Experimental Marine Biology and Ecology, vol. 89, pp. 239-258, 1984.

89. Coutteau, M. Dravers, P. Leger, P. Sorgeloos, "Manipulated yeast diets and dried algae as a partial substitute for live algae in the juvenile rearing of the Manila clam Tapes philippinarum and the Pacific oyster Crassostrea gigas," Special Publication of the European Aquatic Society Ghent, Belgium, vol. 18, pp. 523-531, 1993.

90. J. A. Nell, W. A. O'Connor, "The evaluation of fresh algae and stored algal concentrates as a food source for Sydney rock oyster, Saccostrea commercialis (Iredale and Roughley) larvae," Aquaculture, vol. 99, pp. 277-284, 1991.

91. McCausland, M. R. Brown, S. M. Barrett, J. A. Diemar, M. P. Heasman, "Evaluation of live and pasted microalgae as supplementary food for juvenile Pacific oysters (Crassostrea gigas)," Aquaculture and Research, vol. 174, pp. 323-342, 1999.

92. Heasman, J. Diemar, W. O'Connor, T. Sushames, L. Foulkes, "Development of extended shelf-life micro-algae concentrate diets harvested by centrifugation for bivalve molluscs—a summary," Aquaculture and Research, vol. 31, pp. 637–659, 2000.

93. Muller-Feuga, R. Robert, C. Cahu, J. Robin, P. Divemach, "Use of microalgae in aquaculture." In: Støttrup, J. A., McEvoy, L. A., (Eds), Live Feeds in Marine Aquaculture. Blackwell, Oxford, 2003, pp. 253-299.

94. Montaini, G. C. Zittelli, M. R. Tredici, E. M. Grima, J. M. F. Sevilla, J. A. S. Perez, "Long-term preservation of Tetraselmis suecica: influence of storage on viability

and fatty acid profile," Aquaculture, vol. 134, pp. 81-90, 1995.

95. Csordas, J.-K. Wang, "An integrated photobioreactor and foam fractionation unit for the growth and harvest of Chaetoceros spp. in open systems," Aquacultural Engineering, vol. 30, pp. 15-30, 2004.

96. Millamena, E. J. Aujero, I. G. Borlongan, "Techniques on algae harvesting and preservation for use in culture as larval food," Aquaculture Engineering, vol. 9, pp. 295-304, 1990.

97. Poelman, N. de Pauw, B. Jeurissen, "Potential of electrolytic flocculation for recovery of micro-algae," Resources, Conservation and Recycling, vol. 19, pp. 1-10, 1997.

98. Rossingol, L. Vandanjon, P. Jaouen, F. Quéméneur, "Membrane technology for the continuous separation microalgae/culture medium: compared performances of cross-flow microfiltration and ultrafiltration," Aquaculture Engineering, vol. 20, pp. 191-208, 1999.

99. Sandbank, "The utilization of microalgae as feed for fish," Ergeb Limnology, vol. 11, pp. 108-120, 1978.

100. M. Knuckey, M. R. Brown, R. Robert, D. M. F. Frampton, "Production of microalgal concentrates by flocculation and their assessment as aquaculture feeds," Aquacultural Engineering, vol. 35, pp. 300-313, 2006.

101. J. O. M. Karlson, M. Toner, "Long-term storage of tissues by cryopreservation: critical issues," Biomaterials, vol. 17, pp. 243-256, 1996.

102. Mazur, "Freezing of living cells: mechanisms and implications," American Journal of Physiology, vol. 247, pp. 125-142, 1984.

103. M. Fahy, "The relevance of cryoprotectant 'toxicity' to cryobiology," Cryobiology, vol. 23, pp. 1-13, 1986.

104. M. Fahy, T. H. Lilley, H. Lindsell, M. J. Douglas, H. T. Meryman, "Cryoprotectant toxicity and cryoprotectant toxicity reduction: in search of molecular mechanisms," Cryobiology, vol. 27, pp. 247- 268, 1990.

105. Taylor, R. L. Fletcher, "Cryopreservation of eukaryotic algae—a review of methodologies," Journal of Applied Phycology, vol. 10, pp. 481–501, 1999.

106. W. Barclay, S. Zeller, "Nutritional enhancement of n-3 and n-6 fatty acids in rotifers and Artemia nauplii by feeding spray-dried Schizochytrium sp.," Journal of the World Aquaculture Society, vol. 27, pp. 314-322, 1996

107. Langdon, E. Önal, "Replacement of living microalgae with spray-dried diets for the marine mussel Mytilus galloprovincialis," Aquaculture, vol. 180, pp. 283-294, 1999

108. Dunstan, J. K. Volkman, S. W. Jeffrey, S. M. Barret, "Biochemical composition of microalgae from the green algal classes Chlorophyceae and Prasinophyceae," Journal of Experimental Marine Biology and Ecology, vol. 161, pp. 115-134, 1992.

109. J. Dhont, G. van Stappen, Live Feeds in Marine Aquaculture, Blackwell Science, pp. 65- 121. 2003.

110. Elston, "Mollusc diseases, guide for the shellfish farmer," Washington Sea Grant Program, Seattle, WA, p. 73, 1990.

111. Douillet, C. J. Langdon, "Effects of marine bacteria on the culture of axenic oyster Crassostrea gigas (Thunberg) larvae," Biological Bulletin, vol. 184, pp. 36-51, 1993.

112. Douillet, C. J. Langdon, "Use of a probiotic for the culture of larvae of the Pacific oyster (Crassostrea gigas Thunberg)," Aquaculture, vol. 119, pp. 25-40, 1994.

113. J. Donaldson, Proceedings of US-Asia Workshop, Honolulu, HA, January 28-31, 1991, The Oceanic Institute, HA, pp. 229-236.

114. Zhou, W. Liu, W. Qu, C. K. Tseng, "Application of Spirulina mixed feed in the breeding of bay scallop,"Bioresource Technology, vol. 38, pp. 229-232, 1991.

115. P. D. Nichols, D. G. Holdsworth, J. K. Volkman, M. Daintith, S. Allanson, "High incorporation of essential fatty acids by the rotifer Brachionus plicatilis fed on the prymnesiophyte alga Pavlova lutheri," Australian Journal of Marine and Freshwater Research, vol. 40, pp. 645-655, 1989.

116. Gara, R. J. Shields, L. McEvoy, "Feeding strategies to achieve correct metamorphosis of Atlantic halibut, Hippoglossus hippoglossus L., using enriched Artemia," Aquaculture Research, vol. 29, pp. 935-948, 1998.

117. Rodríguez, J. A. Pérez, P. Badía, M. S. Izquierdo, H. Fernández-Palacios, A. Lorenzo Hernández, "The n-3 highly unsaturated fatty acid requirements of gilthead seabream (Sparus aurata L.) larvae when using an appropriate DHA/EPA ratio in the diet," Aquaculture, vol. 169, pp. 9-23, 1998.

118. M. R. Brown, S. Skabo, B. Wilkinson, "The enrichment and retention of ascorbic acid in rotifers fed with microalgal diets," Aquaculture Nutrition, vol. 4, pp. 151-156, 1998.

119. Merchie, P. Lavens, P. Dhert, M. Dehasque, H. Nelis, A. de Leenheer, P. Sorgeloos, "Variation of ascorbic acid content in different live food organisms," Aquaculture, Nvol. 134, pp. 325-337, 1995.

120. Rønnestad, S. Helland, Ø. Lie, "Feeding Artemia to larvae of Atlantic halibut (Hippoglossus hippoglossus L.) results in lower larval vitamin A content compared with feeding copepods," Aquaculture, vol. 165, pp. 159-164, 1998.

121. S. Tamaru, R. Murashige, C.-S. Lee, "The paradox of using background phytoplankton during the larval culture of striped mullet, Mugil cephalus L.," Aquaculture, vol. 119, pp. 167-174, 1994.

CHAPTER 2

MICROBIAL PROTEIN: FUTURE SUSTAINABLE FOOD SUPPLY ROUTE WITH LOW ENVIRONMENTAL FOOTPRINT

Silvio Matassa, Nico Boon, Ilje Pikaar and Willy Verstraete

INTRODUCTION

From the times when our ancestors decided to settle, growing crops and domesticating animals became consolidated practices allowing constant feed and food production. As human civilization proceeded, new strategies of securing food supply have continuously been discovered, consolidated and improved. The major driver of such process was the need to provide resilience towards the changing elements of nature, continuously threatening food supply (Berglund, **2003**).

The current anthropogenic pressure on earth's finite resources and the concomitant dynamics of climate change, generate serious concerns about the resilience of the contemporary agricultural feed/food chains (Godfray *et al.*, **2010**). In view of the still growing world population towards 10 billion in 2050 (Ezeh *et al.*, **2012**), it has been calculated that the world will need to produce about 70% more food calories than in 2006 (Ranganathan, **2013**). Therefore, there is a need to find reliable alternative solutions, able to strengthen future food security while minimizing the impact on the global sustainability.

Citation: Silvio Matassa, Nico Boon, Ilje Pikaar and Willy Verstraete, Microbial protein: future sustainable food supply route with low environmental footprint, doi:10.1111/1751-7915.12369

Microorganisms have always been central in basic food processing techniques, for instance converting fibres into edible food when fermenting dough to produce bread, or milk into cheese, allowing its long-term preservation (Caplice and Fitzgerald, **1999**). They have been often used as direct food source, as it is the case for yeast or algae. The latter, together with bacteria, constitute the microbial actors involved in processing food. They can also be used directly as feed or food source (Anupama and Ravindra, **2000**). The term 'microbe' is used here in the broad connotation of bacteria, fungi, yeast and algae.

In the early 1960s, when public awareness grew in respect to the impeding global demographic boom, the need to search for alternatives to sustainably feed a growing population corresponded in major efforts to develop alternative feed and food sources (Goldberg, **1985**). Several attempts were made to develop and bring to practice the production of high-quality protein additives from microorganisms, known as microbial protein (MP), or single cell protein (SCP), mainly by using abundant and low cost hydrocarbon substrates such as methanol and methane (Goldberg, **1985**). The Imperial Chemical Industries (ICI) were the first to bring to full scale production and commercialization a MP product called Pruteen®, produced from methanol oxidation by means of*Methylophilus methylotrphus* (Westlake, **1986**). Besides industrially developed hydrocarbon-based MP, researchers investigated a whole range of other possibilities to produce MP, including the use of natural or artificial light, molecular hydrogen and many different organic substrates such as by-products from the sugar industry as well as other food processing residues or even food wastes (Anupama and Ravindra, **2000**). Despite being well accepted and successful in many feed trials with livestock, the actual and definitive breakthrough of MP in the animal feed market was hampered by the low prices achieved by more conventional protein sources such as soybean and fishmeal in the late 1970s as well as the fairly underdeveloped state of fermentation technology. Concomitantly, the rising oil prices in the subsequent decades led to the end of the ICI enterprise because of the relatively high costs of MP production and the consequent competitive disadvantage towards other cheaper more 'natural' alternatives (Øverland *et al.*, **2010**).

In recent years, however, research and development around MP is regaining momentum both in the scientific and industrial domains. The steep increase in the prices of fishmeal (from about $500 per ton in the 1990s to $1500 to $2500 in recent years), together with the environmental pressure of soybean production on land and water use in the tropical areas of the globe justify the re-examination of the microbial alternative (Kupferschmidt, **2015**).

In the present article, we align the possibilities offered as well as the challenges to be faced by the use of MP production as a biotechnological tool to help securing nutritive protein supply in the years to come. Threatened by forthcoming population growth, climate change and agricultural unsustainability, mankind must seek, once more, new forms of adaptation to safeguard itself.

MICROBIAL PROTEIN: FEED, FOOD AND FURTHER

MP as feed

The main driver leading to the renaissance of MP as a source of feed is indubitably the aquaculture sector. Fish farming currently provides about 50% of world's fish food supply, and it is projected to grow further, becoming a key sector in the supply of high-quality protein for the global population. In this context, scientific research and the industrial applications have found in MP a powerful ally. Aquaculture accounts nowadays for more than 73% of the global fishmeal consumption, with wild fish capture clearly unable to provide enough high-quality feed for such a fast growing sector (The World Bank, **2013**). Production of MP from natural gas has recently received a great deal of attention, with innovative fermentation processes allowing high volumetric productivities (3–4 kg MP dry matter (DM) per m^3 reactor volume per hour) by continuous cultures of *Methylococcus capsulatus*, marketed under the name of FeedKind® (Unibio, **2016**). The latter level of productivity has a physical footprint which is a factor 1000, or more, smaller than any conventional vegetable protein production system (Matassa *et al.*, **2015a**). Besides achieving feasible industrial scale production and costs competitiveness with fishmeal, the final MP product is comparable to fishmeal in terms of essential amino acid profile and overall nutritive value (Øverland *et al.*, **2010**). Being tested in numerous feed trials with different fish species, resulting in promising perspectives, full-scale production is currently ongoing, with a production of up to 80 000 ton DM/year foreseen in the near future (see Table 2).

In addition to aquaculture, the MP product has also been successfully tested in feed trials with terrestrial animals including major livestock like ruminants, pigs and chickens, broadening its potential market applications (Øverland *et al.*, **2010**). In this case though, the relatively low price of soybean meal and the abundant and well-established use as main protein additive in livestock production of the latter, still counteract the application of natural gas-based MP as replacement of substantial percentages of feed composed by fishmeal.

An alternative route to produce MP consists of recovering valuable nutrients from various side streams of the food industry, for instance feed and food processing water (Lee *et al.*, **2015**). In this case, the use of heterotrophic microorganisms such as yeast and bacteria allows to convert the organic carbon and the nutrients (N, P) in the waste or processing waters into MP (Anupama and Ravindra, **2000**). Microbial protein produced along this line might constitute a valuable and competitive route to produce a substitute for soy protein for animal feed. Indeed it should be possible to generate such MP at costs which take into account the revenue from the avoidance of the treatment of the mineral nutrients (N, P) present in side (waste) streams. As a matter of fact, dissipation of reactive nitrogen back to atmosphere as dinitrogen gas by means of the conventional nitrification–denitrification pathway comes to a cost of about 2–3 Euro per kg nitrogen-N, while capture of phosphorous in the wastewater line costs of the order of 7 Euro per kg P (Levlin and Hultman, **2003**). Note that the market value of proteinaceous nitrogen from vegetable sources is at the current price of some 1.1–1.6 Euro per kg dry weight protein (see Table 1) corresponding to some 6 Euro per kg proteinaceous N. When this microbial proteinaceous N is converted to high-value animal protein, one can attain an end-value of the same order and even up to 14 Euro per kg dry weight protein, in case of fish.

Table 1. Production volumes and price of various animal and vegetable protein sources

Protein source	Production volume (Mton DM/y)	Farm gate price ($/kg DM)	Average protein content (% DW)	Price per unit protein ($/kg protein DM)	Ref
Animal					
Fish	66.7	2.07	15–20	10–14	Waite *et al.* (**2014**)
Pork	108.5	1.54	20	7.7	Waite *et al.* (**2014**)
Chicken	92.7	1.43	31	4.6	Waite *et al.* (**2014**)
Beef	62.7	2.70	25	10.8	Waite *et al.* (**2014**)
Vegetable					
Soybean	320.2	0.37	35	1.1	Indexmundi (**2016a**); USDA (**2015**)
Wheat	712.7	0.19	12	1.6	FAO, **2015**; Indexmundi (**2016b**)

An example of the implementation of such approach to food processing water is the study published by Lee *et al.* (**2015**), where the effluent of a brewery is used as feedstock for the production of SCP-MP. The latter study relates to a technology implemented by Nutrinsic, dealing with a production volume of 5000 ton DM/year from a brewery effluent (see Table 2). Also, at present in Belgium, a first full-scale MP production installation is under construction dealing with the upgrading of potato process waters. It should be in production by 2016 at the level of 5000 ton MP per year (Valpromic NV, pers. comm.). Yet, for the latter bacterial-based MP products, there are so far no clear cut data in terms of their putative market size or market values. Nevertheless, the sector is attracting growing interests from investors dealing with novel aspects of the cyclic economy (Nutrinsic, **2014**).

Table 2. Overview of current production volumes and market sizes for different microbial protein. Hyphens indicate that values were not available

Organisms	Production volume (ton DM/y)	Production costs (Euro/kg DM)	Global market value (Billion Euro)	Yearly growth (% per year)	Remarks	Ref
Yeast	3 000 000	–	9.2	7.9	Mostly commercialized as baker's yeast and for ethanol fermentation. Global market value projected to 2019	Kellershohn and Russell (**2015**)
Algae (microalgae)	9000	4–25	2.4	10	Besides feed and food, derivatives are also used	Enzing *et al.* (**2014**)
Mycoprotein (Quorn®)	25 000	–	0.214	20	Investments for a plant of 22000 tons per year were done in 2015	Beer (**2015**)
Bacteria (Profloc®)	5000	1–1.1	–	–		Nutrinsic (**2015**)
Bacteria (FeedKind®)	80 000	–	–	–	Commercial production foreseen on 2016	Byrne (**2016**)
Valpromic	5000	–	–	–		Personal communication

MP as food

Microbial protein is an alternative source of high-quality protein able to replace animal protein like fishmeal in livestock nutrition and aquaculture. Going one step higher in the food chain, MP is meeting the FAO/WHO requirements in terms of essential amino acid scoring pattern for human nutrition (Fig. 1) and therefore, also humans could benefit greatly from the use of MP directly as food.

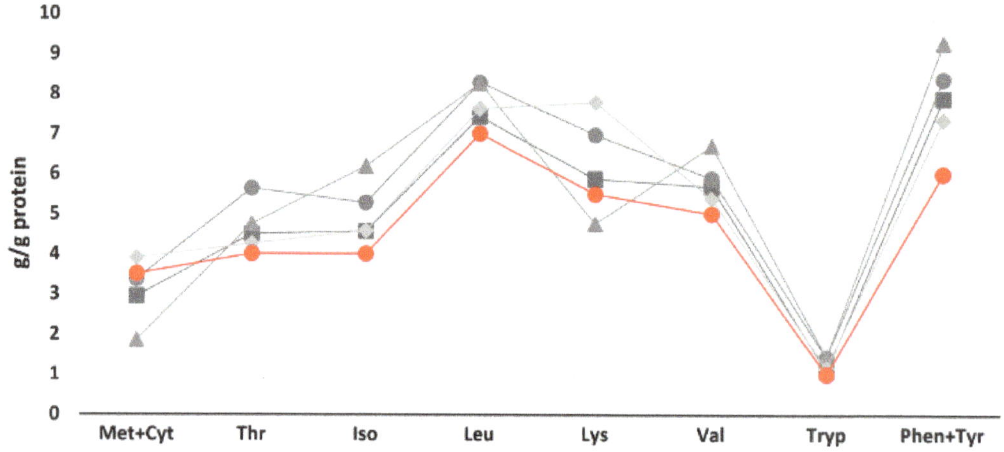

Figure 1. Essential amino acid scoring pattern of microbial protein from bacteria (Pseudomonas/Methylophilus spp.) (-■-), yeast (Candida spp.) (-●-), algae (Spirulina maxima (▲), compared with the high-quality animal protein from fishmeal (◆) as well as to the FAO/WHO standard (-●-) for amino acid scoring pattern for human nutrition.

Algae are reported to have supported the life of ancient populations living close to the sea for millennia, providing a constant source of protein and vitamins. Algae and microalgae are currently used as food and food supplements in food industry (Anupama and Ravindra, **2000**; Becker, **2007**), with a global production achieving 9000 ton DM/year (see Table 2) with a market value estimated about 2.4 billion Euro with a projected yearly growth of 10%.

Yeast, being at the base of food processing since the first bread was baked, or grapes fermented, can also be used as direct food source, as it was the case e.g. with the massive campaign of yeast production and supply, first to the army, then to the whole population during World War II (Khachatourians and Arora, **2002**). Currently, yeast is a major player in the microbial derived production of products for food as well as for other applications. Baker's yeast and alcohols fermentation are the two main processes employing yeast, with a projected global market value for 2019 of up to 9.2 billion Euro and an annual growth forecast of 7.9% (see Table 2).

Fungi are also a suitable alternative and have also made their way as human food. Quorn™ is the most successful example of the so called mycoprotein, which is commercialized and sold in some 15 countries worldwide (Wiebe, 2004). Mycoproteins are particularly suited to reproduce the taste and consistency of meat; this explains their success as alternative to conventional animal-based products. Currently, mycoprotein production supporting Quorn™ products manufacturing amounts to 25 000 ton DM/year, with a global market value of about 214 million Euro, prospected to grow with 20% annually in the coming years.

Added Value Applications

Besides being rich in nutritive protein, microorganisms offer the possibility of producing a broad variety of added-value products, suitable for both animal and human nutrition (Vandamme and Revuelta, **2016**). Table 3 summarizes the average amount of protein producible by algae, fungi and bacteria, as well as other possible added-value products already investigated or produced from microorganisms.

Table 3. Overview of different microorganism for MP and added-value product formation

Microorganism	Average crude protein content (% CDW)	Nutritional value	Added value by-products (% CDW)	Remarks	Ref.
Algae	40–60	Compares favourably to egg, soy and what protein. Cell wall digestibility is an issue	• Microbial oil (50–70%) • Carbohydrates (up to 70%) • Vitamins • …	Triacylglycerides (TAG) can replace partly vegetable oils in food products. Poly unsaturated fatty acids (PUFA) are of interest for health applications	Draaisma*et al.* (**2013**); Harunet *al.* (**2010**)
Fungi (Filamentous and Yeast)	30–70	Amino acids and digestibility of mycoprotein is similar to egg and milk	• Carbohydrates • Pullulan • Xylitol • Astaxanthin • …	High unsaturated/saturated fatty acids and low fat content makes them highly suitable for human nutrition	Thrane (**2007**)

Bacteria	50–83	Amino acids and digestibility is similar to those of fishmeal	• Internal storage polymers (PHB) • Ectoine • Lipids • Extracellular polysaccharides • Growth media and vitamins • ...		Strong*et al.* (**2015**)

Certain microalgae and cyanobacteria are primary producers of microbial oil, suitable as substitutes for vegetable oil in food supplements. Particularly, the high concentration of fatty acids can replace fatty acids otherwise derived from rape seed, soy, sunflower oil and palm oil. The purification of omega-3 fatty acid can offer even higher value applications, e.g. for clinical purposes, eicosapentanoic acid and decosahexaenoic acid, normally obtained from fish oil, can be also concentrated and purified from naturally omega-3-accumulating microalgae. Vitamins such as vitamin B12 and provitamin A are also important high-value products obtainable from edible algae, conferring additional nutritional benefits in livestock production. Carbohydrates, which can be accumulated up to 70% of the cell dry weight by many algal species are also of nutritional value, but the major research effort so far was directed towards the use of algae for biofuel, biogas or biohydrogen generation (Jones and Mayfield, **2012**; Draaisma *et al*., **2013**).

Fungi, mainly yeast-like fungi are the main agents involved in the saccharification of fibres from corn, as well as fermentation of other organic substrates. While processing corn fibre with yeast like*Aureobasidium*, xylose, arabinose and glucose can be produced at different relative concentrations depending on the pre-treatment of the fibres feedstock. The sugars can then be further fermented in bioethanol, xylitol and pullulan. Xylitol and pullulan find particularly application as food additive for their specific property of flavour-enhancing and binding agents. Besides sugars and sugar-derivatives, yeasts like *Phaffa rhodozyma* (now Xanthophyllomyces rodochrous) can be used to produce valuable carotenoid pigments like astaxanthin, mainly used in aquaculture as feed supplement for salmon (Leathers, **2003**).

Bacteria are a versatile group of microorganisms able to produce a large array of added-value bio-products. Biopolymers such as polyhydroxyalkanes (PHA) are named to be biological alternatives to petroleum-based chemicals to produce plastics. Yet, so far large-scale applications are to the best of our knowledge, not industrially established. Recently, other applications for PHA/polyhydroxybuthyrate (PHB) such as in the medical field are emerging. Of interest for aquaculture is the ongoing research on the prebiotic effects of PHB when used as feed supplement, offering an interesting alternative to antibiotics (Defoirdt *et al*., **2007**; De Schryver *et al*., **2010**). Another interesting niche product which can be derived from bacteria is osmo-protectants such as glutamate and ectoine (Lentzen and Schwarz, **2006**). The latter is a high-value cyclic imino acid used in cosmetic formulation, but which has found application also in aquaculture as highly active protectant against oxidative stress. Bacteria can also produce relevant amounts of lipids, commonly employed in biofuel production. High-quality membrane-derived lipids can also be employed as human health supplement, being already tested as effective in reducing plasma cholesterol during animal tests (Strong *et al*., **2015**).

Forthcoming challenges

The extensive use of MP products as partial replacement of conventional protein feed additives such as soybean and fishmeal can offer the opportunity of decreasing part of the environmental pressure that these products exert on land and water use. A recent report of the British Carbon Trust evaluated the environmental impact of FeedKind® protein, a bacterial MP feed additive produced from natural gas (see *MP as feed*). The report evaluated two FeedKind® commercial products in terms of greenhouse gas, land and water use, comparing them with soybean and fishmeal (Cumberlege *et al*., **2016**). In terms of freshwater consumption, the report shows an average value of about 29 m^3 per ton MP produced. A more detailed analysis shows that this 29 m^3 is for about 80% determined by the vegetable oil used as binding agent to produce a MP-pelletized product. If the latter major contribution is excluded by producing a simple straightforward protein powder, the actual freshwater requirement comes down to the order of 1 m^3 per ton MP. From Fig. 2, it can be derived that this value about water foot print is about 20 and 140 times lower than fishmeal and soybean meal respectively.

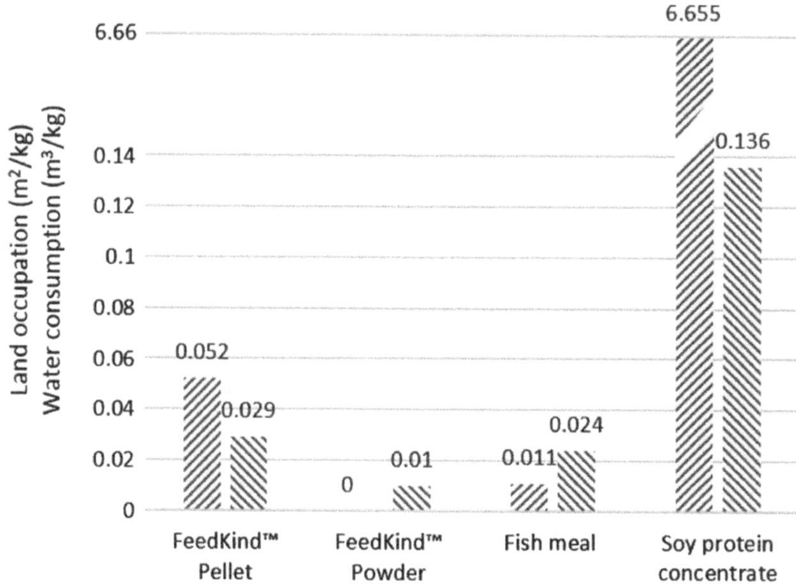

Figure 2. Land ( ) and freshwater ( ) requirements of MP compared with fishmeal and soy protein concentrate. The values are normalized to the protein content of each product.

The same trend is observable for the required land. The value of 52 m^2 per ton MP is in fact due to vegetable oil for the production of the pelletized form, whereas no arable land is required in case of the powdered MP. Compared with the 6655 m^2 land per ton protein required for the production of soybean meal concentrate, the value of quasi zero land foot print reveals how the land footprint of MP production is a major benefit in respect to conventional agricultural-based protein production. Fishmeal of course requires minimal amount of land for its processing, yet the dramatic impact of wild fish capture on ocean ecosystems is well known and documented (Pauly *et al.*, **2005**).

Finally, the above-mentioned report also analyses the carbon footprint of FeedKind®. The value of 5.8 ton CO_{2eq}/ton MP is mainly due to the natural gas necessary for the metabolism of the bacteria involved in the biological fermentation process. This value would be as low as 1.7 ton CO_{2eq}/ton MP in case biogas and renewable energy is used in place of fossil fuels to power the reactor-based production and downstream processing of the final MP product. For fishmeal and soybean meal concentrate, the report indicates values of 2.6 and 0.8 ton CO_{2eq}/ton protein respectively. Nevertheless, if the spared agricultural land in case of MP production would be accounted for its recovered carbon capture potential, the overall benefit in avoided carbon emissions from MP production could possibly be much higher.

In this context, an interesting alternative to natural gas-based MP is represented by autotrophic microorganisms such as algae or hydrogen-oxidizing bacteria. If algae offer the great advantage of being able to use sun light to fix carbon dioxide into biomass, the main drawbacks of such process are the high land footprint required together with the technical challenges of downstream processing of poorly concentrated algal biomass (Majid *et al.*, **2014**). On the other hand, the hydrogen gas needed to fix carbon dioxide into bacterial biomass by means of hydrogen-oxidizing bacteria is more expensive resource, but the land footprint and the biomass concentrations achievable with modern fermentation technologies outscore those of the algal platform. In case the production of the latter is connected to hydrogen generated by means of renewable energies (solar, wind, etc.), this allows an elegant platform for MP production and concomitant carbon dioxide capture (Matassa *et al.*, **2015b**).

To the scientific community, it is evident how a dedicated industrial production of MP can represent a key biotechnological tool to curb down the environmental impact of the current feed and food chain assuring the necessary amounts of nutritive protein for mankind. Clearly, significant efforts are warranted to bring this to practice at relevant scales. A key feature is to deal with the aspects of public awareness. At present, the mere economic market rules justify the application of MP in feed for livestock only in some niche applications such as aquaculture. Yet, if the externalized environmental costs of the current feed/food production system would be taken into account and made clear to the broader public (including decision makers in political institutions), the MP route would result in a more rational alternative, able to offer immediate advantages in terms of water and land use, with direct consequences on increased carbon capture potential of ecosystems restored by better agricultural land use (Galloway and Leach, **2016**). An important aspect, in this sense, relates to nutrients flows, and principally the excessive input in our biosphere of reactive nitrogen species (NH_4^+, NO_2^-, NO_3^-) produced by fixing atmospheric N_2 gas by means of the Haber–Bosch process. Compared with the proposed sustainable boundary of 35 Mton N_2 fixed per year, the current 121 Mton actually converted into reactive nitrogen surpass the sustainability boundary of almost 3.5 times (Rockstrom *et al.*, **2009**). Moreover, if the economic benefit in agricultural production ranges between 20 and 80 billion Euro per year, the annual costs (including damages to both environment and human health) of N pollution by agriculture have been estimated in the range of 35–230 billion Euro per year (Van Grinsven *et al.*, **2013**). It has been recently demonstrated how the high nitrogen inefficiency of the soil–plant system could be mitigated by MP production (Matassa *et al.*, **2015a**), decreasing significantly the impact of eutrophication, nitrous oxide emissions and ecosystems disturbance due to unbalanced anthropogenic nitrogen inputs.

Besides the awareness of the overall environmental benefit of MP production for feed and food, the development of higher value by-products will allow boosting and bolstering the MP biotech platform. Thus, a more powerful penetration into the market of microbial-based product as replacement of chemically derived ones, as discussed above, will be possible. This will play in favour of establishing a public mindset more open and prone to acceptance towards microbial derived products.

Obviously, barriers must be overcome in order to allow a widespread adoption of the MP biotechnology. Besides the official legal recognition of some MP products as feed and food (Øverland *et al.*, **2010**), further openings are warranted in terms of used carbon and nutrient sources recovery and their up-cycling into edible MP products as part of the cyclic economy. This will impact drastically on how efficiently our current society makes use of its precious primary resources.

CONCLUSIONS

- Microbial protein qualifies as an excellent source of nutritive proteins, but other cellular components can also be of increasing importance, driving new developments for microbial-based by-products. In the context of the need to generate new biotechnological processes and products, this line of research and development, particularly in countries where new opportunities are warranted, should be explored with great care (Timmis *et al.*, **2014**).
- The current conventional agricultural based supply route for nutritive animal proteins have a high environmental impact; they should be reconsidered and their externalized environmental costs should be assessed and benchmarked to those of MP.
- Upgrading various valuable nutrients (N, P) and nutritive resources (organic carbon), the production of heterotrophic microbes as food and feed certainly has gaining renewed industrial interest, particularly in the context of the cyclic economy.
- The overall transition of the public mindset towards widespread acceptance and appreciation of MP as main supply route for feed and food needs to be prepared in the near future with care and foresight particularly in terms of quality and regulatory issues.

REFERENCES

1. Anupama, and Ravindra, P. (2000) Value-added food: single cell protein. Biotechnol Adv 18: 459–479.
2. Becker, E.W. (2007) Micro-algae as a source of protein. Biotechnol Adv 25: 207–210.
3. Beer, E. (2015) "No success model" for new protein ingredients [WWW Document]. URL http://www.foodnavigator.com/Market-Trends/No-success-model-for-new-protein-ingredients
4. Berglund, E. (2003) Human impact and climate changes—synchronous events and a causal link? Quat. Int. 105: 7–12.
5. Byrne, J. (2016) Calysta says gas to fishmeal replacement protein on path to commercialization [WWW Document]. URL http://www.feednavigator.com/R-D/Calysta-says-gas-to-fishmeal-replacement-protein-on-path-to-commercialization
6. Caplice, E., and Fitzgerald, G.F. (1999) Food fermentations: role of microorganisms in food production and preservation. Int J Food Microbiol 50: 131–149. doi:10.1016/S0168-1605(99)00082-3.
7. Cumberlege, T., Blenkinsopp, T. and Clark, J. (2016) Assessment of environmental impact of FeedKind protein Carbon Trust, URL https://www.carbontrust.com/media/672719/calysta-feedkind.pdf.
8. De Schryver, P., Sinha, A.K., Kunwar, P.S., Baruah, K., Verstraete, W., Boon, N., et al. (2010) Poly-beta-hydroxybutyrate (PHB) increases growth performance and intestinal bacterial range-weighted richness in juvenile European sea bass, Dicentrarchus labrax. Appl Microbiol Biotechnol 86: 1535–1541. doi:10.1007/s00253-009-2414-9.
9. Defoirdt, T., Boon, N., Sorgeloos, P., Verstraete, W., and Bossier, P. (2007) Alternatives to antibiotics to control bacterial infections: luminescent vibriosis in aquaculture as an example. Trends Biotechnol 25: 472–479. doi:10.1016/j.tibtech.2007.08.001.
10. Draaisma, R.B., Wijffels, R.H., Slegers, P.M.E., Brentner, L.B., Roy, A., and Barbosa, M.J. (2013) Food commodities from microalgae. Curr Opin Biotechnol 24: 169–177. doi:10.1016/j.copbio.2012.09.012.
11. Enzing, C., Ploeg, M., Barbosa, M. and Sijtsma, L. (2014) Microalgae-based products for the food and feed sector: an outlook for Europe. doi:10.2791/3339.
12. Ezeh, A.C., Bongaarts, J., and Mberu, B. (2012) Global population trends and policy options. Lancet 380: 142–148. doi:10.1016/S0140-6736(12)60696-5.
13. FAO (2015) World food situation [WWW Document]. URL http://www.fao.org/worldfoodsituation/csdb/en/
14. Galloway, J.N., and Leach, A.M. (2016) Sustainability: your feet's too big. Nat Geosci 9: 97–98.
15. Godfray, H.C.J., Beddington, J.R., Crute, I.R., Haddad, L., Lawrence, D., Muir, J.F., et al. (2010) Food security: the challenge of feeding 9 billion people. Science 327: 812–818.
16. Goldberg, I. (1985) Single Cell Protein, Biotechnology Monographs. Springer-Verlag, Berlin.
17. Harper, A. (1981) Amino Acid Scoring Patterns [WWW Document]. URL http://www.fao.org/3/contents/aa7e1ca5-4634-51bf-a465-5bf12d5cec2d/M3013E00.HTM
18. Harun, R., Singh, M., Forde, G.M., and Danquah, M.K. (2010) Bioprocess engineering of microalgae to produce a variety of consumer products. Renew Sustain Energy Rev 14: 1037–1047. doi:10.1016/j.rser.2009.11.004.
19. Indexmundi (2016a) Soybeans [WWW Document]. URL http://www.indexmundi.com/commodities/?commodity=soybeans.
20. Indexmundi (2016b) Wheat [WWW Document]. URL http://www.indexmundi.com/commodities/?commodity=wheat.

21. Jones, C.S., and Mayfield, S.P. (2012) Algae biofuels: versatility for the future of bioenergy. Curr Opin Biotechnol 23: 346–351. doi: 10.1016/j.copbio.2011.10.013.

22. Kellershohn, J. and Russell, I. (2015) Yeast biotechnology. In Advances in Food Biotechnology. John Wiley & Sons, pp. 303–310. doi:10.1002/9781118864463.ch18.

23. Khachatourians, G.G. and Arora, D.K. (2002) Applied Mycology and Biotechnology, Volume 2. Amsterdam: Agriculture and Food Production.

24. Kupferschmidt, K. (2015) Why insects could be the ideal animal feed [WWW Document]. Science. doi:10.1126/science.aad4709.

25. Leathers, T.D. (2003) Bioconversions of maize residues to value-added coproducts using yeast-like fungi. FEMS Yeast Res 3: 133–140. doi:10.1016/S1567-1356(03)00003-5.

26. Lee, J.Z., Logan, A., Terry, S., and Spear, J.R. (2015) Microbial response to single-cell protein production and brewery wastewater treatment. Microb Biotechnol 8: 65–76. doi:10.1111/1751-7915.12128.

27. Lentzen, G., and Schwarz, T. (2006) Extremolytes: natural compounds from extremophiles for versatile applications. Appl Microbiol Biotechnol 72: 623–634. doi:10.1007/s00253-006-0553-9.

28. Levlin, E. and Hultman, B. (2003) Phosphorus recovery from phosphate rich side-streams in wastewater treatment plants. In Polish Swedish Seminar, Gdansk March. pp. 47–56.

29. Majid, M., Shafqat, S., Inam, H., Hashmi, U. and Kazi, A.G. (2014) Biomass and Bioenergy: Processing and Properties. Hakeem, R.K., Jawaid, M. and Rashid, U. (eds). Cham, Switzerland: Springer International Publishing, pp. 207–224. doi:10.1007/978-3-319-07641-6_13.

30. Matassa, S., Batstone, D.J., Huelsen, T., Schnoor, J.L., and Verstraete, W. (2015a) Can direct conversion of used nitrogen to new feed and protein help feed the world? Environ Sci Technol 49: 5247–5254. doi:10.1021/es505432w.

31. Matassa, S., Boon, N., and Verstraete, W. (2015b) Resource recovery from used water: the manufacturing abilities of hydrogen-oxidizing bacteria. Water Res 68: 467–478. doi:10.1016/j.watres.2014.10.028.

32. Nutrinsic (2014) Nutrinsic Corporation Raises $12 [WWW Document]. URL http://www.marketwired.com/press-release/nutrinsic-corporation-raises-127-million-to-fuel-growth-1938044.htm

33. Nutrinsic (2015) Nutrinsic Announces Grand Opening of First US ProFloc(TM) Facility. [WWW Document]. URL http://www.marketwired.com/press-release/-2020797.htm

34. Øverland, M., Tauson, A.-H., Shearer, K., and Skrede, A. (2010) Evaluation of methane-utilising bacteria products as feed ingredients for monogastric animals. Arch. Anim. Nutr. 64: 171–189. doi:10.1080/17450391003691534.

35. Pauly, D., Watson, R., and Alder, J. (2005) Global trends in world fisheries: impacts on marine ecosystems and food security. Philos Trans R Soc Lond B Biol Sci 360: 5–12.

36. Ranganathan, J. (2013) The Global Food Challenge Explained in 18 Graphics [WWW Document]. World Resour. Inst. URL http://www.wri.org/blog/2013/12/global-food-challenge-explained-18-graphics

37. Rockstrom, J., Steffen, W., Noone, K., Persson, A., Chapin, F.S., Lambin, E.F., et al. (2009) A safe operating space for humanity. Nature 461: 472–475.

38. Strong, P.J., Xie, S., and Clarke, W.P. (2015) Methane as a resource: can the methanotrophs add value? Environ Sci Technol 49: 4001–4018. doi:10.1021/es504242n.

39. Tacon, G.J.A. (1987) The nutrition and feeding of farmed fish and shrimp - a training manual 2. Nutrient sources and composition. FAO. URL http://www.fao.org/3/contents/d66b3e1f-c059-50fa-9ba2-717e9940b7f1/AB470E00.htm

40. The World Bank (2013) FISH TO 2030 Prospects for Fisheries and Aquaculture. doi:83177-GLB.

41. Thrane, U. (2007) Fungal protein for food. In Food Mycology, Mycology. CRC Press, pp. 353–360. doi:10.1201/9781420020984.ch18.

42. Timmis, K., de Lorenzo, V., Verstraete, W., Garcia, J.L., Ramos, J.L., Santos, H., et al. (2014) Pipelines for New Chemicals: a strategy to create new value chains and stimulate innovation-based economic revival in Southern European countries. Environ Microbiol 16: 9–18. doi:10.1111/1462-2920.12337.

43. Unibio (2016) Introduction _ Unibio [WWW Document]. URL http://www.unibio.dk/technology/introduction.

44. USDA (2015) World agricultural supply and demand estimates. United States Dep. Agric. doi:WASDE-525.

45. Van Grinsven, H.J.M., Holland, M., Jacobsen, B.H., Klimont, Z., Sutton, M., and Willems, W.J. (2013) Costs and benefits of nitrogen for europe and implications for mitigation. Environ Sci Technol 47: 3571–3579.

46. Vandamme, E.J. and Revuelta, J.L. (eds.) (2016) Industrial Biotechnology of Vitamins, Biopigments, and Antioxidants. Weinheim, Germany: Wiley-VCH, pp. 560.

47. Waite, R., Beveridge, M., Brummett, R.E., Castine, S., Chaiyawannakarn, N., Kaushik, S., et al. (2014) Improving productivity and environmental performance of aquaculture. pp. 1–60. doi:10.5657/FAS.2014.0001.

48. Westlake, R. (1986) Large-scale continuous production of single cell protein. Chemie Ing. Tech. 58: 934–937. doi:10.1002/cite.330581203.

49. Wiebe, M.G. (2004) QuornTM Myco-protein - Overview of a successful fungal product. Mycologist 18: 17–20. doi:10.1017/S0269915X04001089.

CHAPTER 3

BIOMOLECULES FROM MICROALGAE FOR ANIMAL FEED AND AQUACULTURE

Zahira Yaakob, Ehsan Ali, Afifi Zainal, Masita Mohamad and Mohd Sobri Takriff

Keywords: Microalgae Carotenoid Biomolecules Nutraceutical Aquaculture Animal feed

INTRODUCTION

Nutritional composition of algae is well documented and it mainly contains proteins, carbohydrates, lipids and trace nutrients, including vitamins, antioxidants, and trace elements. These all algal components have the characteristics to be a natural supplement in human and animal feed to replace synthetic components or meet the increasing demands of such ingredients. Increased consumption of animal proteins is creating a towering demand of high quality fish and animal feed to generate healthy proteins for humans. Fish meal is the protein source traditionally used in aquaculture diets but it is considered as limited and expensive. The nutritional contents of algae are rapidly gaining importance as a renewable source to substitute the conventional ingredients in the aquaculture/animal feed. Algae is defined as heterogeneous assemblage of organisms that range in size from tiny cells to giant seaweeds, thus algae mostly are photosynthetic species which includes eukaryotes and prokaryotic Cyanobacteria (blue-green algae) and live in aquatic habitats. Algae can be differentiated (by the body size and structure) either to microalgae (with algal bodies that need microscope to be observed) or to macroalgae (large enough to be seen with unaided eye) [1]. Algae culturing has been adopted since the late of 1800s and early 1900s, while microalgae mass culturing begin to develop since the achievement of Allen and Nelson in 1910 by cultivating *Chlorella* sp. for aquaculture purposes in Berlin, Germany [2]. To date, it seems that microalgae have the potential to play pivotal roles in order to remedy the energy, environment and food crisis prevailing in the world. This review encompasses the roles of algae in biomolecules production including astaxanthin, lutein, beta carotene, chlorophyll, phycobiliprotein, Polyunsaturated Fatty Acids (PUFAs), beta-1,3-glucan, and pharmaceutical and nutraceutical compounds. This review also emphasizes on the utilization of these biomolecules in livestock, poultry and aquaculture (and especially shrimp farming) feed.

REVIEW

Production of Value Added Products from Algae

The global carotenoids market is projected to reach US$1.2 billion by the year 2015, because of the rising consumer awareness about health benefits offered by various carotenoids, and the shift towards healthy and natural food products (Global Industry Analysts, Inchttp://www.strategyr.com/Carotenoids_Market_Report.asp/). According to a recently published report, global carotenoids market tends to reach US$1.3 billion by 2017 (http://www.prweb.com/pdfdownload/8849957.pdf).

Citation: Zahira Yaakob, Ehsan Ali, Afifi Zainal, Masita Mohamad and Mohd Sobri Takriff, biomolecules from microalgae for animal feed and aquaculture, DOI: 10.1186/2241-5793-21-6

The nutraceuticals boom has also integrated carotenoid mainly on the claim of their proven antioxidant properties; carotenoids are usually considered important for industrial use in food products, cosmetics as vitamin supplements, health food products, and as feed additives for poultry, livestock, fish, and crustaceans [3]. In contrast, carotenoid in fish oil comes from microalgae, therefore it is logical to cultivate microalgae for carotenoids production [4]. Del Campo *et al.* [3] have reported that structure of more than 600 different carotenoids have been derived from a 40-carbon polyene chain. This paper will focus on carotenoids such as astaxanthin, lutein, beta-carotene and other high-valued molecules which are produced by microalgae (or in some case by macro algae) and Cyanobacteria and can be industrialized. Moreover, microalgae play a key role in high grade animal nutrition food from aquaculture to farm animals. Comprehensive nutritional and toxicological evaluations have demonstrated suitability of algal biomass as a valuable feed supplement or substitute for conventional animal feed sources [5]. Therefore, the usage of microalgae biomass for feed productions will be discussed in this review which might be helpful to encourage the trends of mass cultivation of microalgae for commercial and beneficial purposes.

Astaxanthin

Astaxanthin, a carotenoid equipped with two asymmetric carbon located at the 3 and 3′ position of the benzenoid rings on either end of the molecule, has been proposed as super vitamin E due to its natural antioxidant activities [6, 7, 8]. As a natural pigment, the major market for astaxanthin is as pigmentation source in aquaculture [9]. Natural astaxanthin can be produced by *Haematococcus* microalgae, *Chlorella zofingiensis*, *Chlorococcum* sp. and the yeast *Xanthophyllomyces dendrorhous* [10, 11, 12, 13]. Studies regarding effects on lipid metabolism and antioxidant defense mechanisms revealed that astaxanthin is responsible for lowering the plasma lipid concentrations and enhancing antioxidant defense in apolipoprotein E knockout mice [14]. The increasing growth of salmonid farming has created an enormous demand for natural astaxanthin as pigment. Out of total carotenoid (astaxanthin, cantaxanthin), the astaxanthin was determined as the most important carotenoid in salmons and rainbow trouts [15]. Salmonid are unable to synthesize astaxanthin *de novo*, therefore carotenoid pigments must be supplied in their artificial aquaculture diet [16]. The carotenoid pigment astaxanthin has important applications and is routinely being used in the nutraceutical, cosmetics, food and feed industries [17]. In aquaculture feed, it has been used to culture salmon, shrimp, ornamental fish and sea bream. A recent report showed that*Haematococcus pluvialis* especially in dose of 3 g kg^{-1} feed administration may effectively enhance the antioxidant system and some biochemical parameters in rainbow trout [18]. Astaxanthin is not responsible only to provide pigmentation but also has been attributed with extraordinary potential for protecting the organism against a wide range of diseases, and has considerable potential and promising applications in human health [13]. Natural astaxanthin was also found effective to treat the patients with functional dyspepsia, caused by *Helicobacter pylori* [7]. UV radiation from sunlight is known as the most potent environmental risk factor in skin cancer pathogenesis, astaxanthin has been demonstrated as a protective tool against UVA-induced DNA alterations in human skin fibroblasts (1BR-3) [19].

Epidemiological studies have demonstrated a correlation between increased carotenoid intake and reduced incidence of coronary heart disease, certain cancer types, macular degeneration, and increased resistance to viral, bacterial, fungal and parasitic infection [20]. The price for astaxanthin varies in the market and depends on the percentage contents in algal source; however the overall maintained price of 5% astaxanthin is about US$1900 kg^{-1} (http://www.herbridge.com/bencandy.php?fid-55-id-23446-page-1.htm).

Lutein

Lutein is another important carotenoid which can be found naturally in algae. It is prominent in human serum and foods [21]. Lutein has been used for the pigmentation of animal tissues and products, and it is considered important for the natural coloration of foods, drugs and cosmetics. It has been documented that it plays an active role in delaying chronic diseases, stimulating the immune response, hampering the development of cataracts and the progression of early atherosclerosis, and preventing blindness or decrease in vision caused by age-related macular degeneration [22, 23, 24]. Global lutein market in 2004 accounted for US$139 million and was considered as the fastest projected growth in individual carotenoids sales [3]. The recent market value of lutein is~US$233 million (2010), and is expected to reach US$309 million by 2018 with a compound annual growth rate of 3.6% (http://www.companiesandmarkets.com/Market/Food-and-Drink/Market-Research/The-Global-Market-for-Carotenoids/RPT988273). French marigold (*Tagetes patula*) is currently the most widely applied source for lutein production in the world. In the USA, two lutein containing products, Aztec Marigold and Tagetes have been commercialized. Lutein is usually produced from three different marigold flowers (*Tagetes patula* L.) (marigold orange, marigold yellow, and marigold red) but the marigold orange (MGO) variety contains the maximum amount of lutein [25]. However, mass plantation of marigold occupies a large land area and it is easily influenced by seasonal and climatic change [24]. Although an established commercial system for the production of lutein from microalgae does not exist yet, the basis for outdoor production of lutein-rich cells of strains of *Muriellopsis* sp. and *Scenedesmus* sp. at a pilot scale have already been set up [3]. Recently, there has been remarkable interest in dietary carotenoids with respect to their antioxidant properties and their ability to reduce the chances of some chronic diseases which occurred because of the free radicals. Possibly, carotenoids protect cells from oxidative stress by quenching singlet oxygen damage with a variety of mechanisms. Therefore, algae derived carotenoids could be a leading natural resource in the search of potential functional ingredients [26]. The microalgae biomass from chlorophycean is also considered as the major source of lutein, especially *Chlorella* sp., *Scenedesmus* sp. and*Muriellopsis* sp. [3, 24]. Laboratory studies showed that heterotrophically cultivated *Chlorella* sp. produce considerable amount of lutein [27]. The ability of heterotrophic growth in fermenters makes *Chlorella* sp. a potential alternative resource for commercial production of lutein [24]. Studies showed that more lutein will be produced by *Chlorella* sp. using Basal Medium compared to Kuhl Medium because of the high concentration of EDTA and minerals (500 mg l^{-1}: 9.3 mg l^{-1}) [22]. On the other hand, Wu *et al.* [24] have modeled lutein production by heterotrophic *Chlorella* sp. in batch and fed-batch cultures as a potential alternative source for commercial production.

Beta-carotene

Beta-carotene is a pigment of increasing demand and a wide variety of market applications: as food coloring agent, pro-vitamin A (retinol) in food and animal feed, additive to cosmetics and multivitamin preparations, and as a health food product under the antioxidant category. The market value of beta-carotene was estimated about US$253 million in 2009 [3]. According to a report on the global market for carotenoids (September 2001), the market value of beta-carotene was estimated at around US$261 million in 2010 and this market is expected to grow to US$334 million by 2018 at a compound annual growth rate of 3% (http://www.companiesandmarkets.com/News/Chemicals/Beta-carotene-the-Largest-Carotenoid-Sector-is-Forecast-to-be-Worth-US-334-Mn-by-2018/NI2439).

Because of limited supply from natural sources, a number of companies have introduced synthetically produced carotenoids to meet the high demands in the global market. Beta-carotene, a provitamin A is the only carotenoid which has the potential to form two molecules of vitamin A (retinol) [28]. Natural production of beta-carotene can be produced by the microalgae genus *Dunaliella*, strains that have shown to change their color from green to orange may produce and accumulate large amount of cellular beta-carotene [29].*Dunaliella* sp. occurs in oceans, brine lakes, salt marshes, salt lagoon and salt water ditches near sea, predominantly in water bodies containing more than 2 M salt and high level of magnesium. *Dunaliella* sp. is recognized as being the most halotolerant eukaryotic photosynthetic organism, showing a remarkable degree of adaptation to variety of salt concentrations from as low as 0.1 M to salt saturation about 4 M [30]. The unicellular halotolerant alga *Dunaliella bardawil* was shown to contain high concentration of beta-carotene, composed of the all-*trans* and the 9-*cis* isomers [28]. Beta-carotene is a structural analogue of vitamin A and its topical application can effectively be used to treat skin melasma [31]. It has been reported that a daily dose of *Dunaliella* sp. beta-carotene exerts a protective effect against exercise induced asthma probably through *in vivo* antioxidative effect [32]. Age-related macular degeneration is one of the most common eye diseases causing severe or permanent loss of vision, an antioxidant therapy using combinations of high dosage antioxidant vitamins C, E, beta-carotene, and zinc are employed [33, 34].

The price for natural beta-carotene depends on demand varying from US$300 to US$3000 kg^{-1}. Purified beta-carotene is sold in vegetable oil form ranging from 1% to 20% of concentration depending on type of food products. Naturally produced purified beta-carotene is usually accompanied with other *Dunaliella* sp. carotenoid (carotenoid mix) which includes lutein, neoxanthin, zeaxanthin, violaxanthin, cryptoxanthin, and α-carotene. Variety of natural beta-carotene can be found for sale as health food and supplement under vitamin section. In powder form, natural beta-carotene is used for colorization and pro-vitamin A for animal (cattle and poultry) and aquaculture (shrimp and fish) feed [30].

Chlorophyll

Chlorophyll is a green pigment and is an extremely important biomolecule, critical in photosynthesis, which allows plants to absorb energy from light for carbon dioxide fixation. The most green leaf material contains about 0.3% chlorophyll and on extraction using conventional methods yields about only 5%. Cheap sources, like grass and lucerne (alfalfa), are usually selected as chlorophyll source [35]. In more esoteric extractions, leaves of nettles and elder are preferred, whilst algae and silk worm droppings have also been used commercially [36].

Structurally, chlorophyll is essentially two parts: a substituted porphyrin ring known as an excellent chelating ligand and a long carbon chain phytol [37]. Dietary chlorophyll is predominantly composed of lipophilic derivatives including chlorophyll a and b. The structural modifications of the chlorophyll pigments mainly the de-esterification of phytol ring may significantly increase (65-fold) its transfer from the food matrix to the intestinal epithelial cells during digestion [38]. Sodium/copper derivatives of chlorophyll are frequently used as food additives or in beverages [39]. Sodium copper chlorophyllin is an approved food color in the USA (http://www.accessdata.fda.gov/scripts/cdrh/cfdocs/cfCFR/CFRSearch.cfm?fr=73.125). Besides that, it repairs cells, increases hemoglobin in blood and promotes the cell growth [5]. Regarding the cytotoxity of chlorophyll derivatives, the application of pheophorbide and pheophytin against tumor cell revealed that the cellular uptake and inhibition of myeloma cell multiplicity was greater for pheophorbide than pheophytin, hence it indicates that chlorophyll derivates may play a role in cancer prevention [28]. Patients with chronic relapsing pancreatitis can be treated with chlorophyll-α without any unfavorable side-effects, such as of allergic, or photosensitive, or hepatotoxic nature [40].

Chlorophyll has also been investigated as potential source of pigments in cosmetics; the brown and red algae are mostly used in the cosmetic industries [5]. Furthermore, in food industry, chlorophyll is used as natural pigment ingredient in processed foods. Because of its strong green pigment and consumer demand for natural foods, chlorophyll is gaining importance as food additive. This in turn is encouraging food processors to switch from artificial pigments to chlorophyll-based natural coloring. However, a downstream process needs to be developed to purify chlorophyll a and b from algae [5].

Phycobiliproteins

Phycobiliprotein, a high-potential molecule, has been utilized commercially as natural dyes and has a variety of applications in pharmaceutical industry. It is classified according to UV visible absorption maxima as phycocyanin (blue pigment), phycoerythrins (red pigment), and allophycocyanin (pale-blue pigment). The annual market of phycocyanin was around US$5-10 million [41]. The Cyanobacterium *Spirulina* sp. is a source of c-phycocyanins and allophycocyanins and contains relatively low nucleic acid content; is composed of 55%-70% protein, 6%-9% fat, and 15%-20% carbohydrate, and is rich in minerals, vitamins, fibres, and pigments [42]. Another type of algae that can be a source for phycobiliprotein is *Aphanizomenon flos-aquae*[28, 43]. *Arthrospira* sp. is a group of Cyanobacteria characterized by loosely spiral shaped trichomes arranged in an open helix enclosed in a thin mucilaginous sheath, usually found in alkaline, brackish and saline waters. They usually become predominant species and form massive blooms [44]. The phycobiliprotein may be utilized for labeling antibodies in clinical tests (such as immunofluoresscence or flow cytometry) by binding to biologically active molecules such as immunoglobulin, biotin or protein [28, 45]. Scientific studies have shown that the synergistic action of a wide spectrum of antioxidants is better than the activity of a single antioxidant, and that antioxidants from natural sources (primarily foods) have a higher bioavailability and therefore higher protective efficacy than synthetic antioxidants [43]. It is reported that phycocyanin acts as antioxidant against free radicals and selenium and phycocyanin have been reported to show potent cancer chemopreventive activities [46, 47]. It is proven that phycocyanin is 20 times more efficient than ascorbic acid against hemolysis induced by peroxyl radicals in human erythrocytes [28].

Polyunsaturated fatty acids (PUFAs)

Polyunsaturated Fatty Acids (PUFAs) include docosahexaenoic acid (DHA), eicosapentaenoic acid (EPA), arachidonic acid (AA), γ-linolenic acid (GLA) and have been widely recognized as beneficial towards human health [48]. According to a new technical market research report "Global Market for Oleochemical Fatty Acids" (CHM062A) from BCC Research (http://www.bccresearch.com), the naturally produced fatty acid market was valued at US$7.2 billion in 2011 and should reach US$6.9 billion in 2012. The total market value of fatty acids expected to reach US$13 billion in 2017 (http://www.slideshare.net/bccresearch).

Natural PUFAs can be obtained from fish or extracted from fish oil. However, there is report on the possibility of accumulated toxin in the PUFAs derived from fish sources, furthermore the unpleasant smell, taste and poor oxidative stability has limits the application of fish oil as food additive [4]. Fish oil derived PUFAs have low oxidative stability and the oxidative stability of DHA containing algae oils varies widely depending on some physical parameters. However, EDTA can be used to inhibit oxidation of foods containing long-chain PUFA from either fish or algae and fortified with iron [49]. Moreover, a recent study has shown that PUFAs (450 mg day^{-1}) are not being met by the diet in the majority of the population; the current mean intake of PUFAs by adults is estimated to be about 282 mg day^{-1} (EPA and DHA contributing about 244 mg day^{-1}) [48]. It is undeniable that microalgae can be the alternative natural source for PUFAs; PUFAs found in fish are actually originated from microalgae. However, DHA is the only algal PUFA commercially available as an omega-3 fatty acid and has been documented as a major structural fatty acid in the grey matter of the brain [50]. DHA is important for correct brain and eye development in infants and has been shown to support cardiovascular health in adults [4, 51]. DHA helps body to fight against diseases and recent studies explore the significant role of PUFAs in the immune system through peroxisome proliferator-activated receptors (PPAR-α and PPAR-γ), and favorable effects of n-3 PUFA supplementation on allergic diseases and novel therapeutic strategies to treat eosinophilic disorders have also been reported [52].

Marine microalgae have significantly higher DHA contents compared to fresh water microalgae [5]. DHA is found in a limited selection of foods such as fatty fish and organic meat; it also occurs naturally in breast milk and its concentration in breast milk might be influenced by environmental factors, however it is not available in cow's milk [53]. Since 1990, a number of health and nutrition organizations have specifically recommended the inclusion of DHA in infant formula for preterm and full term infants. DHA is the characteristic PUFA of the marine dinoflagellates. *Crypthecodinium cohnii* is a non-photosynthetic, marine dinoflagellate producing DHA predominantly [54]. Nearly 30-50% of its constituent fatty acids are C22:6 (n-3) fatty acid and no other PUFAs are present in excess of 1%. It is, thus easy to separate DHA from the fatty acid mixtures. For this reason,*Crypthecodinium cohnii* represents a promising source for the commercial production of DHA [55]. The world wholesale market for infant formula was estimated to be about US$10 billion per year [56]. A number of food items containing PUFAs (including DHA) are being sold in market [4]. Some algae-related heterotroph strains that have high DHA such as *Schizochytrium mangrove*, reported to have main component of DHA in a range of 33-39% of total fatty acid while *Amphidium caryerea* and *Thrautocytrium aureum* contain 17% and 16.1%, respectively [5].

Nannochloropsis sp. has been proposed as a source of PUFAs due to its high contents of EPA [57]. Thus,*Nannochloropsis sp.* production has been used to address the rotifer requirements of 8 million gilt-head sea bream fingerlings production annually [58]. Being suitable for aquaculture industry, EPA too has been used as nutritional supplement. The diatoms *Phaeodactylum tricornutum* and *Nitzschia laevis* are algae strains to be utilized as sources of EPA [4, 59]. *Chaetoceros muelleri* var. *subsalsum* and *Isochrysis galbana* have record percentage of DHA and EPA; they present the highest output of EPA (3.5% and 4.8% on dry basis, respectively) using vertical plate glass reactor [58, 60]. EPA is an n-3 PUFA that plays an important role in the regulation of biological functions and prevention and treatment of a number of human diseases such as heart and inflammatory diseases [61]. The eicosanoids are hormone-like substances including prostaglandins (PG), thromboxanes (TX) and leukotrienes (LT). AA and EPA are precursors of eicosanoid compounds. However, the eicosanoids from these two fatty acids are different structurally and functionally, and are sometimes even antagonistic in their effects. A balanced uptake of EPA/AA can prevent eicosanoid dysfunctions and may be effective in treating a number of illnesses and metabolic disorder [59]. *Porphyridium cruentum*, a red alga, has been used for massive cultures due to its relatively high AA and EPA concentration. Nevertheless,*Porphyridium cruentum* biomass is high in protein (34.1%), carbohydrates (32.1%) and essential minerals [62]. AA is potentially used for infant formulas (either for full-term or preterm infants) and as nutritional supplement [4].

Aside from phycobiliprotein, *Arthrospira* sp. is known to be the richest algal source of GLA. GLA is important to the infant formulas for full-term infant and has the potential to lower low-density lipoproteins in hypercholesterolemic patients, alleviate the premenstrual syndrome and treat atopical eczema [4, 44]. GLA may attenuate the signs and symptoms of inflammatory diseases such as rheumatoid arthritis and atopic dermatitis [63, 64]. From both animal and human trials, it has been revealed that dietary supplementation with GLA can modulate inflammation [63].

Beta-1,3-glucan

Beta-1,3-glucanis considered responsible to initiate host defense reactions in response to pathogen surface molecules. The most important substance other than oil in *Chlorella* sp. is beta-1,3-glucan. The world annual sales of *Chlorella* sp. are in excess of US$ 38 billion [65]. Initially, beta-1,3-glucan was named as Zymosan in the 1940s in Norway. It was found in the cell wall of baker's yeast. Further studies have been conducted by Riggi & Di Luuzio in 1961 [66] in Italy, and since then the bio-active compound was renamed as beta-1,3-glucan. Further studies on beta-1,3-glucan has been done in Japan around 1970s. Beta-1,3-glucan is important to human health. Feeding of immunostimulant beta-1,3-glucan to healthy fish may raise the non-specific and specific immunity level and the protection against bacterial infection compared with the control [67]. *Chlorella*sp. is a major source of beta-1,3-glucan [68]. Beta-1,3-glucan had significant effect as a prophylactic treatment to reduce the anthrax infection in mice and the same type of treatment can also inhibit the growth of cancer cells *in vivo* by involving stimulation of three important cytokines [69, 70]. Positive effects were also found in patients after cardiopulmonary bypass, and inhibition of antiviral activity has been found in HIV-infected patients, moreover beta-1,3-glucans are routinely used in patients for tumor immunotherapy [71]. The use of beta-1,3-glucan is of special interest in the cancer patient undergoing chemotherapy and/or radiation treatment, since beta-glucans have shown a remarkable ability to accelerate hematopoetic recovery in both sublethally and lethally irradiated mice. It can also stimulate recovery of the bone marrow following chemotherapy, which is vital to restricting tumor growth and preventing infectious complications during treatment. Amparyup *et al.* [72] showed that pattern recognition proteins [including lipopolysaccharide and beta-1,3-glucan binding protein (LGBP)] could enhance the phenoloxidase activity in shrimp.

Pharmaceutical and nutraceutical compounds

Algae have the potential to be a source of compounds having antibiotic and anticancer activity. A wide variety of Cyanobacteria produce compounds including sulfolipids that were active against herpes virus, pneumonia virus and HIV [73, 74]. Cyanobacteria can produce tolytoxin (6-hydroxy-7-O-methylscytophycin B), an anti-fungal antibiotic and highly toxic to mice [75, 76]. The marine cyanobacterium *Lyngbya majuscula* produces aurilides B and C having cytotoxicity against human lung tumor and mouse neuroblastoma cell line [77]. Moreover the macroalgae *Laphocladia* sp. is capable of producing lophocladine alkaloids that exhibit cytotoxicity to NCI-H460 human lung tumor and MDA-MB-435 breast cancer cell lines [78]. Therefore, once the algal sources of useful medicinal compound are identified, the algae can be cultivated in mass scale and the material can be extracted, purified, and marketed [1]. Some microalgae species are established in the skin care market, the main ones being *Arthrospira* sp. and *Chlorella* sp. A number of cosmeticians have even invested in their own microalgae production (http://www.fao.org/docrep/012/i1704e/i1704e01.pdf). Two examples of commercially available products and their properties claimed by their companies are a) a protein-rich extract from *Arthrospira* sp. that repairs the signs of early skin aging, exerts a tightening effect and prevents stria formation [4] (Protulines, Exsymol S.A.M., Monaco) and b) an extract from *Chlorella vulgaris*that stimulates collagen synthesis in skin, thereby supporting tissue regeneration and wrinkle reduction (Dermochlorella, Codif, St. Malo, France http://www.coptis.com/site/desc.php?RefFab=1CDF). Recently, two new products have been launched by Pentapharm (Basel, Switzerland): a compound from *Nannochloropsis oculata* is used for excellent skin-tightening properties (short and long-term effects) (Pepha-Tight) and an ingredient from *D. salina* is known to have the ability to markedly stimulate cell proliferation and turnover to positively influence the energy metabolism of skin (Pepha-Ctive) [4].

Livestock feed

Spirulina has been frequently used in feed supplements due to its excellent nutrient compounds and digestibility because it has small amount of carbohydrates. The livestock feed is another useful product which may be obtained from the algae, since many of algae species have been examined for their biochemical compositions to evaluate their suitability as a substitute or primary livestock feed [79]. Spirulina has a unique blend of nutrients containing nutrients that include B-complex vitamins, minerals, proteins, gamma-linolenic acid and super anti-oxidants (such as beta-carotene, vitamin E, trace elements) and a number of unexplored bioactive compounds [80]. Compared to other microorganisms, *Spirulina* sp. can be cultivated in high saline water and alkaline conditions which give an advantage to function as a feedstock for livestock feed [81]. Recently, studies regarding growth and body conformation responses of genetically divergent Australian sheep to Spirulina (*Arthrospira platensis*) supplementation revealed that Spirulina can increase live weight, growth and body conformation significantly [82]. Feeding lipid-encapsulated algae supplements may increase n-3 content in milk fat without adversely affecting milk fat yield [83].

An algal supplementation level of about 10 g kg^{-1} of dry matter intake proved to be effective in reducing milk fat content and in modifying the milk fatty acid composition toward increased conjugated linolenic acid (CLA) and DEA concentrations [84]. A related study demonstrated that dietary supplementation with fish meal or n-3 PUFAs in early-lactating dairy cows significantly increased milk yield with no change in milk composition [85]. In addition, some sea weeds like red algae (e.g. *Porphyra* sp., *Gracilaria* sp. and *Kappaphycus* sp.) and brown algae (particularly *Laminaria* sp., *Undaria* sp. and *Hizikia fusiforme*) were used in human food exhibiting potential to be used for livestock feed. *Porphyra* sp., *Gracilaria* sp., *Grateloupia* sp., and*Kappaphycus* sp. have also been genetically transformed for enhanced production of respective products [86]. The annual *Porphyra* sp. harvest worldwide has been estimated to be worth of US$2.5 billion. *Porphyra*sp. primarily contains vitamin A, vitamin B, vitamin C, beta-carotene, as well as essential minerals including iodine. Besides that, a study on *Laminaria digitata* suggested that algae supplemented feed has increased the pig weight up to 10% on a daily basis [1, 5].

Poultry feed

There are more chickens in the world than any other bird; in fact, more than 50 billion chickens are reared annually as a source of food, for both their meat and eggs (http://www.ciwf.org.uk/farm_animals/poultry/). Comprehensive analyses and nutritional studies have demonstrated that algae contain proteins and these proteins are of high quality and comparable to conventional vegetable proteins [87]. Ginzberg *et al.* [88] studied the role of algae *Porphyridium* sp. as feed supplement on metabolism of chicken. It was found that cholesterol of egg yolk was reduced about 10% and color of egg yolk became darker, indicating the production of higher carotenoids [5, 89]. In poultry rations, algae up to a level of 5-10% can be used safely as partial replacement for conventional proteins. Prolonged feeding of algae at higher concentrations may cause adverse effects. The yellow color of broiler skin and shanks as well as of egg yolk is the most important characteristic that can be influenced by feeding algae. Moreover, the Institute für Getreideverarbeitung (Bergholz-Rehbrücke, Germany) produces a natural feed with the algae *Chlorella* sp. and *Arthrospira* sp. called Algrow [90].

Aquaculture feed

Demand for fish for human consumption has risen due to the environmental concerns over Open Ocean fishing [91]. The world market for farmed fish and aquatic edible plants was valued in 2008 at US$106 billion. Products from fisheries and aquaculture combined are supplying the world with 142 million tonnes of protein every year (http://www.blueeconomy.eu/m/articles/view/Fish-without-Feed-Part-1-The-Market).

According to another report by United Nations Food and Agriculture Organization (http://www.fao.org/docrep/016/i2727e/i2727e.pdf), in the last three decades (1980-2010), world food fish production of aquaculture has increased by almost 12 times, at an average annual rate of 8.8% and the world aquaculture production attained another all-time high in 2010 at 60 million tonnes (excluding aquatic plants and non-food products), with an estimated total value of US$119 billion (http://aquabounty.com/aquaculture-market/). Non-toxic marine microalgae, including the stramenopiles *Isochrysis* sp., *Pavlova* sp. and*Nannochloropsis* sp. as well as various diatoms, represent the primary food source at some stages in the life cycle of most cultivated marine animals [1]. Therefore, algal biomass will be in high demand for the fish-food and aquaculture markets in the future and provide ample revenues for the algae industry [91].

A challenge in fish nutrition is to generate end products with high levels of health-promoting long chains of omega-3 fatty acids for the consumer, while reducing the use of fish oils. Omega-3 oils have a high market demand in both the human nutraceuticals and animal feed industries. This growing trend is an additional driving force for the marketing of non-marine-based omega-3 oils and alternative feed ingredients. Algae is a rich source of high quality protein, vitamins, micronutrients (trace elements), and carotenoids, Long-chain PUFAs, especially of n-3 and n-6 series such as eicosapentaenoic (EPA), docosahexaenoic (DHA), and arachidonic (AA) are considered pharmacologically important for dietetics and therapeutics [90]. Algal meal is a rich source of high-quality protein, vitamins, micronutrients (trace elements) and carotenoids, which can be used directly in aqua feeds [85]. Furthermore, fish phosphoglycerides generally contain 50% of their total fatty acids as n-3 PUFA with a ratio of 22:6 (n-3):20:5 (n-3) of about 2:1 and it can be seen most clearly in the phosphoglycerides of fish eggs [59]. In addition, better growth weight and protein efficiencies ratio of tilapia was observed when supplied with algae as nutritional source in feed. Also, *Phorphidium valderianum* was successfully used as feed for aquaculture (based on its nutritional performance and non-toxic properties) [5]. The diatom *Thalassiosira pseudonana* is widely cultivated to feed variety of mollusks, including the Pacific oyster *Crassostrea gigas* and rock scallops [1]. The long chain PUFAs are highly susceptible to oxidation because of their unsaturated nature; the addition of natural antioxidants to the highly unsaturated fish oil may protect it from oxidation [92].

Shrimp feed

The total global production of farmed shrimp reached more than 1.6 million tonnes in 2003, representing a value of nearly US$9,000 million (http://fisherymanagement.wikia.com/wiki/Shrimp_farm). Shrimp farming production reached 737,200 tonnes in 1998, an increase of 12% from 1997. This increased production was mainly reported in subtropical regions of America (28%, 457 hatcheries) and SE Asia (72%, 3,718 hatcheries) [59]. According to Shrimp Production Review by Food and Resource Economics Department, University of Florida USA, the shrimp farming was estimated to be about 4 million metric tonnes in 2013, with 10.3% projected annual growth rate (mainly attributed to SE Asian countries and China) (http://www.gaalliance.org/update/GOAL11/DiegoValderrama.pdf). However, disease outbreaks in recent years have affected farmed Atlantic salmon in Chile, oysters in Europe, and marine shrimp farming in several countries in Asia, South America and Africa, resulting in partial or sometimes total loss of production (http://www.fao.org/docrep/016/i2727e/i2727e.pdf). In 2010, aquaculture in China suffered production losses of 1.7 million tons caused by natural disasters, diseases and pollution. Disease outbreaks virtually wiped out marine shrimp farming production in Mozambique in 2011 (http://www.fao.org/docrep/016/i2727e/i2727e.pdf). Microalgae species *Hypnea cervicornis* and *Cryptonemia crenulata* particularly rich in protein were tested in shrimp diets. Amount of algae in fish meal resulted in significant increase in shrimp growth rates [5]. The requirement for DHA in marine fish and shrimp nutrition has been established via feeding diets containing DHA and AA; the findings revealed that shrimp rose on diets rich in DHA and AA showed significant improvement in immune parameters, such as total hemocyte count, phenoloxidase activity, superoxide dismutase activity, and bactericidal activity [93]. Recently, it has been reported that the addition of 0.5% Spirulina meal in a complete diet for shrimp (*L. vannamei* juveniles), with 14% of Peruvian fishmeal, has proved as a nutritionally efficient feeding attractant [94]. Through short-term feeding studies, two marine algal products (MAP) named MAP3 and MAP8 were found suitable as substitute of protein source in the feeds of three prominent farmed species [89]. The studies regarding the utilization of fish oil and algae as a source of PUFAs are going side by side but the trends are more inclined towards algal products. Menhaden fish oil can also be used as a source of dietary fatty acids which influence the composition of Pacific white shrimp muscle tissue by exhibiting significantly higher shrimp muscle DHA level [95]. As fish meal content, DHA is required for the production of ecdysone for molting, growth and egg production. The mechanism involves the biological membranes rich in di-22:6 (n-3) phosphoglycerides which is relatively constant in the face of changing environmental variables like temperature, pressure, and salinity [59].

CONCLUSION

Microalgae have multiple potential to produce high-valued products and there is an urgent need of awareness makes these biomolecules popular in the world to meet the increasing demands with respect to population. Most of these biomolecules are not produced in the animal/human body but termed as essential; therefore, it is highly recommended to make these biomolecules available for food and feed purposes. Although the production of biomolecules from algae requires some awareness regarding technology and market value at the industrial and farmer level, it has been proved that these biomolecules can be produced using algae as a potential source at small or large scale. The demanding market of these biomolecules is quite convincing to stimulate research interests leading to production programs by the related organizations. Algae culturing got attention because of its biofuel producing characteristics but the awareness regarding cultivation technologies may also lead to set trends for biomolecule production as a pharmaceutical or nutritional product. The production of biomolecules from algae and their subsequent utilization in animal and aquaculture feed may support the objectives to establish industry from microalgae at domestic or industrial scale. Some of the biomolecules can be produced as a byproduct while producing biofuel from algae as a main targeted product. Eighty-two percent of the world fish stocks are overexploited, depleted or endangered, while demand for fish protein is exploding. With the world population expected to reach 9 billion by 2050, demand for critical sources of protein continues to outstrip supply.

Consent

Written informed consent was obtained from the patient for the publication of this report and any accompanying images.

ABBREVIATIONS

AA: Arachidonic acid

CLA: Conjugated linolenic acid

DEA: Drug Enforcement agency

DHA: Docosahexaenoic acid

DNA: Deoxyribonucleic acid

EDTA: Ethylene diamine tetra acetic acid

EPA: Eicosapentaenoic acid

GLA: (Gamma) γ-linolenic acid

HIV: Human immunodeficiency virus

LGBP: Lipopolysaccharide and beta-1,3-glucan binding protei

LT: Leukotrienes

M: Molar

MAP: Marine algal products

MGO: Marigold Orange

PG: Prostaglandins

PPAR: Peroxisome Proliferator-activated receptors

SE: South East

Sp: Specie

TX: Thromboxanes

UV: Ultraviolet

UVA: Ultraviolet A.

DECLARATIONS

ACKNOWLEDGEMENTS

Authors, acknowledge the University Kebangsaan Malaysia for providing infrastructure and funding to facilitate the preparation of this review report.

Competing interests

The author's declare that they have no competing interests.

Authors' contributions

ZY: As a team leader has worked for concept development leading to a comprehensive report on algal biomolecules. EA: Has worked for writing, editing, and presenting the facts in the form of a review. AZ: Has worked for literature review, writing, and designing a review report. MM: Was also involved in literature review and extracting facts and figures for this report. MS: Was involved in supervision for designing manuscript. All authors read and approved the final manuscript.

REFERENCES

1. Graham JE, Wilcox LW, Graham LE: *Algae*. City: Pearson Education; 2009.
2. Preisig HR, Andersen RA: Historical review of algal culturing techniques. In *Algal Culturing Techniques*. Edited by: Andersen RA. London: Elsevier Academic Press; 2005:1–12.
3. Del Campo JA, García-González M, Guerrero MG: Outdoor cultivation of microalgae for carotenoid production: current state and perspectives. *Appl Microbiol Biot* 2007, 74: 1163–1174. 10.1007/s00253-007-0844-9
4. Spolaore P, Joannis-Cassan C, Duran E, Isambert A: Commercial applications of microalgae. *J Biosci Bioeng* 2006, 101:87–96. 10.1263/jbb.101.87
5. Harun R, Singh M, Forde GM, Danquah MK: Bioprocess engineering of microalgae to produce a variety of consumer products. *Renew Sust Energ Rev* 2010, 14: 1037–1047. 10.1016/j.rser.2009.11.004
6. Naguib YMA: Antioxidant activities of astaxanthin and related carotenoids. *J Ag Food Chem* 2000, 48: 1150–1154. 10.1021/jf991106k
7. Kupcinskas L, Lafolie P, Lignell Å, Kiudelis G, Jonaitis L, Adamonis K, Andersen LP, Wadström T: Efficacy of the natural antioxidant astaxanthin in the

treatment of functional dyspepsia in patients with or without Helicobacter pylori infection: A prospective, randomized, double blind, and placebo-controlled study. *Phytomedicine* 2008, 15: 391–399. 10.1016/j.phymed.2008.04.004

8. Nakagawa K, Kiko T, Miyazawa T, Burdeos GC, Kimura F, Satoh A, Miyazawa T: Antioxidant effect of astaxanthin on phospholipid peroxidation in human erythrocytes. *Bri J Nutr* 2011, 105: 1563–1571. 10.1017/S0007114510005398
9. Canales-Gómez E, Correa G, Viana MT: Effect of commercial carotene pigments (astaxanthin, cantaxanthin and β-carotene) in juvenile abalone Haliotis rufescens diets on the color of the shell or nacre. *Vet Mexico* 2010, 41: 191–200.
10. Del Campo JA, Rodríguez H, Moreno J, Vargs MA, Rivas J, Guerrero MG: Accumulation of astaxanthin and lutein in Chlorella zofingiensis (Chlorophyta). *Appl Microbiol Biot* 2004, 64: 848–854. 10.1007/s00253-003-1510-5
11. Zhao LY, Chen G, Zhao G, Hu X: Optimization of microwave-assisted extraction of astaxanthin from Haematococcus pluvialis by response surface methodology and antioxidant activities of the extracts. *Separ Sci Technol* 2009, 44:243–262. 10.1080/01496390802282321
12. Rodríguez-Sáiz M, de la Fuente JL, Barredo JL: Xanthophyllomyces dendrorhous for the industrial production of astaxanthin. *Appl Microbiol Biot* 2010, 88: 645–658. 10.1007/s00253-010-2814-x
13. Yuan J-P, Peng J, Yin K, Wang J-H: Potential health-promoting effects of astaxanthin: a high-value carotenoid mostly from microalgae. *Mol Nutr Food Res* 2011, 55: 150–165. 10.1002/mnfr.201000414
14. Yang Y, Seo JM, Nguyen A, Pham TX, Park HJ, Park Y, Kim B, Bruno RS, Lee J: Astaxanthin-rich extract from the green alga Haematococcus pluvialis lowers plasma lipid concentrations and enhances antioxidant defense in apolipoprotein E knockout mice. *J Nutr* 2011, 141: 1611–1617. 10.3945/jn.111.142109
15. Tolasa S, Cakli S, Ostermeyer U: Determination of astaxanthin and canthaxanthin in salmonid. *Eur Food Res Technol*2005, 221: 787–791. 10.1007/s00217-005-0071-5
16. Dominguez A, Ferreira M, Coutinho P, Fábregas J, Otero A: Delivery of astaxanthin from Haematococcus pluvialis to the aquaculture food chain. *Aquaculture* 2005, 250: 424–430. 10.1016/j.aquaculture.2005.08.005
17. Guerin M, Huntley ME, Olaizola M: Haematococcus astaxanthin: applications for human health and nutrition. *Trends Biotechnol* 2003, 21: 210–216. 10.1016/S0167-7799(03)00078-7
18. Sheikhzadeh N, Tayefi-Nasrabadi H, Oushani AK, Enferadi MHN: Effects of Haematococcus pluvialis supplementation on antioxidant system and metabolism in rainbow trout (Oncorhynchus mykiss). *Fish Physiol Biochem* 2012, 38:413–419. 10.1007/s10695-011-9519-7
19. Lyons NM, O'Brien NM: Modulatory effects of an algal extract containing astaxanthin on UVA-irradiated cells in culture. *J Dermatol Sci* 2002, 30: 73–84. 10.1016/S0923-1811(02)00063-4
20. Cysewski GR, Lorenz RT: Industrial production of microalgal cell-mass and secondary products - species of high Potential: *Haematococcus* . In *Handbook of Microalgal Culture Biotechnology and Applied Phycology*. Edited by: Richmond A. Oxford: Blackwell Publishing Ltd; 2004:281–288.
21. Landrum JT, Bone RA: Lutein, zeaxanthin, and the macular pigment. *Arch Biochem Biophys* 2001, 385: 28–40. 10.1006/abbi.2000.2171
22. Shi X-M, Chen F, Yuan J-P, Chen H: Heterotrophic production of lutein by selected Chlorella strains. *J Appl Phycol*1997, 9: 445–450. 10.1023/A:1007938215655
23. Zhang X-W, Shi X-M, Chen F: A kinetic model for lutein production by the green microalga Chlorella protothecoides in heterotrophic culture. *J Ind Microbiol Biot* 1999, 23: 503–507. 10.1038/sj.jim.2900760
24. Wu Z-Y, Shi C-L, Shi X-M: Modeling of lutein production by heterotrophic Chlorella in batch and fed-batch cultures.*World J Microb Biot* 2007, 23: 1233–1238. 10.1007/s11274-007-9354-2
25. Bhattacharyya S, Datta S, Mallick B, Dhar P, Ghosh S: Lutein content and in vitro antioxidant activity of different cultivars of Indian marigold flower (Tagetes patula L.) extracts. *J Agr Food Chem* 2010, 58: 8259–8264. 10.1021/jf101262e
26. Christaki E, Bonos E, Giannenas I, Florou-Paneri P: Functional properties of carotenoids originating from algae. *J Sci Food Agr* 2012, 93: 5–11.
27. Wei D, Chen F, Chen G, Zhang XW, Liu LJ, Zhang H: Enhanced production of lutein in heterotrophic Chlorella protothecoides by oxidative stress. *Sci China Ser C* 2008, 51: 1088–1093. 10.1007/s11427-008-0145-2
28. Becker W: Microalgae in human and animal nutrition. In *Handbook of Microalgal Culture Biotechnology and Applied Phycology*. Edited by: Richmond A. Oxford: Blackwell Publishing Ltd; 2004:312–351.
29. Lers A, Biener Y, Zamir A: Photoinduction of massive β-carotene accumulation by the alga Dunaliella bardawil: Kinetics and dependence on gene activation. *Plant Physiol* 1990, 93: 389–395. 10.1104/pp.93.2.389
30. Ben-Amotz A: Industrial production of microalagal cell-mass and secondary products - major industrial species*Dunaliella* . In *Handbook of Microalgal Culture Biotechnology and Applied Phycology*. Edited by: Richmond A. UK: Blackwell Publishing Ltd; 2004:273–280.
31. Kar HK: Efficacy of beta-carotene topical application in melasma: an open clinical trial. *Indian J Dermatol Ve* 2002, 68:320–322.
32. Neuman I, Nahum H, Ben-Amotz A: Prevention of exercise-induced asthma by a natural isomer mixture of β-carotene. *Ann Allerg Asthma Im* 1999, 82: 549–553. 10.1016/S1081-1206(10)63165-1

33. Hr T, Tikellis G, Robman LD, McCarty CA, McNeil JJ: Vitamin E supplementation and macular degeneration: randomized controlled trial. *BMJ* 2002, 325: 11–14. 10.1136/bmj.325.7354.11

34. Hall NF: Prevention of age related, macular degeneration. *BMJ* 2002, 325: 1–2. 10.1136/bmj.325.7354.1

35. Humphrey AM: Chlorophyll. *Food Chem* 1980, 5: 57–67. 10.1016/0308-8146(80)90064-3

36. Humphrey AM: Chlorophyll as a color and functional ingredient. *J Food Sci* 2004, 69: C422-C425. 10.1111/j.1750-3841.2004.tb18019.x

37. Ferruzzi MG, Blakeslee J: Digestion, absorption, and cancer preventative activity of dietary chlorophyll derivatives.*Nutr Res* 2007, 27: 1–12. 10.1016/j.nutres.2006.12.003

38. Gandul-Rojas B, Gallardo-Guerrero L, Mínguez-Mosquera IM: Influence of the chlorophyll pigment structure on its transfer from an oily food matrix to intestinal epithelium cells. *J Agr Food Chem* 2009, 57: 5306–5314. 10.1021/jf900426h

39. Scotter MJ, Castle L, Roberts D: Method development and HPLC analysis of retail foods and beverages for copper chlorophyll (E141[i]) and chlorophyllin (E141[ii]) food colouring materials. *Food Addit Contam* 2005, 22: 1163–1175. 10.1080/02652030500306885

40. Yoshida A, Yokono O, Oda T: Therapeutic effect of chlorophyll-a in the treatment of patients with chronic pancreatitis. *Gastroenterol Jpn* 1980, 15: 49–61.

41. Sekar S, Chandramohan M: Phycobiliproteins as a commodity: trends in applied research, patents and commercialization. *J Appl Phycol* 2008, 20: 113–136. 10.1007/s10811-007-9188-1

42. Boussiba S, Richmond AE: Isolation and characterization of phycocyanins from blue green algae Spirulina platensis.*Arch Microbiol* 1979, 120: 155–159. 10.1007/BF00409102

43. Benedetti S, Benvenuti F, Pagliarani S, Francogli S, Scoglio S, Canestrani F: Antioxidant properties of a novel phycocyanin extract from the blue-green alga Aphanizomenon flos-aquae. *Life Sci* 2004, 75: 2353–2362. 10.1016/j.lfs.2004.06.004

44. Hu Q: Industrial production of microalgal cell-mass and secondary products - major industrial species *Arthrospira (Spirulina) platensis* . In *Handbook of Microalgal Culture Biotechnology and Applied Phycology*. Edited by: Richmond A. Oxford: Blackwell Publishing Ltd; 2004:264–272.

45. Shapiro HM, Glazer AN, Christenson L, Williams JM, Strom TB: Immunofluorescence measurement in a flow cytometer using low-power helium-neon laser excitation. *Cytometry* 1983, 4: 276–279. 10.1002/cyto.990040314

46. Patel A, Mishra S, Ghosh PK: Antioxidant potential of C-phycocyanin isolated from cyanobacterial species Lyngbya, Phormidium and Spirulina spp. *Indian J Biochem Bio* 2006, 43: 25–31.

47. Chen T, Wong YS: In vitro antioxidant and antiproliferative activities of selenium-containing phycocyanin from selenium-enriched Spirulina platensis. *J Agr Food Chem* 2008, 56: 4352–4358. 10.1021/jf073399k

48. Givens DI, Gibbs RA: Very long chain n-3 polyunsaturated fatty acids in the food chain in the UK and the potential of animal-derived foods to increase intake. *Nutrition Bull* 2006, 31: 104–110. 10.1111/j.1467-3010.2006.00554.x

49. Frankel EN, Satué-Gracia T, Meyer AS, German BJ: Oxidative stability of fish and algae oils containing long-chain polyunsaturated fatty acids in bulk and in oil-in-water emulsions. *J Agri Food Chem* 2002, 50: 2094–2099. 10.1021/jf0111458

50. Bradbury J: Docosahexaenoic Acid (DHA): an ancient nutrient for the modern human brain. *Nutrients* 2011, 3: 529–554. 10.3390/nu3050529

51. Cottin SC, Sanders TA, Hall WL: The differential effects of EPA and DHA on cardiovascular risk factors. *P Nutr Soc*2011, 70: 215–231. 10.1017/S0029665111000061

52. Tanigai T, Ueki S, Kihara J, Kamada R, Yamauchi Y, Sokal A, Takeda M, Ito W, Kayaba H, Adachi T, Ohta K, Chihara J:Docosahexaenoic Acid exerts anti-inflammatory action on human eosinophils through peroxisome proliferator-activated receptor-independent mechanisms. *Int Arch Allergy Imm* 2012, 158: 375–386. 10.1159/000332965

53. Thiombiano-Coulibaly N, Rocquelin G, Eymard-Duvernay S, Kiffer-Nunes J, Tapsoba S, Traoré SA: Seasonal and environmental effects on breast milk fatty acids in Burkina Faso and the need to improve the omega 3 PUFA content. *Acta Paediatr* 2003, 92: 1388–1393. 10.1111/j.1651-2227.2003.tb00820.x

54. Da Silva TL, Mendes A, Mendes RL, Calado V, Alves SS, Vasconcelos JMT, Reis A: Effect of n-dodecane on Crypthecodinium cohnii fermentations and DHA production. *J Ind Microbiol Biot* 2006, 33: 408–416. 10.1007/s10295-006-0081-8

55. Jiang Y, Chen F, Liang S-Z: Production potential of docosahexaenoic acid by the heterotrophic marine dinoflagellate Crypthecodinium cohnii. *Process Biochem* 1999, 34: 633–637. 10.1016/S0032-9592(98)00134-4

56. Milledge JJ: Commercial application of microalgae other than as biofuels: a brief review. *Rev Environ Sci Biotechnol*2011, 10: 31–41. 10.1007/s11157-010-9214-7

57. Zittelli GC, Rodolfi L, Tredici MR: Industrial production of microalgal cell-mass and secondary products - species of high potential: mass cultivation of *Nannochloropsis* in Closed Systems. In *Handbook of Microalgal Culture Biotechnology and Applied Phycology*. Edited by: Richmond A. Oxford: Blackwell Publishing Ltd; 2004:298–303.

58 Zamora O, Richmond A: Microalgae production for aquaculture. In *Handbook of Microalgal Culture Biotechnology and Applied Phycology*. Edited by: Richmond A. Oxford: Blackwell Publishing Ltd; 2004:365–379.

59. Hemaiswarya S, Raja R, Kumar RR, Ganesan V, Anbazhagan C: Microalgae: a sustainable feed source for aquaculture.*World J Microb Biot* 2011, 27: 1737–1746. 10.1007/s11274-010-0632-z

60. Zhang CW, Richmond A: Sustainable, high-yielding outdoor mass cultures of Chaetoceros muelleri var. subsalsum and Isochrysis galbana in vertical plate

reactors. *Mar Biotechnol* 2003, 5: 302–310. 10.1007/s10126-002-0075-0

61. Wen Z-Y, Chen F: Heterotrophic production of eicosapentaenoic acid by microalgae. *Biotechnol Adv* 2003, 21: 273–294. 10.1016/S0734-9750(03)00051-X
62. Arad S, Richmond A: Industrial production of microalgal cell-mass and secondary products - Species of high potential: *Porphyridium* sp. In *Handbook of Microalgal Culture Biotechnology and Applied Phycology*. Edited by: Richmond A. Oxford: Blackwell Publishing Ltd; 2004:289–297.
63. Johnson MM, Swan DD, Surette ME, Stegner J, Chilton T, Fonteh AN, Chilton FH: Dietary supplementation with g-linolenic acid alters fatty acid content and eicosanoid production in healthy humans. *J Nutr* 1997, 127: 1435–1444.
64. Fan Y-Y, Chapkin RS: Importance of dietary γ-linolenic acid in human health and nutrition. *J Nutr* 1998, 128: 1411–1414.
65. Yamaguchi K: Recent advances in microalgal bioscience in Japan, with special reference to utilization of biomass and metabolites: a review. *J Appl Phycol* 1997, 8: 487–502.
66. Riggi SJ, Di Luzio NR: Identification of a reticuloendothelial stimulating agent in zymosan. *Am J Physiol* 1961, 200:297–300.
67. Sahoo PK, Mukherjee SC: Effect of dietary beta-1,3 glucan on immune responses and disease resistance of healthy and aflatoxin B1-induced immunocompromised rohu (Labeo rohita Hamilton). *Fish Shellfish Immun* 2001, 11: 683–695. 10.1006/fsim.2001.0345
68. Iwamoto H: Industrial production of microalgal cell-mass and secondary products - major industrial species*Chlorella* . In *Handbook of Microalgal Culture Biotechnology and Applied Phycology*. Edited by: Richmond A. Oxford: Blackwell Publishing Ltd; 2004:255–263.
69. Vetvicka V, Yvin J-C: *Int Immunopharmacol*. 2004, 4: 721–730. 10.1016/j.intimp.2004.02.007
70. Blaylock RL: Yeast β-1,3-glucan and its use against anthrax infection and in the treatment of cancer. *J Am Nutraceutical Assoc* 2002, 5: 3–4.
71. Vetvicka V, Terayama K, Mandevill R, Brousseau P, Kournikakis B, Ostroff G: Pilot study: orally-administered yeast β1,3-glucan prophylactically protects against anthrax infection and cancer in mice. *J Am Nutraceutical Assoc* 2002, 5: 1–5.
72. Amparyup P, Sutthangkul J, Charoensapsri W, Tassanakajon A: Pattern recognition protein binds to lipopolysaccharide and beta-1,3-glucan and activates shrimp prophenoloxidase system. *J Biol Chem* 2012, 287:10060–10069. 10.1074/jbc.M111.294744
73. Gustafson KR, Cardellina JH 2nd, Fuller RW, Weislow OS, Kiser RF, Snader KM, Patterson GM, Boyd MR: AIDS-antiviral sulfolipids from cyanobacteria (blue-green algae). *J Natl Cancer I* 1989, 81: 1254–1258. 10.1093/jnci/81.16.1254
74. Archer SD, McDonald KA, Jackman AP: Effect of light irradiance on the production of sulfolipids from Anabaena 7120 in a fed-batch photobioreactor. *Appl Biochem Biotechnol* 1997, 67: 139–152. 10.1007/BF02787848
75. Carmeli S, Moore RE, Patterson GML: Tolytoxin and new scytophycins from three species of Scytonema. *J Nat Prod*1990, 53: 1533–1542. 10.1021/np50072a021
76. Patterson GML, Carmeli S: Biological effects of tolytoxin (6-hydroxy-7-O-methyl-scytophycin b), a potent bioactive metabolite from cyanobacteria. *Arch Microbiol* 1992, 157: 406–410. 10.1007/BF00249096
77. Han B, Gross H, Goeger DE, Moobery SL, Gerwick WH: Aurilides B and C, cancer cell toxins from a Papua New Guinea collection of the marine cyanobacterium Lyngbya majuscula. *J Nat Prod* 2006, 69: 572–575. 10.1021/np0503911
78. Gross H, Goeger DE, Hills P, Moobery SL, Ballantine DL, Murray TF, Valeriote FA, Gerwick WH: Lophocladines, bioactive alkaloids from the red alga Lophocladia sp. *J Nat Prod* 2006, 69: 640–644. 10.1021/np050519e
79. Singh J, Gu S: Commercialization potential of microalgae for biofuels production. *Renew Sust Energ Rev* 2010, 14:2596–2610. 10.1016/j.rser.2010.06.014
80. Kulshreshtha A, Zacharia AJ, Jarouliya U, Bhadauriya P, Prasad GBKS, Bisen PS: Spirulina in health care management.*Curr Pharm Biotechno* 2008, 9: 400–405. 10.2174/138920108785915111
81. Costa JA, de Morais MG, Dalcanton F, Reichert CC, Durante AJ: Simultaneous cultivation of Spirulina platensis and the toxigenic cyanobacteria Microcystis aeruginosa. *Z Naturforsch C* 2006, 61: 105–110.
82. Holman B, Kashani A, Malau-Aduli AEO: Growth and body conformation responses of genetically divergent Australian sheep to Spirulina (Arthrospira platensis) supplementation. *Am J Experiment Agric* 2012, 2: 160–173.
83. Stamey JA, Shepherd DM, de Veth MJ, Corl BA: Use of algae or algal oil rich in n-3 fatty acids as a feed supplement for dairy cattle. *J Dairy Sci* 2012, 95: 5269–5275. 10.3168/jds.2012-5412
84. Boeckaert C, Vlaeminck B, Dijkstra J, Issa-Zacharia A, Van Nespen T, Van Straalen W, Fievez V: Effect of dietary starch or micro algae supplementation on rumen fermentation and milk fatty acid composition of dairy cows. *J Dairy Sci*2008, 91: 4714–4727. 10.3168/jds.2008-1178
85. Subhadra B, Grinson G: Algal biorefinery-based industry: an approach to address fuel and food insecurity for a carbon-smart world. *J Sci Food Agr* 2011, 91: 2–13. 10.1002/jsfa.4207
86. Qin S, Jiang P, Tseng C: Transforming kelp into a marine bioreactor. *Trends Biotechnol* 2005, 23: 264–268. 10.1016/j.tibtech.2005.03.010
87. Becker EW: Micro-algae as a source of protein. *Biotechnol Adv* 2007, 25: 207–210. 10.1016/j.biotechadv.2006.11.002

88. Ginzberg A, Cohen M, Sod-Moriah UA, Shany S, Rosenshtrauch A, Arad SM: Chickens fed with biomass of the red microalga Porphyridium sp. have reduced blood cholesterol level and modified fatty acid composition in egg yolk. *J Appl Phycol* 2000, 12: 325–330. 10.1023/A:1008102622276

89. Kiron V, Phromkunthong W, Huntley M, Archibald I, De Scheemaker G: Marine microalgae from biorefinery as a potential feed protein source for Atlantic salmon, common carp and whiteleg shrimp. *Aquacul Nutr* 2012, 18: 521–531. 10.1111/j.1365-2095.2011.00923.x

90. Pulz O, Gross W: Valuable products from biotechnology of microalgae. *Appl Microbiol Biot* 2004, 65: 635–648. 10.1007/s00253-004-1647-x

91. Anemaet IG, Bekker M, Hellingwerf KJ: Algal photosynthesis as the primary driver for a sustainable development in energy, feed, and food production. *Mar Biotechnol* 2010, 12: 619–629. 10.1007/s10126-010-9311-1

92. Keithly JI, Kott RW, Berardinelli JG, Moreaux S, Hatfield PG: Thermogenesis, blood metabolites and hormones, and growth of lambs born to ewes supplemented with algae-derived docosahexaenoic acid. *J Anim Sci* 2011, 89: 4305–4313. 10.2527/jas.2010-3391

93. Nonwachai T, Purivirojkul W, Limsuwan C, Chuchird N, Velasco M, Dhar AK: Growth, nonspecific immune characteristics, and survival upon challenge with Vibrio harveyi in Pacific white shrimp (Litopenaeus vannamei) raised on diets containing algal meal. *Fish Shellfish Immun* 2010, 29: 298–304. 10.1016/j.fsi.2010.04.009

94. Silva-Neto JF, Nunes AJP, Sabry-Neto H, Sá MVC: Spirulina meal has acted as a strong feeding attractant for Litopenaeus vannamei at a very low dietary inclusion level. *Aquac Res* 2012, 43: 430–437. 10.1111/j.1365-2109.2011.02846.x

95. Perez-Velazquez M, González-Félix ML, Navarro-García G, Valenzuela-Escalante E: Nutritional value of various ray fish liver oils to the pacific white shrimp Litopenaeus vannamei. *Lipids* 2008, 43: 1009–1016. 10.1007/s11745-008-3231-0

CHAPTER 4

USE OF PROBIOTICS IN AQUACULTURE

Patricia Martínez Cruz, Ana L. Ibáñez, Oscar A. Monroy Hermosillo, and Hugo C. Ramírez Saad

INTRODUCTION

Aquaculture is the farming of aquatic organisms by intervention in the rearing process to enhance production and private ownership of the stock being cultivated. Compared to fishing, this activity allows a selective increase in the production of species used for human consumption, industry or sport fishing. Due to overfishing of wild populations, aquaculture has become an economic activity of great importance around the world. Aquaculture's contribution to world food production, raw materials for industrial and pharmaceutical use, and aquatic organisms for stocking or ornamental trade has increased dramatically in recent decades. The report World Aquaculture 2012 found that global production of fish from aquaculture grew more than 30 percent between 2006 and 2011, from 47.3 million tons to 63.6 million tons. It also forecasts that by 2012 more than 50 percent of the world's food fish consumption will come from aquaculture, so it is expected to overtaking capture fisheries as a source of edible fish. This growth rate is due to several factors: (1) many fisheries have reached their maximum sustainable exploitation, (2) consumer concerns about security and safety of their food, (3) the market demand for high-quality, healthy, low-calorie, and high-protein aquatic products, and (4) aquatic breeding makes only a minimum contribution to carbon dioxide emission [1, 2].

Aquaculture has a long history, originating at least in the year 475 B.C. in China [3], but became important in the late nineteen-forties, since the methods of aquaculture could be used to restock the waters as a complement to natural spawning. Nowadays, aquaculture is a lucrative industry [2, 4]. However, the intensification of aquaculture practices requires cultivation at high densities, which has caused significant damage to the environment due to discharges of concentrated organic wastes, that deplete dissolved oxygen in ponds, giving rise to toxic metabolites (such as hydrogen sulfide, methane, ammonia, and nitrites), that often are responsible for mortality. Additionally, aquaculture has appropriated of water bodies used for recreational purpose, and sometimes makes a water's waste because this natural resource is not reused in extensive aquaculture systems [5, 6]. Moreover, under these conditions of intensive production, aquatic species are subjected to high-stress conditions, increasing the incidence of diseases and causing a decrease in productivity [7].

Outbreaks of viral, bacterial, and fungal infections have caused devastating economic losses worldwide, that is, China reported disease-associated losses of $750 million in 1993, while India reported $210 million losses from 1995 to 1996. Added to this, significant stock mortality has been reported

Citation: Martínez Cruz, Ana L. Ibáñez, Oscar A. Monroy Hermosillo, and Hugo C. Ramírez Saad, "Use of Probiotics in Aquaculture," ISRN Microbiology, vol. 2012, Article ID 916845, 13 pages, 2012. doi:10.5402/2012/916845

due to poor environmental conditions on farms, unbalanced nutrition, generation of toxins, and genetic factors [8]. In recent decades, prevention and control of animal diseases has focused on the use of chemical additives and veterinary medicines, especially antibiotics, which generate significant risks to public health by promoting the selection, propagation, and persistence of bacterial-resistant strains [9–11].

Resistance to antibiotics is acquired in two ways: chromosomal mutation or acquisition of plasmids. Chromosomal mutations are not transferred laterally to other bacteria but resistance plasmids can be transferred very rapidly, producing a high percentage of pathogenic bacteria that develop mediated plasmid-resistance in a short period of time [12]. This can be an important public health issue as well. For example, transfer of multidrug resistance occurred in Ecuador during the cholera epizootic happened between 1991 and 1994, which originated from shrimp farm workers. Although the original epidemic strain Vibrio cholerae 01 was susceptible to 12 antimicrobial agents tested, on the coast of Ecuador it acquired multidrug resistance due to transfer of resistance genes from other Vibrio species that were pathogenic to shrimp [13].

Probiotic is a relatively new term which is used to name microorganisms that are associated with the beneficial effects for the host. Kozasa made the first empirical application of probiotics in aquaculture [14], considering the benefits exerted by the use of probiotics on humans and poultry. He used spores of Bacillus toyoi as feed additive to increase the growth rate of yellow tail, Seriola quinqueradiata. In 1991, Porubcan [15, 16] documented the use of Bacillus spp, to test its ability to increase productivity of Penaeus monodon farming and to improve water quality by decreasing the concentrations of ammonia and nitrite. In order to avoid or reduce the use of certain antimicrobials, biological control was tested, described as the use of natural enemies to reduce the damage caused by harmful organisms. Strictly speaking, a probiotic should not be classified as a biological control agent because it is not necessarily a natural enemy of the pathogen [17]. However, certain probiotics have the ability to inhibit the growth of pathogenic bacteria. Moriarty determined the ability ofBacillus spp. to decrease the proportion of Vibrio spp. in shrimp ponds, especially in sediments [18]. Further studies have stressed probiotics ability to stimulate appetite, improve absorption of nutrients, and strengthen the host immune system [19, 20]. The aim of this paper is to present an updated concept of probiotics, current uses in aquaculture, commercial presentations, and a final approach to prospect applications in this area.

DEFINITION OF PROBIOTIC

The term "probiotic" comes from Greek pro and bios meaning "prolife" [56], having different meanings over the years. In 1905, Dr. Elie Metchnikoff was the first to describe the positive role played by some bacteria among farmers who consumed pathogen-containing milk and that "reliance on gut microbes for food makes it possible to take steps to change the flora of our bodies and to replace harmful microbes by beneficial microbes" [57]. However, the term probiotic was introduced until 1965 by Lilly and Stillwell [58] as a modification of the original word "probiotika." It was used to describe substances produced by a microorganism that prolong the logarithmic growth phase in other species. It was described as an agent which has the opposite function of antibiotics. Later, Sperti [59] modified the concept of "tissue extracts that stimulate microbial growth."

The first use of the term to describe a microbial feed/food supplement was by Parker in 1974 [60]. He defined it as "organisms and substances that contribute to intestinal microbial balance." Fuller [61] expanded the definition to "live microbial food supplement that benefits the host (human or animal) by improving the microbial balance of the body" and said that it would be effective in a range of extreme temperatures and salinity variations. Afterwards, it was suggested that probiotics were "monocultures or mixed cultures of microorganisms applied to animals or humans, that benefit the host by improving properties of indigenous microflora" [62]. In 1998, Guarner and Schaafsma assumed that probiotics are live microorganisms which, when consumed in adequate amounts, confer health benefits to the host [63]. Gatesoupe in 1999, defined them as "microbial cells administered in a certain way, which reaches the gastrointestinal tract and remain alive with the aim of improving health" [25]. In the same year, studies were carried out on the inhibition of pathogens using probiotics, this expanded the definition to " live microbial supplement which benefits the host by improving its microbial balance" [34].

Knowledge ofprobiotics has increased, currently it is known that these microorganisms have an antimicrobial effect through modifying the intestinal microbiota, secreting antibacterial substances (bacteriocins and organic acids), competing with pathogens to prevent their adhesion to the intestine, competing for nutrients necessary for pathogen survival, and producing an antitoxin effect. Probiotics are also capable of modulating the immune system, regulating allergic response of the body, and reducing proliferation of cancer in mammals. Because of this, when provided at certain concentration and viability, probiotics favorably affect host health [64]. In fact, terms such as "friendly bacteria," "friendly," or "healthy" are commonly used to describe probiotics [19].

For many years, studies focused on microorganisms characteristic from intestinal microbiota, and the term "probiotic" was mainly restricted to gram-positive lactic acid bacteria [65], particularly representative of the genera Bifidobacterium, Lactobacillus, and Streptococcus [66]. In contrast to terrestrial animals, gastrointestinal microbiota of aquatic species is particularly dependent on the external environment due to the flow of water passing through the digestive tract. Thus, the majority of bacteria are transient in the intestine, due to constant intake of water and food, together with microorganisms present in them. Although in the gastrointestinal tract (GIT) of aquatic animals have been reported potentially pathogenic bacteria such as Salmonella, Listeria, andEscherichia coli, probiotic bacteria and other microorganisms have also been identified. These include gram-positive bacteria such as Bacillus, Carnobacterium, Enterococcus, and several species of Lactobacillus; gram-negative, facultative anaerobic such as Vibrio and Pseudomonas, as well as certain fungi, yeasts, and algae of the genera Debaryomyces, Saccharomyces, and Tetraselmis, respectively [20, 67, 68]. Due to the increasing interest of probiotics in aquaculture, Moriarty [18] proposed extending the definition of these to "living microbial additives that benefit the health of hydrobionts and therefore increase productivity."

A more general and common concept of probiotic is "one or more microorganisms with beneficial effects for the host, able to persist in the digestive tract because of its tolerance to acid and bile salts" [20]. Although the use of probiotics in aquaculture is relatively recent, interest in them has increased due to their potential in disease control [19]; however, gradually other applications have been proposed (summarized in Table 1), that will be addressed in detail later.

Table 1: Different applications of probiotics in aquaculture.

Application	Identity of the probiotic	Applied to aquatic species	Reference
Growth promoter	Bacillus sp. S11	Penaeus monodon	[21]
	Bacillus sp.	Catfish	[22]
	Carnobacterium divergens	Gadus morhua	[23]
	Alteromonas CA2	Crassostrea gigas	[24]
	Lactobacillus helveticus	Scophthalmus maximus	[25]
	Lactobacillus lactis AR21	Brachionus plicatilis	[26]
	Streptococcus thermophilus	Scophthalmus maximus	[25]
	Streptomyces	Xiphophorus helleri	[27]
	L. casei	Poeciliopsis gracilis	[28]
	Bacillus NL 110, Vibrio NE 17	Macrobrachium rosenbergii	[29]
	Bacillus coagulans	Cyprinus carpio koi	[30]
Pathogen inhibition	Bacillus sp.	Penaeids	[18]
	Enterococcus faecium SF 68	Anguilla anguilla	[31]
	L. rhamnosus ATCC53103	Oncorhynchus mykiss	[32]
	Micrococcus luteus A1-6	Oncorhynchus mykiss	[33]
	Pseudomonas fluorescens	Oncorhynchus mykiss	[34]
	P. fluorescens AH2	Oncorhynchus mykiss	[35]
	Pseudomonassp.	Oncorhynchus mykiss	[36]
	Roseobacter sp. BS. 107	Scallop larvae	[37]
	Saccharomyces cerevisiae, S. exiguous, Phaffia rhodozyma	Litopenaeus vannamei	[38]
	Vibrio alginolyticus	Salmonids	[39]
	V. fluvialis	Oncorhynchus mykiss	[33]
	Tetraselmis suecica	Salmo salar	[40]
	Carnobacterium sp. Hg4-03	Hepialus gonggaensis larvae	[41]
	Lactobacillus acidophilus	Clarias gariepinus	[42]
	Bacillus spp., Enterococcussp.	Farfantepenaeus brasiliensis	[43]
	Lactococcus lactis	Epinephelus coioides	[44]

Nutrient digestibility	L. helveticus	Scophthalmus maximus	[25]
	Bacillus NL 110, Vibrio NE 17	Macrobrachium rosenbergii	[29]
	Carnobacterium sp. Hg4-03	Hepialus gonggaensis larvae	[41]
	Lactobacillus acidophilus	Clarias gariepinus	[45]
	Shewanella putrefaciens Pdp11	Solea senegalensis	[46]
	Bacillus sp. 48	Penaeus monodon	[19]
	Bacillus NL 110, Vibrio sp. NE 17	Macrobrachium rosenbergii	[29]
Water quality	Lactobacillus acidophilus	Clarias gariepinus	[45]
	B. coagulans SC8168	Pennaeus vannamei	[47]
	Bacillus sp., Saccharomyces sp.	Penaeus monodon	[48]
	Lactobacillus delbrueckii	Dicentrarchus labrax	[49]
	Alteromonas sp.	Sparus auratus	[50]
Stress tolerance	B. subtilis, L. acidophilus, S. cerevisiae	Paralichthys olivaceus	[51]
	L. casei	Poecilopsis gracilis	[28]
	Pediococcus acidilactici	Litopenaeus stylirostris	[52]
	Shewanella putrefaciens Pdp11	Makimaki	[46]
Reproduction improvement	Bacillus subtilis	Poecilia reticulata, Xiphophorus maculatus	[53]
	L. rhamnosus	Danio rerio	[54]
	L. acidophilus, L. casei, Enterococcus faecium, Bifidobacterium thermophilum	Xiphophorus helleri	[55]

COMMERCIAL PREPARATIONS

The interest in probiotics as an environmentally friendly alternative is increasing and its application is both empirical and scientific. According to Soccol et al. [69], the global market for probiotic ingredients, supplements and foods, reached US $15,900 million in 2008 and is projected to increase to US $19,600 million in 2013, representing an annual growth rate of 4.3%. At present, there are several commercial preparations of probiotics that contain one or more live microorganisms, which have been introduced to improve the cultivation of aquatic organisms. Probiotics can be used as a food additive added directly to the culture tank or mixed with food.

Apart from laboratory preparation of bacteria, some commercially available products are now available. One of the first evaluations of commercial products focussed on a bacterial preparation called Biostart that is derived from Bacillus isolates. It was used during the production of cultured catfish studying the effect of inoculum concentration [22]. In 1998, Moriarty reported that the use of commercial probiotic strains of Bacillus spp. increased the quality and viability of pond-raised shrimp [18]. Meanwhile, Chang and Liu [31] evaluated the effect of Enterococcus faecium SF68 and Bacillus toyoi isolates present in Cernivet LBC and Toyocerin, respectively, to decrease the mortality of the European eel because of the edwardsielosis, ensuring greater efficiency with E. faecium SF68. It is relevant to note that E. faecium has long been known as a probiotic for humans, whereas B. toyoi has been used with terrestrial animals. Moreover, a B. subtilis strain combined with hydrolytic enzymes to produce Biogen, was used to supplement the feed of Oreochromis niloticus, obtaining significant increases in productivity [70].

The lactic acid-producing bacteria have been the focus of much interest. The human probiotic, Lactobacillus rhamnosus ATCC (American Type Culture Collection, Rockville, MD, USA), was used in rainbow trout for 51 days to reduce mortality by Aeromonas salmonicida, the causative agent of the fish disease "furunculosis" (one of the major fish diseases in many parts of world). Mortality was reduced from 52.6 to 18.9% when 10^9 cells g^{-1}were administered with feed, when probiotic dose was increased to 10^{12} cells g^{-1} of feed the mortality reached 46.3% [32]. Apparently, increasing dosage does not necessarily improve protection. Abasali and Mohamad [55] increased the gonadosomatic index and the production of fingerlings in females of reproductive age, using mixed cultures consisting of L. acidophilus, L. casei, E. faecium, and B. thermophilum (Primalac).

Studies with Penaeus vannamei showed that using mixed cultures of probiotics increases survival, feed conversion, and the final production of farmed shrimp [71]. Meanwhile, Taoka et al. [72] used Alchem Poseidon and Alchem Korea CO and Wonju Korea CO, which have mixed cultures of bacteria (Bacillus subtilis, Lactobacillus acidophilus, and Clostridium butyricum) and yeast (Saccharomyces cerevisiae), enhanced nonspecific immune parameters of tilapia Oreochromis niloticus such as lysozyme activity, migration onf neutrophils, and plasma bactericidal activity, resulting in improvement of resistance to Edwardsiella tardainfection. Previous studies in humans and land animals using prebiotics (nondigestible ingredients of the diet that stimulates the growth of microorganisms) showed their ability to stimulate the activity of probiotic bacteria in the colon [73, 74]. Some commercial aquaculture products included prebiotics in their formulation, such as mannans, glucans, and yucca extract that further increase the beneficial effects of the product [75, 76].

Currently, commercial products are available in liquid or powder presentations, and various technologies have been developed for improvement. on the case of fermentation processes, the interest has been focused on optimizing the fermentation conditions to increase the viability and functionality of probiotics, improving performance [77]. Generally, the production is carried out in batch cultures due to the difficulty of industrial scale operation of continuous systems [69]. More recently, systems have been developed for immobilization of probiotics, especially using microencapsulation. Microbial cells at high density are encapsulated in a colloidal matrix using alginate, chitosan, carboxymethylcellulose, or pectin to physically and chemically protect the microorganisms. The methods commonly used for microencapsulation of probiotics are the emulsion, extrusion, spray drying, and adhesion to starch [78]. Focused on the application to aquaculture, Rosas et al. [79] have effectively encapsulated cells of Shewanella putrefaciens in calcium alginate, demonstrating the survival of encapsulated probiotic cells through the gastrointestinal tract of sole (Solea senegalensis). Encapsulation in alginate matrices protects bacteria from low pH and digestive enzymes; this protection helps to release the probiotic into the intestine without any significant damage [80]. Currently, the lyophilized commercial preparations have advantages for storage and transport. However, conditions for reconstitution of these preparations such as temperature, degree of hydration, and osmolarity of the solution are vital to ensure the viability of bacteria [81]. It is important to emphasize that these products must provide a health benefit to the host; for this, it is necessary that contained microorganisms have the ability to survive storage conditions, and after that in the digestive tract of aquatic species, remaining viable and stable, and finally improving production [20]. According to the opinion of the producers, these preparations are safe to use and effective in preserving the health of aquatic animals [19].

APPLICATIONS OF PROBIOTICS IN AQUACULTURE

The need for sustainable aquaculture has promoted research into the use of probiotics on aquatic organisms. The initial interest was focused on their use as growth promoters and to improve the health of animals; however, new areas have been found, such as their effect on reproduction or stress tolerance, although this requires a more scientific development.

Growth Promoter

Probiotics have been used in aquaculture to increase the growth of cultivated species, in reality it is not known whether these products increase the appetite, or if, by their nature, improve digestibility. Some people are inclined to think that it could be both factors; furthermore, it would be important to determine whether probiotics actually taste good for aquaculture species [20]. According to Balcázar et al. [13], probiotic microorganisms are able to colonize gastrointestinal tract when administered over a long period of time because they have a higher multiplication rate than the rate of expulsion, so as probiotics constantly added to fish cultures, they adhere to the intestinal mucosa of them, developing and exercising their multiple benefits. This also depends on factors such as hydrobionts species, body temperature, enzyme levels, genetic resistance, and water quality.

The effect of probiotics has been tested on phytoplankton (microalgae), which forms the basis of aquatic food chains, due to its nutrient-producing photosynthetic machinery that in most cases, higher organisms are unable to synthesize such is the case of polyunsaturated fatty acids and vitamins. Within groups of microalgae used in aquaculture are distinguished central diatoms as Chaetoceros spp., which have proven to be a good live food; however, production has limitations due to the complexity of their nutritional requirements [82]. Gómez et al. [83] assessed the growth of Vibrio alginolyticus C7b probiotic in the presence of the microalgae Chaetoceros muelleri, proving that these organisms can be grown together to achieve high density and fed to shrimp.

Rotifers are indispensable as the first live feed for larvae of most cultured aquatic species, due to their small size they are more accessible to larvae, for example, the nauplii of brine shrimp, which is a very common live feed. Planas et al. [84] used lactic acid bacteria to increase the growth of the rotifer Brachionus plicatilis and obtained best results with the addition of Lactococcus casei ssp. casei, Pediococcus acidilactici, and Lactobacillus lactis ssp.lactis.

The use of probiotics as growth promoters of edible fishes has been reported. Diet of Nile tilapia (Oreochromis niloticus) was amended with a probiotic Streptococcus strain, increasing significantly the content of crude protein and crude lipid in the fish, also weight has increased from 0.154 g to 6.164 g in 9 weeks of culture [85]. Due to the commercial importance of this species, the effect of supplementing diet with probiotics produced an increase of 115.3% when commercial formulation was used at a concentration of 2% [70].

Examples of growth improvement of ornamental fishes include swordtail (Xiphophorus helleri, X. maculatus) and guppy, (Poecilia reticulate, P. sphenops), their feed was supplemented with Bacillus subtilis and Streptomyces, finding significative increases in growth and survival of Xiphophorus and Poecilia after 90 and 50 days of administration, respectively [27, 86].

Probiotics also have been tested successfully in shellfish culture. Macey and Coyne [87] isolated two yeasts and one bacterial strain (designated SS1, AY1, and SY9, resp.) from the digestive tract of abalone (Haliotis midae). A diet was formulated with a mixture of the three putative probiotics. Each probiont was added to the feed to achieve a final concentration of approximately 10^7 cells g^{-1} of dry feed. The growth rate of small (20 mm) and large (67 mm) abalone was improved by 8% and 34%, respectively, in eight-month cultures. Furthermore, abalones supplemented with probiotics had a survival rate of 62% to the pathogenic bacterium Vibrio anguillarum compared to 25% survival of untreated animals.

Inhibition of Pathogens

Antibiotics were used for a long time in aquaculture to prevent diseases in the crop. However, this caused various problems such as the presence of antibiotic residues in animal tissues, the generation of bacterial resistance mechanisms, as well as an imbalance in the gastrointestinal microbiota of aquatic species, which affected their health [88]. In fact, the European Union has regulated the use of antibiotics in organisms for human consumption [89]. Today, consumers demand natural products, free of additives such as antibiotics; moreover, there is a tendency for preventing diseases rather than treating them. Thus, the use of probiotics is a viable alternative for the inhibition of pathogens and disease control in aquaculture species.

Probiotic microorganisms have the ability to release chemical substances with bactericidal or bacteriostatic effect on pathogenic bacteria that are in the intestine of the host, thus constituting a barrier against the proliferation of opportunistic pathogens. In general, the antibacterial effect is due to one or more of the following factors: production of antibiotics, bacteriocins, siderophores, enzymes (lysozymes, proteases) and/or hydrogen peroxide, as well as alteration of the intestinal pH due to the generation of organic acids [65].

Taoka et al. [51] showed that viable probiotics administered to tilapia Oreochromis niloticus, increased non-specific immune response, determined by parameters such as lysozyme activity, neutrophile migration, and bactericidal activity, which improved the resistance of fish to infection by Edwardsiella tarda. In turn, Robertson et al. [90] isolated a strain of Carnobacterium sp. from salmon bowel and administered alive to rainbow trout and Atlantic salmon, demonstrating in vitro antagonism against known fish pathogens:Aeromonas hydrophila, A. salmonicida, Flavobacterium psychrophilum, Photobacterium damselae, and Vibriospecies. There is also evidence on the effect of dead probiotic cultures consisting on a mixture of Vibrio fluvialisA3-47S, Aeromonas hydrophila A3-51, and Carnobacterium BA211, in the control of furunculosis in rainbow trout. For this specific case, the number of leukocytes was greater than with live cells, in fact, the data suggest that cellular immunity more than humoral factors was involved in the benefits of these preparations of inactivated bacterial cells [91]. In the case of shrimp, studies have focused on the evaluation of probiotics such as Bacillus cereus, Paenibacillus polymyxa, and Pseudomonas sp. PS-102 as biocontrol agents against pathogens of various Vibrio species [92, 93].

Probiotic strains isolated from the gastrointestinal tract of clownfish (Amphiprion percula) have been used to inactivate several pathogens such as Aeromonas hydrophila and Vibrio alginolyticus among others. It has been observed that probiotics in vivo generate a density such that allow the production of antimicrobial metabolites therefore, the bacteria isolated from adult clownfish have the potential to colonize the intestinal mucus and therefore can be used as prophylactic agent and/or therapeutic [94, 95]. Furthermore, it has been found that concentrations of 10^6 to 10^8 cells g^{-1} of probiotic promote the development of healthy microbiota in the gastrointestinal tract of ornamental fishes from the genera Poecilia and Xiphophorus, decreasing the amount of heterotrophic microorganisms [86].

Gómez et al. [96] reported the use of Vibrio alginolyticus strains as probiotics to increase survival and growth of white shrimp (Litopenaeus vannamei), also by using probiotics in Ecuadorian shrimp hatcheries, production increased by 35%, while with the use of antimicrobials it decreased by 94%.

Improvement in Nutrient Digestion

A study has suggested that probiotics have a beneficial effect on the digestive processes of aquatic animals because probiotic strains synthesize extracellular enzymes such as proteases, amylases, and lipases as well as provide growth factors such as vitamins, fatty acids, and aminoacids [13]. Therefore, nutrients are absorbed more efficiently when the feed is supplemented with probiotics [70].

In this sense, probiotics have been used in edible fishes, as in the case of larvae of European bass (Dicentrarchus labrax). It has been reported that the probiotic yeast Debaryomyces hansenii HF1 has the ability to produce spermine and spermidine, two polyamines involved in the differentiation and maturation of the gastrointestinal tract in mammals. In addition this yeast secretes amylase and trypsin, enzymes that aid digestion in sea bass larvae [97]. Studies in juvenile common dentex Dentex dentex L. showed that when diet is supplemented with 0.5 g of Bacillus cereus strain E kg^{-1} of food, increased fish growth due to more efficient use of the food [98]. In the case of rainbow trout, similar results were obtained using B. subtilis, B. licheniformis, and Enterococcus faecium, when these probiotics were provided for 10 weeks along with the diet of fish [99]. In some trials, diet of European sea bass larvae (Dicentrarchus labrax) has supplemented with probiotic yeast (Saccharomyces cerevisiae strain X2180), assessing fish growth and activity and expression of antioxidative key enzymes (catalase, glutathione peroxidase, and superoxide dismutase), finding differences in enzyme activity and gene expression patterns between probiotic supplemented and nonsupplemented treatments, which was attributed to the presence of the yeast [100].

In white shrimp Litopenaeus vannamei Boone and Fenneropenaeus indicus, various strains of Bacillus have been used as probiotics to increase apparent digestibility of dry matter, crude protein, and phosphorus. Results showed higher sizes when the diet is supplemented with 50 g of probiotic kg^{-1} of food [101]. Other research has suggested the importance of managing the probiotic in all ontogenetic stages of the shrimp to generate a constant effect on the production of digestive enzymes [102].

In guppies (Poecilia reticulata, P. sphenops), and swordtail (Xiphophorus helleri, X. maculatus), the effect of incorporating Bacillus subtilis, isolated from the intestine of Cirrhinus mrigala into their diet has been evaluated. The results show an increase in the length and weight of the ornamental fishes as well as the specific activity of proteases and amylases in the digestive tract [86]. According to Moriarty [103], Bacillus secretes a wide range of exoenzymes that complement the activities of the fish and increases enzymatic digestion. In fact, the bacteria isolated from the digestive tract of aquatic animals have shown chitinases, proteases, cellulases, lipases, and trypsin [67].

Improvement of Water Quality

In several studies, water quality was recorded during the addition of probiotic strains especially of the gram-positive genus Bacillus. Probably since this bacterial group is more efficient than gram-negative in transforming organic matter to CO_2. It is suggested that maintaining high levels of probiotics in production ponds, fish farmers can minimize the accumulation of dissolved and particulate organic carbon during the growing season. In addition, this can balance the production of phytoplankton [13]. However, this hypothesis could not be confirmed on tests carried out during cultivation of shrimps or channel catfish, using one or more species ofBacillus, Nitrobacter, Pseudomonas, Enterobacter, Cellulomonas, and Rhodopseudomonas. Thus published evidence for improving water quality is limited, except for the nitrification [65].

In the case of edible fish, trout production farms generate high concentrations of nitrogen ranging from 0.05–3.3 mg L^{-1} of total Kjeldahl nitrogen and up to 6.4 mg L^{-1} after 7 months of monitoring [104]. For tilapia production in recirculating systems, concentrations of total ammonia (NH_4 + NH_3) increased from 4.73 to 14.87 mg L^{-1} in a 21-day-experiment, while the nitrite concentration increased from 3.75 to 9.77 mg L^{-1} [105]. Due to the high concentrations of produced nitrogen compounds, especially the highly toxic total ammonia; the use of probiotics is recommended, as they may improve water quality. Haroun et al. supplemented the food of Nile tilapia Oreochromis niloticus L. with a commercial probiotic made from Bacillus licheniformis and B. subtilis in 17 weeks of culture. Assessment of water quality parameters showed an acceptable range for fish cultivation: 5.7–6.3 mg L^{-1} for dissolved oxygen concentration, 0.36–0.42 mg L^{-1} for ammonia concentration, and pH between 6.3 and 8.2 [70].

Lalloo et al. [106] isolated several strains of Bacillus from Cyprinus carpio and carried out tests to improve water quality in ornamental fish culture and to inhibit the growth of Aeromonas hydrophila. Three out of nine isolates showed high capacity to inhibit the pathogen in 78 of relative incidence rate; moreover, concentrations of ammonia, nitrate, and phosphate were lowered in rates of 74%, 76% and 72%, respectively. Contradictorily, Queiroz and Boyd [22] tested a commercial probiotic in catfish (Ictalurus punctatus), noting a survival and net fish production significantly higher when the probiotic was applied. However, very little differences were significant for the determined water quality variables (ammonia, chemical oxygen demand, nitrate, soluble reactive phosphorus, and dissolved oxygen) between the treated and control ponds. Taoka et al. [51] studied effects of commercial probiotics formulated from mixed cultures of bacteria and yeast on survival of Japanese flounder Paralichthys olivaceus, and water quality in a closed recirculating system. The probiotics-treated groups showed significantly greater survival rate as compared to the control group at the end of rearing experiment (50 days of culture), and water quality parameters were significantly lower in probiotics diet groups (from 0.24 ± 0.22 to 0.12 ± 0.10 mg L^{-1} of NH_4, from 0.15 ± 0.08 to 0.08 ± 0.08 mg L^{-1} of NO_2, and from 13.0 ± 3.9 to 10.2 ± 3.0 mg L^{-1} of PO_4). Meanwhile, Wang et al. showed that a commercial product made fromBacillus sp., Saccharomyces cerevisiae, Nitrosomonas sp., and Nitrobacter sp. had the ability to increase the beneficial bacterial microbiota of Penaeus vannamei shrimp, further reducing the concentrations of inorganic nitrogen from 3.74 to 1.79 mg L^{-1} and phosphate from 0.1105 to 0.0364 mg L^{-1} [107].

Stress Tolerance

Aquaculture practices demand intensive productions in shorter times, causing stress in crop species. For example, it has been reported that chronic stress in zebra fish, Danio rerio, induces a general depression on the synthesis of muscle protein [108]. As a result, it was sought to increase stress tolerance by using probiotics. One of the firsts formal reports on this field studied the supplementation of Lactobacillus delbrueckii ssp. delbrueckiiin the diet of European sea bass (Dicentrarchus labrax), at time intervals of 25 to 59 days. In addition to evaluating the growth improvement, hormone cortisol was quantified in fish tissue as stress marker, since it is directly involved in the animal›s response to stress. Cortisol levels obtained in the treated fishes were significantly lower than those in the control (3.6 ± 0.36 ng g^{-1} and 5.1 ± 0.47 ng g^{-1}, resp.) [49]. Another way to assess stress in fish involve subjecting them to heat shock, as in the case of Japanese flounder (Paralichthys olivaceus) grown in a recirculating system [72]. The stress tests were carried out until half the population died, thus calculated the mean lethal time (LT_{50}) in the absence and with addition of a commercial probiotic containing Bacillus subtilis, Lactobacillus acidophilus, Clostridium butyricum, and Saccharomyces cerevisiae. The group treated with probiotics showed greater tolerance in the stress test than the control group, the LT_{50} was 40 and 25 min, respectively.

Lactate and plasma glucose levels are considered appropriate indicators of stress as they increase as a secondary response during periods of stress to cover high energy requirements induced by this situation. Therefore, Varela et al. have conducted studies on the gilt-head bream (Sparus auratus); the glycogen and triglycerides reserves in the livers of the control group were significantly decreased in relation to concentrations obtained when the fish feed was supplemented with the probiotic Alteromonas sp. Strain Pdp 11 [50].

Castex et al. [52] evaluated the effect of Pediococcus acidilactici MA 18/5 in the antioxidative response of the shrimp Litopenaeus stylirostris to oxidative stress. The results showed high activities of antioxidant enzymes; superoxide dismutase (283.7 and 153.7 mol min^{-1} in control and treated groups, resp.) and catalase (3.08 and 1.34 mol min^{-1} for the control and probiotic-treated, resp.).

The results obtained so far raise the possibility of preparing the fish in advance with probiotic treatment, for conventional aquacultures practices that create stress in animals, such as transport, change in water temperature, and periodic manipulations [46].

Effect on Reproduction of Aquatic Species

According to Izquierdo et al. [109], breeding aquaculture species have high nutritional requirements, thus reproductive capacity depends on appropriate concentrations of lipids, proteins, fatty acids, vitamins C and E, and carotenoids. Furthermore, the relationship of these components influences reproduction in various processes such as fertility, fertilization, birth and development of larvae. At present, for most cultured fish species, there are commercially available "broodstock diets" that just are larger-sized diets. In practice, many fish hatcheries improve the nutrition of their broodstock by feeding them solely on fresh fish byproducts or in combination with commercial diets. The most common fresh organisms used to feed broodstock fishes include squid, cuttlefish, mussels, krill, and small crustaceans. The use of these unprocessed fish products often do not provide adequate levels of nutrients needed by broodstock fishes, increasing the risk of pathogens transmission to the parents and offspring, including parasites, bacteria, and viruses. Therefore, probiotics added to food or water were used in order to prevent infections and to explore their effect on reproduction.

The pioneer study on the effect of probiotic supplementation on reproductive performance of fish was carried out by Ghosh et al. [53], using a strain of B. subtilis isolated from intestine of Cirrhinus mrigala, incorporated at different concentrations to four species of ornamental fishes: Poecilia reticulata, P. sphenops, Xiphophorus helleri, and X. maculates, in a one-year experiment. The results showed that using B. subtilis concentrations of 10^6–10^8 cells g^{-1} of food, produced increases in the gonadosomatic index, fecundity, viability, and production of fry from the females of all four species. Furthermore, these authors proposed that complex B vitamins synthesized by the probiotic, especially thiamine (vitamin B1) and vitamin B12, contribute to reduce the number of dead or deformed alevins.

Abasali and Mohamad [55] carried out similar studies with X. helleri, using a commercial probiotic containingLactobacillus acidophilus, L. casei, Enterococcus faecium, and Bifidobacterium thermophilum. The total production of alevin per female and the relative fecundity were evaluated. Their results showed significant differences between the control and probiotic-treated groups; in the first parameter 105 and 150 alevins were found on average, respectively. While in the second, 28 females were fecundated in the control and 41 for the probiotic treatment.

SAFETY CONSIDERATIONS OF PROBIOTIC

Traditionally, probiotics used in food industry have been deemed safe, in fact, no human risks have been determined, remaining as the best proof of its safety [110]. Theoretically, probiotics may be responsible for four types of side effects in susceptible individuals: systemic infections, deleterious metabolic activities, excessive immune stimulation, and gene transfer. However, no hard evidence has been found.

In practice, there are few reports of bacteremia in humans, where isolation of probiotic bacteria from infections seems to be the result of an opportunistic infection caused by skin lesions, cancer, chronic illness, or a drug-induced abnormality. These conditions lead to a decreased intestinal barrier that promotes the passage of the bacteria through the mucosal epithelium. Subsequently, these microorganisms are transported to the mesenteric lymph nodes and other organs, leading to bacteremia that may progress to septicemia [111]. All reported cases of bacteremia occurred in patients with chronic illness or a weakened immune system [69, 112].

In the case of some lactic acid bacteria that are regarded as probiotics, the resistance to antibiotics can be linked to genes on chromosomes, plasmids, or transposons. However, there is insufficient information about the circumstances in which these genetic elements could be mobilized [113]. Moreover, it is recognized that some enterococci may possess virulence characteristics and have the ability to transfer antibiotic resistance elements, so it is recommended that no reference should be made to these organisms as probiotics for human consumption, unless the producer can demonstrate that this strain cannot acquire or transfer antibiotic resistance or induce infection [112].

Regarding safety of aquaculture products, in Asia and more recently in Latin America cultures of Penaeus monodon have been reported with bacterial white spot syndrome (BWSS), in farms with recurrent use of probiotics based on Bacillus subtilis. The spots are similar to those generated in the white spot viral syndrome (WSS), which is a deadly disease that spreads rapidly and causes mass mortalities in shrimp cultures [114, 115]. Wang et al. [116] demonstrated in 2000 that BWSS is a nonsystemic infection in P. monodon and lesions usually disappear after molting, under this condition cultures are still active and grow normally without significant mortality. However, it is of great concern because most farmers cannot distinguish BWSS from WSS. Farmers are advised in case of suspicion to send samples to the laboratory for confirmatory diagnosis.

Furthermore, since some aquaculture products are consumed raw or half cooked, has raised the question of whether residual probiotics may cause any infection in the final consumer. Shakibazadeh et al. [117] assessed the potential risk to humans caused by the use of probiotic Shewanella algae in shrimp farms. Studies were performed in mice that were given up to 10^{36} cfu to reach the LD_{50} value for S. algae, proving the safety of using probiotics in mice. Based on their results, these authors state that the use of S. algae is safe for the consumer of shrimps, such as for workers in the processing plants and farms.

Because there was no international consensus to ensure efficiency and safety of probiotics, FAO and WHO recognized the need to create guidelines for a systematic approach for the evaluation of probiotics in food, in order to substantiate their health claims. A working group with experts in the field was formed in order to recommend criteria and methodology for the evaluation of probiotics, based on scientific evidence [118]. As a result the "Guide for the Evaluation of Probiotics in Food" was presented, providing guidelines on the evaluation of health and nutrition properties of probiotics in food. The working group stated that no pathogenic or virulent properties were found in lactobacilli, bifidobacteria, or lactococci, although they acknowledged that under certain conditions, some strains of lactobacilli have been associated with rare cases of bacteremia. However, its incidence does not increase with raising the use of lactobacillus in probiotics. It was also mentioned that enterococci may possess virulence characteristics; therefore, it is not recommended as a probiotic for human consumption [113]. Although the guide is not focused on aquaculture products, it creates a precedent for conducting studies to evaluate the safety of probiotics in this area.

To date, the use of animal models including mice, rats, and fish has not revealed specific determinants of virulence or pathogenicity of the studied probiotic microorganisms, suggesting the overall safety of them [112]. However, it is important to continue research using three approaches: (i) analyzing the intrinsic properties of probiotic strains, (ii) studying its pharmacokinetics (survival, activity in the intestine, dose response, and recovery from mucosa), and (iii) understanding the interactions between the microorganism and the host [69].

CONCLUDING REMARKS

The current global food crisis and increasing production costs has put pressure on governments and the international community to ensure sufficient food supply for a growing population. Thus, aquaculture is presented as a way to meet the growing demand for fresh water food or seafood, and to meet current challenges relating to the ongoing globalization of trade, intensification and diversification of aquaculture, progress in technological innovations for food production, changes in ecological systems and human behavior, including a greater awareness to protect biodiversity, public health, and the environment. These challenges will lead to increased attention for improving aquaculture practices, and will become an important alternative to overexploitation and modification of aquatic ecosystems caused by capture fisheries. The use of probiotics can potentiate the benefits of this activity because, as presented in this paper, it offers viable alternatives for the generation of a higher-quality livestock product in terms of size, production time, and health.

In the near future, it is necessary to conduct studies relating to probiotics resistance to antibiotics, and the chances of transmission of genetic elements to other microorganisms in the fish GIT, and thus to humans when consuming the aquaculture product.

On the other hand, there is a need to strengthen studies of microbial ecology in aquaculture systems, correlating microbial communities (microorganisms on water and in the GIT of aquatic species) with animal growth and its relationship to the water quality. Nowadays, a variety of probiotic strains present in the GIT of aquatic animals and nitrifying bacteria from biofilters have been isolated and characterized using biochemical, morphological, and molecular techniques [68, 119–121]. The development of molecular techniques such as PCR, FISH (fluorescent in situ hybridization), DGGE (denaturing gradient gel electrophoresis), and generation of genomic libraries have started to unveil the diversity present in aquaculture systems. Currently, next-generation sequencing methodologies offer great potential for phylogenetic identification of probiotic microorganisms without using conventional cultivation techniques.

CONFLICT OF INTERESTS

The authors declare that there is no conflict of interests between authors and the commercial probiotics and trademarks mentioned in the paper. Mentioning of those products is related to the scientific literature cited in the text with the mere purpose of illustrating the wide range of probiotic applications.

REFERENCES

1. FAO, The State of World Fisheries and Aquaculture, 2012.
2. M. Timmons, J. Ebeling, F. Wheaton, S. Summerfelt, and B. Vinci, Recirculating Aquaculture Systems, Cayuga Aqua Ventures, 2nd edition, 2002.
3. C. E. Boyd and C. S. Tucker, Pond Aquaculture Water Quality Management, Kluwer, Norwell, Mass, USA, 1998.
4. D. Cressey, "Aquaculture: future fish," Nature, vol. 458, no. 7237, pp. 398–400, 2009.
5. P. G. Amaya and D. F. Castellano, Pesca, Acuicultura e Investigación en México, Edited by Cedrssa, 2006.
6. Y. B. Wang and Z. R. Xu, "Probiotics treatment as method of biocontrol in aquaculture," Feed Research, vol. 12, pp. 42–45, 2004.
7. M. G. Bondad, R. P. Subasinghe, J. R. Arthur et al., "Disease and health management in Asian aquaculture," Veterinary Parasitology, vol. 132, no. 3-4, pp. 249–272, 2005.
8. N. Kautsky, P. Rönnbäck, M. Tedengren, and M. Troell, "Ecosystem perspectives on management of disease in shrimp pond farming," Aquaculture, vol. 191, no. 1–3, pp. 145–161, 2000.
9. F. A. O. /OIE/WHO, "Antimicrobial use in aquaculture and antimicrobial resistance," Report of a Joint. Expert Consultation on Antimicrobial Use in Aquaculture And Antimicrobial Resistance, 2006.
10. K. Nomoto, "Prevention of infections by probiotics," Journal of Bioscience and Bioengineering, vol. 100, no. 6, pp. 583–592, 2005.
11. WHO, "Antimicrobial resistance. Fact sheet N° 194," 2012, http://www.who.int/mediacentre/factsheets/fs194/es/index.html.
12. C. S. Lewin, "Mechanisms of resistance development in aquatic microorganisms," in Chemothe-Rapy in Aquaculture: From Theory To Reality, C. Michel and D. Alderman, Eds., pp. 288–301, Office International des Epizooties, Paris, France, 1992.
13. J. L. Balcázar, I. D. Blas, I. Ruiz-Z, D. Cunningham, D. Vendrell, and J. L. Múzquiz, "The role of probiotics in aquaculture," Veterinary Microbiology, vol. 114, no. 3-4, pp. 173–186, 2006.
14. M. Kozasa, "Toyocerin (Bacillus toyoi) as growth promotor for animal feeding," Microbiologie Aliments Nutrition, vol. 4, no. 2, pp. 121–135, 1986.

15. R. S. Porubcan, "Reduction of ammonia nitrogen and nitrite in tanks of Penaeus monodon using floating biofilters containing processed diatomaceous earth media pre-inoculated with nitrifying bacteria," inProceedings of the Program and Abastracts of the 22nd Annual Conference and Exposition, World Aquaculture Society, Puerto Rico, Spain, June 1991.

16. R. S. Porubcan, "Reduction in chemical oxygen demand and improvement in Penaeus monodon yield in ponds inoculated with aerobic Bacillus bacteria," in Proceedings of the Program and Abastracts of the 22nd Annual Conference and Exposition, World Aquaculture Society, Puerto Rico, Spain, June 1991.

17. G. B. Gómez, A. Roque, and J. F. Turnbull, "The use and selection of probiotic bacteria for use in the culture of larval aquatic organisms," Aquaculture, vol. 191, no. 1–3, pp. 259–270, 2000. ··

18. D. J. W. Moriarty, "Control of luminous Vibrio species in penaeid aquaculture ponds," Aquaculture, vol. 164, no. 1–4, pp. 351–358, 1998.··

19. Y. B. Wang, J. R. Li, and J. Lin, "Probiotics in aquaculture: challenges and outlook," Aquaculture, vol. 281, no. 1–4, pp. 1–4, 2008.··

20. A. Irianto and B. Austin, "Probiotics in aquaculture," Journal of Fish Diseases, vol. 25, no. 11, pp. 633–642, 2002.··

21. S. Rengpipat, W. Phianphak, S. Piyatiratitivorakul, and P. Menasveta, "Effects of a probiotic bacterium on black tiger shrimp Penaeus monodon survival and growth," Aquaculture, vol. 167, no. 3-4, pp. 301–313, 1998.··

22. J. F. Queiroz and C. E. Boyd, "Effects of a bacterial inoculum in channel catfish ponds," Journal of the World Aquaculture Society, vol. 29, no. 1, pp. 67–73, 1998.·

23. A. Gildberg, H. Mikkelsen, E. Sandaker, and E. Ringø, "Probiotic effect of lactic acid bacteria in the feed on growth and survival of fry of Atlantic cod (Gadus morhua)," Hydrobiologia, vol. 352, no. 1–3, pp. 279–285, 1997.·

24. P. A. Douillet and C. J. Langdon, "Use of a probiotic for the culture of larvae of the Pacific oyster(Crassostrea gigas Thunberg)," Aquaculture, vol. 119, no. 1, pp. 25–40, 1994.··

25. F. J. Gatesoupe, "The use of probiotics in aquaculture," Aquaculture, vol. 180, no. 1-2, pp. 147–165, 1999. ··

26. A. R. S. Harzeveli, H. VanDuffel, P. Dhert, J. Swing, and P. Sorgeloos, "Use of a potential probioticLactococcus lactis AR21 strain for the enhancement of growth in the rotifer Brachionus plicatilis(Muller)," Aquaculture Research, vol. 29, no. 6, pp. 411–417, 1998.··

27. S. Dharmaraj and K. Dhevendaran, "Evaluation of Streptomyces as a probiotic feed for the growth of ornamental fish Xiphophorus helleri," Food Technology and Biotechnology, vol. 48, no. 4, pp. 497–504, 2010.·

28. L. H. H. Hernandez, T. C. Barrera, J. C. Mejia et al., "Effects of the commercial probiotic Lactobacillus casei on the growth, protein content of skin mucus and stress resistance of juveniles of the Porthole livebearer Poecilopsis gracilis (Poecilidae)," Aquaculture Nutrition, vol. 16, no. 4, pp. 407–411, 2010.··

29. M. Rahiman, J. Yousuf, T. Ambat, and M. Hatha, "Probiotic effect of Bacillus NL110 and Vibrio NE17 on the survival, growth performance and immune response of Macrobrachium rosenbergii (de Man),"Aquaculture Research, vol. 41, no. 9, pp. e120–e134, 2010.··

30. Lin Sh, Y. Guan, L. Luo, and Y. Pan, "Effects of dietary chitosan oligosaccharides and Bacillus coagulanson growth, innate immunity and resistance of koi (Cyprinus carpio koi)," Aquaculture, vol. 342-343, pp. 36–41, 2012.

31. C. I. Chang and W. Y. Liu, "An evaluation of two probiotic bacterial strains, Enterococcus faecium SF68 and Bacillus toyoi, for reducing edwardsiellosis in cultured European eel, Anguilla anguilla L," Journal of Fish Diseases, vol. 25, no. 5, pp. 311–315, 2002.··

32. S. Nikoskelainen, A. Ouwehand, S. Salminen, and G. Bylund, "Protection of rainbow trout (Oncorhynchus mykiss) from furunculosis by Lactobacillus rhamnosus," Aquaculture, vol. 198, no. 3-4, pp. 229–236, 2001.··

33. A. Irianto and B. Austin, "Use of probiotics to control furunculosis in rainbow trout, Oncorhynchus mykiss (Walbaum)," Journal of Fish Diseases, vol. 25, no. 6, pp. 333–342, 2002.··

34. L. Gram, J. Melchiorsen, B. Spanggaard, I. Huber, and T. F. Nielsen, "Inhibition of Vibrio anguillarum byPseudomonas fluorescens AH2, a possible probiotic treatment of fish," Applied and Environmental Microbiology, vol. 65, no. 3, pp. 969–973, 1999.·

35. L. Gram, T. Løvold, J. Nielsen, J. Melchiorsen, and B. Spanggaard, "In vitro antagonism of the probiontPseudomonas fluorescens strain AH2 against Aeromonas salmonicida does not confer protection of salmon against furunculosis," Aquaculture, vol. 199, no. 1-2, pp. 1–11, 2001.··

36. B. Spanggaard, I. Huber, J. Nielsen et al., "The probiotic potential against vibriosis of the indigenous microflora of rainbow trout," Environmental Microbiology, vol. 3, no. 12, pp. 755–765, 2001.··

37. C. Ruiz-Ponte, J. F. Samain, J. L. Sánchez, and J. L. Nicolas, "The benefit of a Roseobacter species on the survival of scallop larvae," Marine Biotechnology, vol. 1, no. 1, pp. 52–59, 1999.·

38. U. Scholz, D. G. Garcia, D. Riccque, S. L. E. Cruz, A. F. Vargas, and J. Latchford, "Enhancement of vibriosis resistance in juvenile Penaeus vannamei by supplementation of diets with different yeast products," Aquaculture, vol. 176, no. 3-4, pp. 271–283, 1999.··

39. B. Austin, L. F. Stuckey, P. A. W. Robertson, I. Effendi, and D. R. W. Griffith, "A probiotic strain of Vibrioalginolyticus effective in reducing diseases caused by Aeromonas salmonicida, Vibrio anguillarum andVibrio ordalii," Journal of Fish Diseases, vol. 18, no. 1, pp. 93–96, 1995.·

40. B. Austin, E. Baudet, and M. Stobie, "Inhibition of bacterial fish pathogens by Tetraselmis suecica,"Journal of Fish Diseases, vol. 90, pp. 389–392, 1992.·

41. Y. Youping, M. Dongdong, L. Shijiang, and W. Zhongkang, "Effects on growth and digestive enzyme activities of the Hepialus gonggaensis larvae caused by

introducing probiotics," World Journal of Microbiology and Biotechnology, vol. 27, no. 3, pp. 529–533, 2011.··

42. A. M. Abdullah, R. Hashim, and P. M. Aliyu, "Evaluating the use of Lactobacillus acidophilus as a biocontrol agent against common pathogenic bacteria and the effects on the haematology parameters and histopathology in African catfish Clarias gariepinus juveniles," Aquaculture Research, vol. 42, no. 2, pp. 196–209, 2011.··
43. S. D. Moreira, S. S. Medeiros, L. L. Pereira, L. A. Romano, W. Wasielesky, and B. E. Cupertino, "The use of probiotics during the nursery rearing or the pink shrimp Farfantepenaeus brasiliensis (Latreille, 1817) in a zero exchange system," Aquaculture Research, pp. 1–10, 2011.·
44. S. Zhang, Y. Sing, M. Long, and Z. Wei, "Does dietary administration of Lactococcus lactis modulate the gut microbiota of grouper, Epinephelus coioides," Journal of the World Aquaculture Society, vol. 43, no. 2, pp. 198–207, 2012.
45. A. Dohail, M. Abdullah, H. Roshada, and M. Aliyu, "Effects of the probiotic, Lactobacillus acidophilus, on the growth performance, haematology parameters and immunoglobulin concentration in African Catfish (Clarias gariepinus, Burchell 1822) fingerling," Aquaculture Research, vol. 40, no. 14, pp. 1642–1652, 2009.··
46. P. A. Tapia, R. P. Díaz, R. J. León et al., "Use or the probiotic Shewanella putrefaciens Pdp11 on the culture of Senegalenses sole (Solea senegalensis, Kaup 1858) and gilthead seabream (Sparaus aurata L.),"Aquaculture International, vol. 21, pp. 1–15, 2012.
47. X. X. Zhou, Y. B. Wang, and W. F. Li, "Effect of probiotic on larvae shrimp (Penaeus vannamei) based on water quality, survival rate and digestive enzyme activities," Aquaculture, vol. 287, no. 3-4, pp. 349–353, 2009.··
48. F. Shishehchian, F. M. Yusoff, and M. Shariff, "The effects of commercial bacterial products on macrobenthos community in shrimp culture ponds," Aquaculture International, vol. 9, no. 5, pp. 429–436, 2001.··
49. O. Carnevali, L. de Vivo, R. Sulpizio et al., "Growth improvement by probiotic in European sea bass juveniles (Dicentrarchus labrax, L.), with particular attention to IGF-1, myostatin and cortisol gene expression," Aquaculture, vol. 258, no. 1–4, pp. 430–438, 2006.··
50. J. L. Varela, I. Ruíz, L. Vargas et al., "Dietary administration of probiotic Pdp11 promotes growth and improves stress tolerance to high stocking density in gilthead seabream Sparus auratus," Aquaculture, vol. 309, no. 1–4, pp. 265–271, 2010.··
51. Y. Taoka, H. Maeda, J. Y. Jo et al., "Growth, stress tolerance and non-specific immune response of Japanese flounder Paralichthys olivaceus to probiotics in a closed recirculating system," Fisheries Science, vol. 72, no. 2, pp. 310–321, 2006.··
52. M. Castex, P. Lemaire, N. Wabete, and L. Chim, "Effect of dietary probiotic Pediococcus acidilactici on antioxidant defences and oxidative stress status of shrimp Litopenaeus stylirostris," Aquaculture, vol. 294, no. 3-4, pp. 306–313, 2009.··
53. S. Ghosh, A. Sinha, and C. Sahu, "Effect of probiotic on reproductive performance in female livebearing ornamental fish," Aquaculture Research, vol. 38, no. 5, pp. 518–526, 2007.··
54. G. Gioacchini, F. Maradonna, F. Lombardo, D. Bizzaro, I. Olivotto, and O. Carnevali, "Increase of fecundity by probiotic administration in zebrafish (Danio rerio)," Reproduction, vol. 140, no. 6, pp. 953–959, 2010.··
55. H. Abasali and S. Mohamad, "Effect of dietary supplementation with probiotic on reproductive performance of female livebearing ornamental fish," Research Journal of Animal Sciences, vol. 4, no. 4, pp. 103–107, 2010.
56. J. Schrezenmeir and M. De Vrese, "Probiotics, prebiotics, and synbiotics—approaching a definition,"American Journal of Clinical Nutrition, vol. 73, pp. 361S–364S, 2001.·
57. E. Metchnikoff, "Lactic acid as inhibiting intestinal putrefaction," in The Prolongation of Life, Optimistic Studies, pp. 161–183, Mitchell Heinemann, London, UK, 1907.
58. D. M. Lilly and R. H. Stillwell, "Probiotics: growth-promoting factors produced by microorganisms,"Science, vol. 147, no. 3659, pp. 747–748, 1965.·
59. G. S. Sperti, Probiotics, Avi Publishing, West Point, Miss, USA, 1971.
60. R. B. Parker, "Probiotics, the other half of the antibiotics story," Animal Nutrition Health, vol. 29, pp. 4–8, 1974.
61. R. Fuller, "Probiotics in man and animals," Journal of Applied Bacteriology, vol. 66, no. 5, pp. 365–378, 1989.·
62. R. Havenaar and I. Huis, The Lactic Acid Bacteria in Health and Disease, Volume 1, Edited by B. J. B. Wood, Elsevier, New York, NY, USA, 1992.
63. F. Guarner and G. J. Schaafsma, "Probiotics," International Journal of Food Microbiology, vol. 39, no. 3, pp. 237–238, 1998.··
64. D. Myers, "Probiotics," Journal of Exotic Pet Medicine, vol. 16, no. 3, pp. 195–197, 2007. ··
65. L. Verschuere, G. Rombaut, P. Sorgeloos, and W. Verstraete, "Probiotic bacteria as biological control agents in aquaculture," Microbiology and Molecular Biology Reviews, vol. 64, no. 4, pp. 655–671, 2000. ··
66. L. Vázquez, R. Hernández, E. Sainz et al., "Cambio en la flora intestinal de ratones por la administración de bifidobacterias y jugo de girasol," Veterinaria México, vol. 27, no. 2, pp. 127–131, 1996.
67. N. G. Vine, W. D. Leukes, and H. Kaiser, "Probiotics in marine larviculture," FEMS Microbiology Reviews, vol. 30, no. 3, pp. 404–427, 2006.··
68. G. Burr, D. Gatlin, and S. Ricke, "Microbial ecology of the gastrointestinal tract of fish and the potential application of prebiotics and probiotics in Finnish aquaculture," Journal of the World Aquaculture Society, vol. 36, no. 4, pp. 425–435, 2005.··
69. C. Soccol, L. Porto, M. Rigon et al., "The potential of probiotics: a review," Food Technology and Biotechnology, vol. 48, no. 4, pp. 413–434, 2010.·

70. E. Haroun, A. Goda, and M. Kabir, "Effect of dietary probiotic Biogen supplementation as a growth promoter on growth performance and feed utilization of Nile tilapia Oreochromis niloticus (L.),"Aquaculture Research, vol. 37, no. 14, pp. 1473–1480, 2006.··

71. Y. B. Wang, Z. R. Xu, and M. S. Xia, "The effectiveness of commercial probiotics in northern white shrimp Penaeus vannamei ponds," Fisheries Science, vol. 71, no. 5, pp. 1036–1041, 2005.··

72. Y. Taoka, H. Maeda, J. Y. Jo et al., "Use of live and dead probiotic cells in tilapia Oreochromis niloticus,"Fisheries Science, vol. 72, no. 4, pp. 755–766, 2006.··

73. C. P. Martínez, L. Mayorga, T. Ponce et al., "Efecto de los fructooligosacáridos en la población bacteriana fecal de un neonato, crecida en cultivo por lote," Revista Mexicana De Ciencias Farmacéuticas, vol. 39, no. 1, pp. 32–37, 2008.

74. K. M. Tuohy, H. M. Probert, C. W. Smejkal, and G. R. Gibson, "Using probiotics and prebiotics to improve gut health," Drug Discovery Today, vol. 8, no. 15, pp. 692–700, 2003.··

75. A. Y. Dakar, S. M. Shalaby, and I. P. Saoud, "Assessing the use of a dietary probiotic/prebiotic as an enhancer of spinefoot rabbitfish Siganus rivulatus survival and growth," Aquaculture Nutrition, vol. 13, no. 6, pp. 407–412, 2007.··

76. Smart Microbials Inc, "Un universo de microbios inteligentes a su servicio. Technical details on the use of commercial probiotic product Aqua BOOSTER," 2012, http://smartmicrobialsinc.com/aqua.php.

77. C. Lacroix and S. Yildirim, "Fermentation technologies for the production of probiotics with high viability and functionality," Current Opinion in Biotechnology, vol. 18, no. 2, pp. 176–183, 2007.··

78. S. Rokka and P. Rantamäki, "Protecting probiotic bacteria by microencapsulation: challenges for industrial applications," European Food Research and Technology, vol. 231, no. 1, pp. 1–12, 2010.··

79. P. Rosas-Ledesma, J. M. León-Rubio, F. J. Alarcón, M. A. Moriñigo, and M. C. Balebona, "Calcium alginate capsules for oral administration of fish probiotic bacteria: aassessment of optimal conditions for encapsulation," Aquaculture Research, vol. 43, pp. 106–116, 2012.··

80. M. A. Moriñigo, V. Sánchez, and T. F. Martínez, "Encapsulation of a bacterial fish probiiotic in alginate beads: protective effect under in vitro simulations of fish gastric conditions," in Abstracts of the International Conference on Fish Diseases and Fish Immunology, pp. 6–9, Iceland University, Reykjavik, Iceland, 2008.

81. J. A. Muller, R. P. Ross, G. F. Fitzgeralk, and C. Stanton, "Manufacture of Probiotic Bacteria," Prebiotics and Probiotics Science and Technology, pp. 125–759, 2009.

82. R. C. Medina and E. B. Cordero, "Crecimiento y composición bioquímica de la diatomea Chaetoceros muelleri Lemmerman, mantenida en cultivo estático con un medio comercial," Ciencia y Mar, pp. 19–25, 2011.

83. G. Gómez, A. Roque, and B. Velasco, "Culture of Vibrio alginolyticus C7b, a potential probiotic bacterium, with the microalga Chaetoceros muelleri," Aquaculture, vol. 211, no. 1–4, pp. 43–48, 2002. ··

84. M. Planas, J. Vázquez, J. Marqués, R. Pérez, M. González, and M. Murado, "Enhancement of rotifer (Brachionus plicatilis) growth by using terrestrial lactic acid bacteria," Aquaculture, vol. 240, no. 1–4, pp. 313–329, 2004.··

85. F. Lara, N. Olvera, M. Guzmán, and M. López, "Use of the bacteria Streptococcus faecium andLactobacillus acidophilus, and the yeast Saccharomyces cerevisiae as growth promoters in Nile tilapia (Oreochromis niloticus)," Aquaculture, vol. 216, no. 1–4, pp. 193–201, 2003.··

86. S. Ghosh, A. Sinha, and C. Sahu, "Dietary probiotic supplementation on growth and health of live-bearing ornamental fishes," Aquaculture Nutrition, vol. 14, no. 4, pp. 289–299, 2008.· ·

87. B. M. Macey and V. E. Coyne, "Improved growth rate and disease resistance in farmed Haliotis midae through probiotic treatment," Aquaculture, vol. 245, no. 1–4, pp. 249–261, 2005.··

88. T. Nakano, Microorganism. En dietary supplements for the health and quality of cultured fish, CAB International, London, UK, 2007.

89. P. J. Ronsón and R. Medina, Probióticos en laAcuicultura, Ciencia y Mar Notas, 2002.

90. P. A. W. Robertson, C. O›Dowd, C. Burrells, P. Williams, and B. Austin, "Use of Carnobacterium sp. as a probiotic for Atlantic salmon (Salmo salar L.) and rainbow trout (Oncorhynchus mykiss, Walbaum),"Aquaculture, vol. 185, no. 3-4, pp. 235–243, 2000.··

91. A. Irianto and B. Austin, "Use of dead probiotic cells to control furunculosis in rainbow trout,Oncorhynchus mykiss (Walbaum)," Journal of Fish Diseases, vol. 26, no. 1, pp. 59–62, 2003.··

92. A. V. Ravi, K. S. Musthafa, G. Jegathammbal, K. Kathiresan, and S. K. Pandian, "Screening and evaluation of probiotics as a biocontrol agent against pathogenic Vibrios in marine aquaculture," Letters in Applied Microbiology, vol. 45, no. 2, pp. 219–223, 2007.··

93. K. K. Vijayan, I. S. Bright Singh, N. S. Jayaprakash et al., "A brackishwater isolate of Pseudomonas PS-102, a potential antagonistic bacterium against pathogenic Vibrios in penaeid and non-penaeid rearing systems," Aquaculture, vol. 251, no. 2–4, pp. 192–200, 2006.··

94. N. G. Vine, W. D. Leukes, and H. Kaiser, "In vitro growth characteristics of five candidate aquaculture probiotics and two fish pathogens grown in fish intestinal mucus," FEMS Microbiology Letters, vol. 231, no. 1, pp. 145–152, 2004.··

95. N. G. Vine, W. D. Leukes, H. Kaiser, S. Daya, J. Baxter, and T. Hecht, "Competition for attachment of aquaculture candidate probiotic and pathogenic bacteria on fish intestinal mucus," Journal of Fish Diseases, vol. 27, no. 6, pp. 319–326, 2004.··

96. R. G. Gómez, J. L. Balcázar, and M. A. Shen, "Probiotics as control agents in aquaculture," Journal of Ocean University of China, vol. 6, no. 1, pp. 76–79, 2007.··

97. D. Tovar, J. Zambonino, C. Cahu, F. Gatesoupe, R. Vázquez, and R. Lesel, "Effect of live yeast incorporation in compound diet on digestive enzyme activity in sea bass (Dicentrarchus labrax) larvae,"Aquaculture, vol. 204, no. 1-2, pp. 113–123, 2002.··

98. M. C. Hidalgo, A. Skalli, E. Abellán, M. Arizcun, and G. Cardenete, "Dietary intake of probiotics and maslinic acid in juvenile dentex (Dentex dentex L.): Effects on growth performance, survival and liver proteolytic activities," Aquaculture Nutrition, vol. 12, no. 4, pp. 256–266, 2006.··

99. D. L. Merrifield, G. Bradley, R. T. M. Baker, and S. J. Davies, "Probiotic applications for rainbow trout (Oncorhynchus mykiss Walbaum) II. Effects on growth performance, feed utilization, intestinal microbiota and related health criteria postantibiotic treatment," Aquaculture Nutrition, vol. 16, no. 5, pp. 496–503, 2010.

100. R. Tovar, D. Mazurais, J. Gatesoupe, P. Quazuquel, C. Cahu, and J. Zambonino, "Dietary probiotic live yeast modulates antioxidant enzyme activities and gene expression of sea bass (Dicentrarchus labrax) larvae," Aquaculture, vol. 300, no. 1–4, pp. 142–147, 2010.··

101. Z. Heizhao, G. Zhixun, Y. Yingying, Z. Wenhui, and J. Zhuojia, "Effect of dietary probiotics on apparent digestibility coefficients of nutrients of white shrimp Litopenaeus vannamei Boone," Aquaculture Research, vol. 35, no. 15, pp. 1441–1447, 2004.··

102. S. Ziaei, M. Habibi, G. Azari, D. Lovett, A. Mirvaghefi, and M. Shakouri, "The effect of Bacillus spp. bacteria used as probiotics on digestive enzyme activity, survival and growth in the Indian white shrimpFenneropenaeus indicus," Aquaculture, vol. 252, no. 2–4, pp. 516–524, 2006.··

103. D. J. W. Moriarty, "Probiotics and bioremediation in aquaculture," Asian Shrimp News, vol. 26, article 3, 1996.

104. V. M. Maillard, G. D. Boardman, J. E. Nyland, and D. D. Kuhn, "Water quality and sludge characterization at raceway-system trout farms," Aquacultural Engineering, vol. 33, no. 4, pp. 271–284, 2005.··

105. G. Rafiee and C. R. Saad, "Nutrient cycle and sludge production during different stages of red tilapia (Oreochromis sp.) growth in a recirculating aquaculture system," Aquaculture, vol. 244, no. 1–4, pp. 109–118, 2005.··

106. R. Lalloo, S. Ramchuran, D. Ramduth, J. Görgens, and N. Gardiner, "Isolation and selection of Bacillusspp. as potential biological agents for enhancement of water quality in culture of ornamental fish,"Journal of Applied Microbiology, vol. 103, no. 5, pp. 1471–1479, 2007.··

107. L. Jiqiu, T. Beiping, M. Kangsen et al., "Comparative study between probiotic bacterium ArthrobacterXE-7 and chloramphenicol on protection of Penaeus chinensis post-larvae from pathogenic Vibrios,"Aquaculture, vol. 253, no. 1–4, pp. 140–147, 2006.··

108. S. Vianello, L. Brazzoduro, V. Dalla, P. Belvedere, and L. Colombo, "Myostatin expression during development and chronic stress in zebrafish (Danio rerio)," Journal of Endocrinology, vol. 176, no. 1, pp. 47–59, 2003.··

109. M. Izquierdo, H. Palacios, and A. Tacon, "Effect of broodstock nutrition on reproductive performance of fish," Aquaculture, vol. 197, no. 1–4, pp. 25–42, 2001.··

110. M. Saxelin, H. Rautelin, S. Salminen, and P. H. Makela, "Safety of commercial products with viableLactobacillus strains," Infectious Diseases in Clinical Practice, vol. 5, no. 5, pp. 331–335, 1996.·

111. N. Ishibashi and S. Yamazaki, "Probiotics and safety," American Journal of Clinical Nutrition, vol. 73, no. 2, pp. 465S–470S, 2001.·

112. S. J. Lahtinen, R. J. Boyle, A. Margolles, R. Frías, and M. Gueimonde, "Safety assessment of probiotics,"Prebiotics and Probiotics Science and Technology, pp. 1193–1235, 2009.

113. FAO/WHO, Guidelines for the Evaluation of Probiotics in Food, 2006.

114. H. Y. Chou, X. Y. Huang, C. H. Wang, H. C. Chiang, and C. F. Lo, "Pathogenicity of a baculovirus infection causing white spot syndrome in cultured penaeid shrimp in Taiwan," Diseases of Aquatic Organisms, vol. 23, no. 3, pp. 165–173, 1995.·

115. Y. G. Wang, M. D. Hassan, M. Shariff, S. M. Zamri, and X. Chen, "Histopathology and cytopathology of white spot syndrome virus (WSSV) in cultured Penaeus monodon from peninsular Malaysia with emphasis on pathogenesis and the mechanism of white spot formation," Diseases of Aquatic Organisms, vol. 39, no. 1, pp. 1–11, 1999.·

116. Y. G. Wang, K. L. Lee, M. Najiah, M. Shariff, and M. D. Hassan, "A new bacterial white spot syndrome (BWSS) in cultured tiger shrimp Penaeus monodon and its comparison with white spot syndrome (WSS) caused by virus," Diseases of Aquatic Organisms, vol. 41, no. 1, pp. 9–18, 2000.·

117. S. Shakibazadeh, C. R. Saad, A. Christianus, M. S. Kamarudin, K. Sijam, and P. Sinaian, "Assessment of possible human risk of probiotic application in shrimp farming," International Food Research Journal, vol. 18, pp. 433–437, 2011.

118. M. Pineiro and C. Stanton, "Probiotic bacteria: Legislative framework—requirements to evidence basis,"Journal of Nutrition, vol. 137, pp. 850S–835S, 2007.·

119. A. D. Schulze, O. A. Abayomi, S. A. Tattersall, and K. M. Miller, "Bacterial diversity in a marine hatchery: balance between pathogenic and potentially probiotic bacterial strains," Aquaculture, vol. 256, no. 1–4, pp. 50–73, 2006.··

120. H. Sugita, H. Mizuki, and S. Itoi, "Diversity of siderophore-producing bacteria isolated from the intestinal tracts of fish along the Japanese coast," Aquaculture Research, vol. 43, no. 4, pp. 481–488, 2012. ..

121. Y. H. Ling, S. Y. Shang, M. R. Long, and Y. J. Dan, "PCR-DGGE analysis of the autochthonous gut microbiota of grouper Ephinephelus coioides following probiotic Bacillus clausii administration,"Aquaculture Research, vol. 43, no. 4, pp. 489–497, 2012.

CHAPTER 5

DEFATTED BIOMASS OF THE MICROALGA, DESMODESMUS SP., CAN REPLACE FISHMEAL IN THE FEEDS FOR ATLANTIC SALMON

Viswanath Kiron, Mette Sørensen, Mark Huntley, Ghana K. Vasanth, Yangyang Gong, Dalia Dahle and Anjana M. Palihawadana

Keywords: Atlantic salmon, microalga, Desmodesmus, feed ingredient, fishmeal replacement, growth, feed performance, intestinal health

INTRODUCTION

Production of farmed aquatic species has increased significantly and therefore the aquafeed industry is actively engaged in finding alternatives for the highly used fishmeal. Such a strategy is imperative for the sustainable growth of the aquaculture industry. The aquafeed sector has been using terrestrial plant-derived ingredients as substitutes for fishmeal and fish oil in the feeds of various farmed fish including those for Atlantic salmon,*Salmo salar*. Terrestrial plant ingredients employed in the feeds of Atlantic salmon include soyprotein concentrate, sunflower expeller, wheat gluten, fava beans, pea protein, maize gluten, horse beans, rapeseed oil, wheat starch, pea starch and tapioca starch (Ytrestøyl et al., 2015). However, some of these ingredients can compromise the intestinal health of carnivorous species like Atlantic salmon (Baeverfjord and Krogdahl, 1996;Knudsen et al., 2007; Bakke, 2011). Hence, it would be ideal to use ingredients from microalgae, which are present in the aquatic food web, in the feeds of Atlantic salmon.

Microalgae are rich in essential nutrients and bioactive molecules, and therefore, are valuable nutrient sources for humans and animals (Becker, 2004). They contain high-quality protein, 50–70% of which is likely to be bioavailable to animals (Becker, 2007). The bioactive components include polyunsaturated fatty acids, sugars, polysaccharides, vitamins, minerals, and pigments (Becker, 2004). Thus, microalgae can be high quality feed components.

In Europe and the United States, efforts are underway to commercially exploit the full potential of microalgae (Enzing et al., 2014; Energy.Gov; Office of Energy Efficiency Renewable Energy, 2016). The co-products from the cultivated microalgae such as *Staurosira* sp. and *Desmodesmus* sp. include "biocrude" oil and residual protein-rich fractions (Huntley et al., 2015). The defatted protein-rich biomass obtained after the extraction of the biofuel could serve as an ingredient in animal feeds. Even though some studies have proved the potential of using microalgae in fish feeds, the protein quality of an alga can be determined only through elaborate feeding studies on animals (Becker, 2004). The effect of current plant-derived fishmeal replacers on farmed animals varies widely. Therefore, feeding studies should evaluate the potential of the microalgae products as fishmeal substitutes by investigating not only the growth of the animals but also their health status.

In the present study, the ability of the defatted biomass of *Desmodesmus* sp. to replace fishmeal in the feeds of Atlantic salmon was evaluated by assessing the growth, feed performance, antioxidant status, and intestinal health of the

Citation: Kiron V, Sørensen M, Huntley M, Vasanth GK, Gong Y, Dahle D and Palihawadana AM (2016) Defatted Biomass of the Microalga,Desmodesmus sp., Can Replace Fishmeal in the Feeds for Atlantic salmon. Front. Mar. Sci. 3:67. doi: 10.3389/fmars.2016.00067

fish. The hypothesis is that the *Desmodesmus* biomass in the feeds of Atlantic salmon could impart good growth and health to the fish.

MATERIALS AND METHODS

Fish and Rearing

The study was approved by the National Animal Research Authority (FDU: Forsøksdyrutvalget ID - 5887) in Norway. The fish rearing and handling procedures were according to the approved protocols of FDU.

Atlantic salmon, *Salmo salar* smolts obtained from Cermaq, Hopen, Bodø, Norway were maintained on commercial feeds at the Research Station of the Nord University, Bodø, Norway for 2 months. Before the start of the feeding experiment, the fish were transferred to the indoor rearing tanks (800 L), which were part of a flow-through seawater system. The tanks had feed traps (Aquatic Eco-Trap, Pentair Aquatic Eco-Systems®, Florida, USA) to collect the uneaten feed. Each of the 18 tanks used for the study was stocked with 25 fish of average weight 167.6 g. The temperature (7.6 ± 0.9°C) and dissolved oxygen saturation (av. 90%) of the rearing water were recorded daily. A 24-h light regime was employed throughout the experimental period.

Experimental Feeds, Design and Feeding

The three types of experimental feeds (one control feed and two microalgal feeds, Table 1) were made to order (Center for Feed Technology, Norwegian University of Life Sciences, Ås, Norway). In the algal feeds, the algal biomass of *Desmodesmus* sp., a co-product from biofuel production (Cellana, San Diego, USA), replaced a portion of the fishmeal. The extruded feed pellets (3 mm size) were stored at 4°C until they were used for feeding the fish.

Table 1. Ingredients and proximate composition of the experimental feeds.

Feed ingredients (g/100 g)	**Experimental feeds**		
	4C	**4L**	**4H**
Fish meal	69	60	51
Algal meal	0	10	20
Wheat	12	12	12
Wheat gluten	5	5	5
Fish oil	13.5	12.5	11.5
Microingredients*	0.5	0.5	0.5
Marker (Yttrium)	0.01	0.01	0.01
Proximate composition (g/100 g dry basis)			
Moisture	3.5	7.6	7.1
Crude protein	53.8	50.4	47.8
Crude lipid	20.1	19.8	19.5
Ash	11.1	11.5	12.2
Energy (MJ/100g)	2.3	2.3	2.3

**Vitamin and mineral mix is a proprietary formulation of Polarfeed, Leknes, Norway.*

The three fish groups in this study were: a control group that received alga-devoid feed (4C), and 2 alga-fed groups that were offered feeds containing the defatted *Desmodesmus* biomass at inclusion levels of 10% (4L) and 20% (4H). Fish in six tanks were randomly allotted to one feed group and they were offered the respective feeds daily for a period of 10 weeks. Automatic feeder (Arvo-Tec, Huutokoski, Finland) of a particular tank was programmed to deliver daily doses of feed (until the end of the study) corresponding to 1.2% of the body mass of fish in the tank. The daily ration was delivered intermittently, between 08:00 and 09:00, and between 14:00 and 15:00. The left-over feeds were collected from the trap, 1 h after each meal. They were then dried and weighed to determine the actual feed intake. The fish in the tanks were visually inspected in the afternoon, after the feeding session, to ascertain their well-being.

Sampling

At the start of the experiment, individual weights and lengths of the fish were recorded. The fish were starved for 24 h and anesthetized with MS-222 (Tricaine methane sulphonate; Argent Chemical Laboratories, Redmond, USA; 80 mg/l), prior to taking the measurements. Similarly, the final weights and lengths were recorded at day 70. Representative samples were collected at day 70 from fish in all tanks—blood (serum), distal intestine, filet and whole body—to assess the different parameters of interest. In addition, the weight of the visceral organs and liver of the sampled fish were recorded. After the termination of the feeding study, the feces for the digestibility determination was collected by stripping the fish, as described by Austreng (1978). Feces from individual fish in one tank of a particular group were pooled during collection. Prior to chemical analysis, freeze-dried feces from two tanks had to be pooled to get enough material for analyses. The sample size is indicated in the figure legends and table footnotes.

STUDY PARAMETERS

Growth and Feed Performance

Initial and final weights, and feed consumption of the fish were used to calculate the growth and feed performance indices. The formulae for calculating the condition factor, specific growth rate (SGR), feed conversion ratio (FCR), protein efficiency ratio (PER) and survival are provided in the footnote of Table 3. Hepatosomatic index [HSI = (liver weight/final weight) × 100] and viscerosomatic index [VSI = (visceral organ weight/final weight) × 100] of the fish from different groups were calculated.

To determine the proximate composition of the whole body and filet, the samples were thawed and homogenized in a grinder (Grindomix GM 200, Retsch, Haan, Germany). Moisture (gravimetry), ash (incineration at 550°C), crude protein (N*6.25, Kjeldahl Autoanalyser, Tecator, Sweden) and total lipid (in fish -ethyl acetate extraction method, Norges Standardiseringsforbund, 1994) contents of the fish and the feeds were determined using standard protocols (Kiron et al., 2012). The energy content in the feeds and feces were determined by bomb calorimetry (C200 calorimeter, IKA, Staufen, Germany). The total lipid and yttrium content in the feeds and feces were determined by Eurofins AS (Moss, Norway) and they were used to calculate the digestibility coefficients. The total lipid content was determined using the Soxhlet method with acid hydrolysis. Yttrium content was analyzed after combustion of the samples at 550°C and boiling with hydrochloric acid and nitric acid. Thereafter the samples were cooled and dissolved in 5% nitric acid. Yttrium was then detected with a spectrometer (Perkin Elmer, Massachusetts, USA).

Histological Studies

The processing and sectioning techniques of the distal intestinal samples ($n = 6$ fish) are described in Vasanth et al. (2015). Approximately 5 µm sections were fixed for the histological studies. The sections were stained with Alcian Blue-Periodic Acid Schiff›s reagent (AB-PAS, pH 1.0) (Bancroft and Gamble, 2007). Photomicrographs were captured employing an Olympus BX61/Camera Color View *III*u (Olympus Europa GmbH, Hamburg, Germany) and processed using the imaging system, photo program Cell P (Soft Imaging System GmbH, Munster, Germany).

Antioxidant Indices

The frozen (−80°C) serum aliquots were thawed prior to conducting the different assays, which were performed using commercially available kits employing a microtiter plate (Nuclon sterile 96 flat, Fisher Scientific, Massachusetts, USA). The respective absorbances were measured at the suggested wavelengths employing a microplate reader (Fluostar Optima; BMG Labtech, Ortenberg, Germany).

Total antioxidant capacity (TAC): Aliquots (20 µL) of the serum samples were used for determining the TAC, employing a kit from Cell Biolabs (STA-360, Cell Biolabs Inc., San Diego, CA, USA). The initial and final absorbances (490 nm) of the samples and standards were measured, and based on the net absorbance values and the uric acid standard curve, the uric acid equivalence (UAE) of the samples was obtained. The µM copper reducing equivalents (CRE), which is proportional to TAC, were calculated by multiplying the UAE concentrations by a factor of 2189 µM Cu^{++}/µM uric acid.

Catalase (CAT) activity: The Cayman Chemicals assay kit 707002 (Cayman Chemicals, Ann Arbor, Michigan, USA) was employed to determine the CAT activity of serum aliquots (20 µL), following the manufacturer›s protocol. The enzyme activity was calculated based on the absorbance at 540 nm and the formaldehyde standard curve. The activity (nmol/min/ml) is expressed as the amount of enzyme that will cause the formation of 1.0 nmol of formaldehyde per minute at 25°C.

Superoxide dismutase (SOD) activity: A commercially available kit (STA-340, Cell Biolabs Inc.) was employed to determine the SOD activity in 20 µL serum samples. The assay was performed following the manufacturer›s instructions. The SOD activity (inhibition %) of the sample is inversely proportional to the colorimetric signal and is calculated using the absorbance values of the sample and blank at 490 nm. The activity is expressed as SOD units U/µL.

GENE EXPRESSION STUDY

The mRNA levels of selected genes - antioxidant-related (*superoxide dismutase1 - sod1; nuclear factor erythroid 2-related factor - nrf2*), mucosa-related (*immunoglobulin T - igt*), inflammation-related (*interleukin 1b - il1b; interleukin 10 - il10; interleukin 17a* and *17d - il17a, il17d; transforming growth factor beta - tgfb; tumor necrosis factor a - tnfa*), antimicrobial (*cathelicidin 1* and *2 - cathl1, cathl2; defensin beta 2 - defb2*) - were assessed in this study.

RNA extraction and cDNA synthesis

The total RNA from the distal intestine was extracted using E-Z 96 Total RNA Kit (OMEGA Bio-Tek Inc., Norcross, GA, USA), following the instructions of the manufacturer. About 100 mg of the tissue was homogenized with Zirconium oxide beads (1.4 mm; Percellys, Tarnos, France) in TRK lysis buffer in a precooled free standing tube (VWR International, Oslo, Norway) at 6000 rpm. The resulting crude mixture was centrifuged (18,000 × g, 20°C) and approximately 300 µl of the supernatant was collected and mixed with equal volume of 70% ethanol. This mixture was added to the assigned wells of the E-Z 96 RNA plate mounted on an E-Z 96®Vacuum Manifold (VAC-03, Omega Bio-tek Inc.). A pressure of 1100 kPa was applied to draw the sample through the well, which had an RNA HiBind® matrix. This step was followed by buffer washes; first with RWF Wash Buffer (500 µl) and then twice with 600 µL of RNA Wash Buffer II. Pressure was applied each time the wells were treated with the wash buffers. Finally, the purified RNA was recovered from the HiBind® matrix by adding 50 µl of RNase-free water (5 Prime GmbH, Hilden, Germany) to each well and applying a final round of pressure.

The cDNA was synthesized as described by Vasanth et al. (2015). The obtained cDNA was 30-fold diluted and used in qPCR.

Real-time PCR (qPCR) and the quantification of gene expression

The primers for the reference and target genes are given in Table 2. The PCR reactions were carried out using LightCycler® 96 Real-Time PCR System (Roche Holding AG, Basel, Switzerland). Each reaction from one fish sample was done in duplicate, and to quantify the mRNA level of a particular gene, reactions were done on samples from 12 fish (2 from each tank). All the reactions were performed using a total volume of 10 µl, containing 5 µl of FastStart Universal SYBR Green Master (ROX) mixed with 1 µl primer mix (200 nM), and 4 µl cDNA (5 ng/µl). Each reaction was done under the following thermal cycling conditions: preincubation at 95°C for 600 s followed by 40 cycles of two step amplification at 95°C (10 s), and 60°C (30 s) and melting at 95°C (10 s), 65°C (60 s) and 97°C (1 s). Specificity of the amplified products was checked by performing melt curve analyses for each sample and the specificity of the primers was confirmed by sequencing the amplified products. For relative quantification, the total RNA from all samples was pooled, and normalized to 1000 ng, from which a 6-point 3-fold serial dilution was prepared (1:1 to 1:243) to make a standard curve. The efficiency of the primers was calculated using the equation $E = (10^{-1/m}-1)$ 100. A normalization factor was computed for each of the samples based on the relative quantities of the two most stable genes, *rpl13* and *rps29* from among the set of four reference genes—*elongation factor 1ab* (*ef1ab*), *ribosomal protein L13* (*rpl13*), *ribosomal protein S29*(*rps29*) and *ubiquitin* (*ubi*), employing geNorm (Vandesompele et al., 2002). The expression levels of all the target genes were then calculated relative to the normalization factor.

Table 2. List of primers for the differentially expressed genes and reference genes.

Gene name	Sequence (5′-3′)	Amplicon size (bp)	PCR efficiency (%)	Reference
sod1	CCACGTCCATGCCTTTGG **F** TCAGCTGCTGCAGTCACGTT **R**	141	95.3	GenBank : AY736282.1
nrf2	TCAACAGGACATCGACCTAAT **F** GGCAGTAGTCAAACACCTCT **R**	83	91.2	GenBank : BT059007.1
cathl1	CTAGCAACAACCTGAACACTG **F** CTTCTTGTCCGAATCTTCTGCATA **R**	88	96.4	GenBank : AY728057.1
cathl2	TGCTTCAAATCAGCTGCTTCC **F** CTGTATTCAAAGTCTCGACTTCCA **R**	83	104.4	GeneBank : AY542961.1
il17d	CTTGTCTCCCTGGGCATACAG **F** CAATATGCCTCGGGTATGAACCT **R**	201	112.7	GenBank : EU689087.1
tgfb	GGTACTGGCTCTGTATAAGCAC **F** CTGCTCCACCTTGTGTTGTC **R**	120	100.9	GenBank : BT059581
igt	CAACACTGACTGGAACAACAAGGT **F** CGTCAGCGGTTCTGTTTTGGA **R**	97	107.7	GenBank : GQ907004
REFERENCE GENES				
ef1ab	TGCCCCTCCAGGATGTCTAC **F** CACGGCCCACAGGTACTG **R**	59	96	GenBank : BG933853
rpl13	CGCTCCAAGCTCATCCTCTTCCC **F** CCATCTTGAGTTCCTCCTCAGTGC **R**	79	96.4	GenBank : BT048949.1
rps29	GGGTCATCAGCAGCTCTATTGG **F** AGTCCAGCTTAACAAAGCCGATG **R**	167	94.5	GenBank : BT043522.1
ubi	AGCTGGCCCAGAAGTACAACTGTG **F** CCACAAAAAGCACCAAGCCAAC **R**	162	92.7	GenBank : AB036060.1

Protein Expression Study

Two-dimensional electrophoresis (2-DE) was performed on the distal intestinal samples (n = 6 fish from 6 tanks) from each of the three groups. The various steps such as protein extraction, two-dimensional electrophoresis and capturing of the gel images are described previously (Vasanth et al., 2015).

The gel images were analyzed using SameSpots (Totallab, Newcastle, UK). The spots were automatically aligned and those with significantly different volumes (with a minimum fold-change of 1.5 and statistical significance at$p < 0.05$) were selected as the spots of interest.

To identify the proteins corresponding to the selected spots, 300 µg of protein was loaded onto a preparative Sypro® Ruby stained gel. The selected spots were excised, and sent to the Tromsø University Proteomics Platform (TUPP), Norway—for performing liquid chromatography and tandem mass spectrometry (LC-MS/MS). The peak list files obtained from TUPP were used to determine the protein identity. Mascot search engine (version 2.5.00) was used to search the NCBInr database or the whole genome sequences fasta files from NCBI. The search gave the matched peptides of Atlantic salmon. At least two unique peptides were considered for the protein inference and contaminants were eliminated from the search list.

Statistical Analysis

GraphPad Prism V6.03 was used to carry out the statistical analyses. The tanks were considered as the experimental units. Normality of the data was tested by KS normality test, and the assumption of equal variance was checked by the Brown-Forsythe test. One-way ANOVA followed by Tukey's multiple comparisons test was employed to detect the significant differences between the means of interest. For non-parametric data, the Kruskal-Wallis test followed by Dunn's multiple comparisons test was employed to understand the difference between the study groups.

RESULTS

Growth and Feed Performance

The experimental feeds (Table 1) were formulated to provide the nutrient and energy needs of Atlantic salmon. At the end of the feeding period, the fish of 4C, 4L, and 4H attained mean weights of 359, 344, 360 g, respectively, compared to the initial weight of 167 g (Table 3). The condition factors of the 4C, 4L, and 4H groups were 1.06, 1.05, and 1.05, respectively.

The viscero- and hepato-somatic indices of the alga-fed groups were not significantly different from the values of the control group (Table 3). The SGR of the alga-fed groups were similar to that of the control group. The FCR was significantly lower for the control group compared to the algae-fed groups. PER also did not vary significantly among the groups.

Table 3. Growth, feed performance, organosomatic indices and survival rate of Atlantic salmon from the different groups.

	4C	4L	4H
Initial weight (g)	166.61 ± 2.10	165.90 ± 1.95	170.30 ± 2.00
Final weight (g)	358.60 ± 3.16[a]	343.90 ± 4.61[b]	359.80 ± 4.57[a]
Condition factor	1.06 ± 0.01	1.05 ± 0.01	1.05 ± 0.02
SGR (% day-1)	1.10 ± 0.01	1.04 ± 0.01	1.07 ± 0.01
FCR (g feed/g gain)	0.81 ± 0.01[a]	0.86 ± 0.01[b]	0.90 ± 0.01[b]
PER	2.29 ± 0.03	2.32 ± 0.03	2.36 ± 0.03
HSI	1.30 ± 0.08	1.29 ± 0.08	1.30 ± 0.05
VSI	8.11 ± 0.26	8.06 ± 0.36	8.29 ± 0.24
Survival (%)	100	99.33 ± 0.67	100

n = 6 tanks, values from 25 fish/tank, mean ± s.e.m. Different superscripts in a row indicate statistically significant differences.
Condition factor, CF = (Final weight/Total length^3) × 100
Specific growth rate, SGR = 100 × ((ln Final mean fish weight in a tank - ln Initial mean fish weight in a tank)/Number of feeding days)
Feed conversion ratio, FCR = (Apparent feed intake in dry basis/(Final biomass-Initial biomass + Biomass of dead fish))
Protein efficiency ratio, PER = ((Final biomass-Initial biomass)/Protein intake)
Survival = 100 × (Final fish number/Initial fish number).

The protein and lipid digestibility of the experimental feeds did not differ significantly. The digestibility of energy in the 10% alga-containing feed was lower compared to that of the control feed (Table 4).

Table 4. Digestibility of the different feeds.

	4C	4L	4H
Protein	84.57 ± 0.33	83.08 ± 0.47	83.61 ± 0.32
Lipid	89.04 ± 1.74	88.30 ± 1.05	93.55 ± 0.16
Energy	81.02 ± 0.75[a]	78.28 ± 0.63[b]	79.70 ± 0.07[ab]

All values are calculated using the dry basis values.
Faeces from 23 fish/tank were pooled and freeze dried. Faeces from 2 tanks of a particular group were pooled prior to analysis; n = 3, mean ± s.e.m. Different superscripts in a row indicate statistically significant differences.

The proximate composition of the whole body of fish from the alga-fed groups was similar to that of the control group (Table 5). In the case of the filet from fish fed 10% of the alga the lipid content was lower compared to the control group. Furthermore, the ash content of the filet of fish fed 20% of the alga was relatively lower compared to those of the control group and the fish fed 10% of the alga. The protein content in the whole body and filet of the alga-fed groups were higher and lower, respectively, compared to the control group, although not statistically significant.

Table 5. Proximate composition of Atlantic salmon (g/100 g dry basis) from the different groups.

Group	Moisture	Protein	Lipid	Ash
WHOLE BODY				
Initial	74.75 ± 0.51	65.89 ± 1.17	27.66 ± 1.30	8.94 ± 0.31
Final				
4C	69.13 ± 0.30	56.44 ± 0.73	32.68 ± 0.91	6.21 ± 0.21
4L	68.77 ± 0.49	57.81 ± 1.01	31.76 ± 0.80	6.62 ± 0.23
4H	68.64 ± 0.32	58.02 ± 0.43	32.46 ± 0.82	6.57 ± 0.21
FILLET				
Initial	73.17 ± 0.67	78.94 ± 2.07	7.48 ± 0.30	6.07 ± 0.37
Final				
4C	73.84 ± 0.22	79.71 ± 1.30	16.73 ± 0.56^{a}	6.76 ± 0.24^{a}
4L	73.08 ± 0.39	75.00 ± 1.97	12.75 ± 0.39^{bc}	6.53 ± 0.08^{a}
4H	73.47 ± 0.25	75.19 ± 1.20	14.84 ± 0.75^{ac}	5.57 ± 0.18^{b}

n = 6 tanks, values from 6 fish/tank; mean ± s.e.m. Different superscripts in a column indicate statistically significant differences.

Antioxidant Status

The total antioxidant capacity as well as the catalase activity of the alga-fed fish was similar to that of the control group (Figure 1). The superoxide dismutase activity of the alga-fed groups was higher than that in the control group, although a statistical significance was not detected (Figure 1). In addition, the mRNA levels of *sod1* and*nrf2* in the different experimental groups were similar (Figure 2).

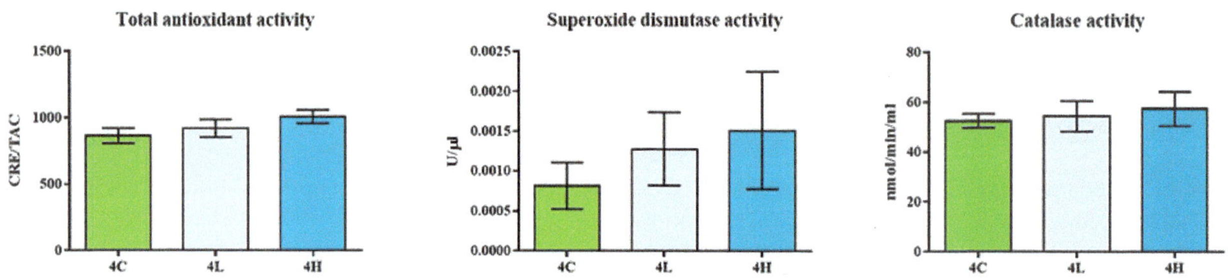

Figure 1. Serum antioxidant capacity of Atlantic salmon from the different groups. Data is generated from 12 fish of each group (*n* = 6 tanks, 2 fish/tank) and the values are expressed as mean ± s.e.m.

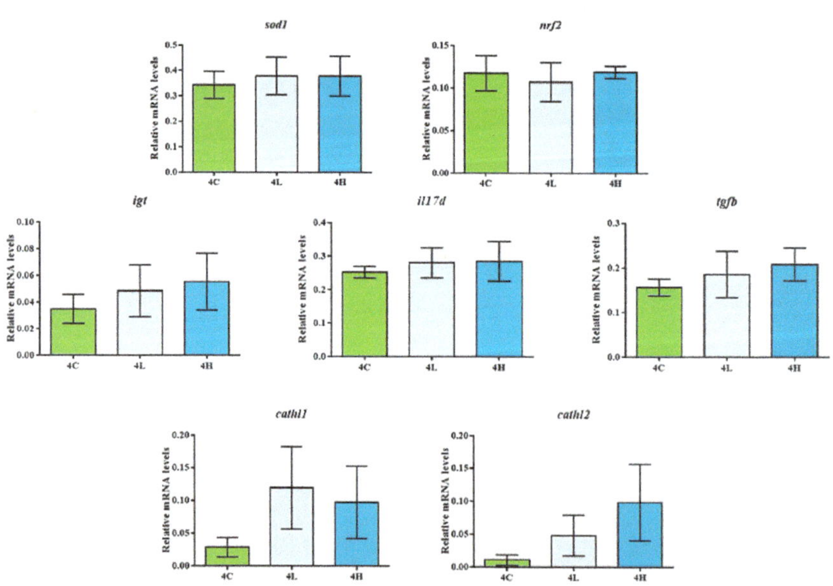

Figure 2. Relative mRNA levels of ***sod*, *nrf2*, *igt*, *il17d*, *tgfb*, *cathl1*, *cathl2*** in the distal intestine of Atlantic salmon from the different groups. The values are presented as mean ± s.e.m.; *n*= 6 tanks, 2 fish/tank.

Intestinal Health

Concerning the expression of the selected genes in the distal intestine of Atlantic salmon, the mRNA levels of the gut mucosa-linked immunoglobulin *igt* in the alga-fed groups were similar to that of the control group (Figure2). The expression of inflammation-related genes, *il1b, il17a, il10, tnfa* were below the detection limit. The mRNA levels of *il17d* in the alga fed groups were similar to the value in the control group (Figure 2). The expression of *tgfb* was similar in all the groups (Figure 2). Among the antimicrobial genes examined, the expression of *defb2* was not detectable. The expression of *cathl1* and *cathl2* were higher in the alga-fed groups compared to the values of the control group (Figure 3). However, a statistical significance was not detected.

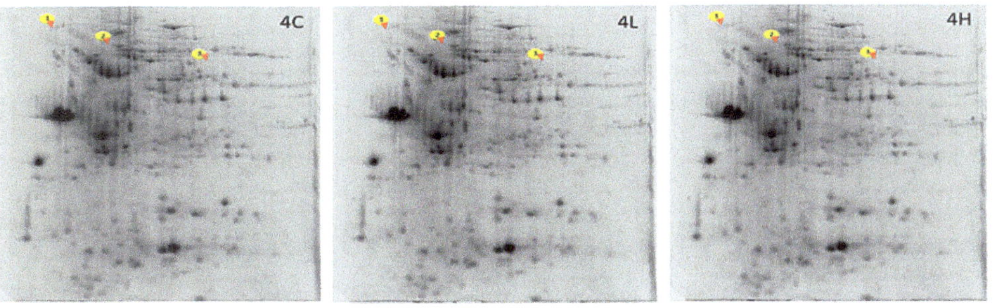

Figure 3. Representative 2-DE gels showing the spots of proteins from the distal intestine of Atlantic salmon. For generating the gels, the intestinal proteins from the distal intestine of Atlantic salmon from the 4C, 4L, and 4H groups were first isoelectrically focused on 17 cm IPG strips (pI 3-10) and then subjected to 12.5% SDS-PAGE. The Sypro®Ruby protein gel stain was used to stain the gels. The spots 1, 2, and 3 corresponds to Ahsg, My11 and Dld, respectively (see Table 6). $n = 6$ tanks, 1 fish/tank.

Analysis of the intestinal proteome of Atlantic salmon revealed that the expressions of Alpha-2-HS-glycoprotein-like (Ahsg), Myosin-11 isoform X1 (My11), Dihydrolipoyl dehydrogenase, mitochondrial-like (Dld) were altered by algal feeding (Figure 3, Table 6; Supplementary Figure 2). Dld was underexpressed in the alga-fed groups (10% − 0.45-fold, $p < 0.05$; 20% − 0.49-fold, $p > 0.05$) compared to the expression in the control group. Ahsg was underexpressed in the alga-fed groups (0.55 and 0.50-fold, $p < 0.05$) compared to the protein expression in the control group. My11 was underexpressed in the 20% alga-fed group (0.6-fold, $p < 0.05$) compared to the expression in the control group.

Table 6. Information of the proteins that were differentially expressed in the distal intestine of Atlantic salmon from different groups.

Spot no.	Protein name	Apparent pI/MW (kDa)	Peptides[a]
1↓	Alpha-2-HS-glycoprotein-like (Ahsg)	3.2/143.5	**YALNQIDDIK** **VVTAVEGDCDVVLR** **KVGGVMSVTAFK** **VGGVMSVTAFK** **ESLFAIMEVGR**
2↓	Myosin-11 isoform X1 (My11)	5.1/99.9	**DVSTLSAQVQDTQELLAEETR** **NSLQEQLEEEVEAK** **KLEEMAGNNELLEESK** **DLEASNTQFEEK** KFDQMLAEEK **ALEEAQDSREELER** **TEMEDLVSSK** **GLEAQVEEMK** **LEVNMQALK** **ELETELEDER** **DLEGQIEMSNK** **ISQLEEELEEESSNMEILNDR** **KSTQQVDQLNNELQTER** **STQQVDQLNNELQTER**
3↓	Dihydrolipoyl dehydrogenase, mitochondrial-precursor (Dld)	7.6/88.6	**GIEISGITLNLEK** **VTAVEFLGHVGGLGIDMEISK** **RPDGQIDVAVEAAAGGK** **NLGLDTVGLELDNR** **VPSIYAIGDVIAGPMLAHK** **EANLAASFGK**

[a] *Unique peptides in bold; ↓ indicates underexpression in the algal groups compared to the control group.*

Algal feeding did not affect the distal intestinal morphology of Atlantic salmon. The normal structure of lamina propria and the appearance of goblet cells are evident from the photomicrographs (Figure 4; Supplementary Figure 1).

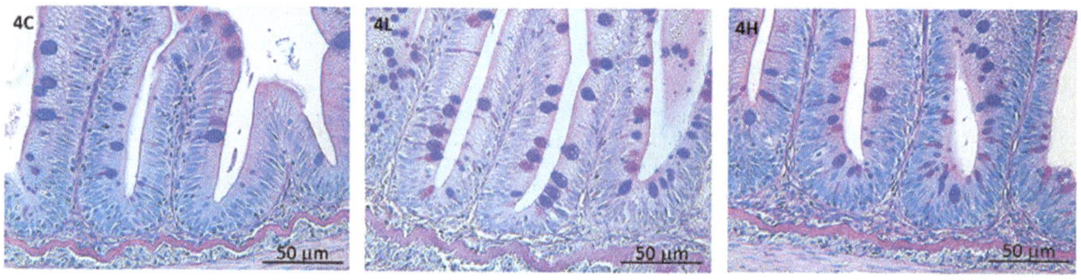

Figure 4. Photomicrographs of the distal intestine of Atlantic salmon from the different groups. Representative images show the normal architecture of the intestine. Scale bar: 50 µm. *n* = 6 tanks, 1 fish/tank.

DISCUSSION

The present study reports the effects of replacement (10 or 20%) of fishmeal with the microalgal biomass of*Desmodesmus* sp. in the feeds of Atlantic salmon. *Desmodesmus*-containing feeds did not significantly alter the specific growth rate and health indices of Atlantic salmon.

Atlantic salmon is a carnivorous fish and requires quality protein for their normal growth and development. As per the National Research Council report, Atlantic salmon should be provided with 36% digestible protein via feeds (National Research Council of the National Academies, 2011). The experimental feeds used in the present study had protein content (dry basis) in the range 478–538 g/kg.

Growth and Feed Conversion Efficiency upon Feeding *Desmodesmus* to Atlantic salmon

Overall, the fish performed well, and the weights of the fish after 70 days were twice that of the initial values. The results indicate that defatted *Desmodesmus* sp. can be included (up to 20%) in the feeds for Atlantic salmon. The final weights of the 4C and 4H groups were similar while that of the 4L group was significantly lower than that of the other groups. Higher FCR of the alga-fed groups indicates that nutrient availability in the alga-containing feeds was poorer than that in the 4C feed, although this was reflected only in the energy digestibility of the 4L group. Different studies have reported the effects of feeding algae to omnivorous fish. Nile tilapia (*Oreochromis niloticus*) larvae/juveniles, when fed with 44% *Spirulina* for 11 weeks, had higher SGR and PER, and good FCR values compared to the corn gluten-based control feed (Hussein et al., 2013). The fry of the same fish fed on 10 g of live *Spirulina* (*Arthrospira platensis*)/kg of feed had optimum growth and feed efficiency (Abdel-Tawwab and Ahmad, 2009). On the contrary, feeding 25% of *Chlorella* spp. or 23% of *Scenedesmus* spp. for 90 days led to poor growth, FCR and PER of Nile tilapia fingerlings (Badwy et al., 2008).

Studies have indicated the ability of carnivorous fish to make use of proteins from microalgae. Rainbow trout (*Oncorhynchus mykiss*) can tolerate up to 68% *Spirulina* (*A. maxima*, 35% algal protein) (Matty and Smith, 1978). On the other hand, an eight-week feeding study on rainbow trout fry indicated a significant reduction in the growth of the fish that received 25 and 50% algae (a consortium of algal biomass consisting of Chlorophyceae and Cyanobacteria) in their diets (Dallaire et al., 2007). The authors concluded that the maximum incorporation level should be 12.5% to avoid negative effects on growth, body lipid and energy content of the fish. The growth and feed performance of longfin yellowtail (*Seriola rivoliana*) was affected upon feeding with the microalga*Schizochytrium limacinum* (Kissinger et al., 2016). Feeding European sea bass (*Dicentrarchus labrax*) with freeze-dried *Isochrysis* sp. did not alter the SGR, FCR, and PER of the fish (Tibaldi et al., 2015). These indices in gilthead sea bream (*Sparus aurata*) juveniles were also not affected by feeding with *Scenedesmus almeriensis*(38%) (Vizcaíno et al., 2014). Similarly, feeding juvenile Atlantic salmon with land animal- or plant-protein based feeds that contained *Spirulina* meal (11%) did not alter the growth, feed performance, and survival of the fish (Burr et al., 2012). On the contrary, *Entomoneis* spp. (5%) in the diets of Atlantic salmon did not alter the growth or PER, but lowered the FCR (Norambuena et al., 2015). In the present feeding study, we did not observe any difference in either the specific growth rate (1.04–1.07) or protein efficiency ratio (2.32–2.36) of salmon that consumed the defatted meal of *Desmodesmus* sp.

Organosomatic indices are indicators of the general well-being of an organism (Dethloff and Schmitt, 2000). The HSI and VSI of sturgeon (*Acipenser baeri*) fed on 60% of an Indian strain of *Spirulina* were 3.07 and 9.49, respectively; the VSI values were significantly higher than that of the control group (Palmegiano et al., 2005). Fifteen percent replacement of fishmeal with *Spirulina pacifica* in the diets of parrot fish (*Oplegnathus fasciatus*) also did not alter the HSI but increased the VSI compared to the control group (Sung-Sam et al., 2013). Furthermore, the HSI value of longfin yellowtail was not altered by feeding with *S. limacinum* (Kissinger et al., 2016). In the present study, the VSI (8.1–8.3) and HSI (1.3) of the algal groups were not significantly different from the values of the control group.

The apparent digestibility coefficients of diets containing microalgae have been evaluated for farmed fishes. A study on European sea bass reported that the ADC values of protein, lipid and energy in the diets containing freeze-dried *Isochrysis* sp. (14%) were 92.6, 87.6, and 85.0, respectively (Tibaldi et al., 2015). The digestibility of protein in fingerling mirror carp and rainbow trout feeds that contained *A. maxima* (68%) were 87.1 and 83.1, respectively (Atack et al., 1979). In Atlantic salmon, the digestibility of protein and lipid of feeds containing*Entomoneis* spp. (5%) were 83 and 87, respectively (Norambuena et al., 2015). The protein, lipid and energy digestibility values of an algal product made from *Schizochytrium* sp. (20%) in the feeds of rainbow trout were 87.3, 53, and 60, respectively (Zhang, 2013). In the present study, the protein, lipid and energy digestibility values in the 20% alga-containing feeds were 84, 94, and 80, respectively. The digestibility of energy of the feeds containing 10% alga was significantly different from that of the control feed. The lower digestibility of energy in the 4L feed coincided with the lower final weights and the filet lipid content of the group. However, the higher inclusion of the alga did not cause a significant reduction in energy digestibility.

The protein, lipid and ash content of the whole body and filet of Atlantic salmon were analyzed. Algal feeding did not significantly alter the proximate composition of the whole body of Atlantic salmon, but the protein content in the whole body and fillet of the alga-fed groups was relatively higher and lower, respectivley. Feeding red sea bream (*Pagrus major*) for 95 days on a diet supplemented with 2% *Spirulina* sp. elevated the protein synthesis of the fish, and the stromal fraction of the muscle protein was significantly higher in the *Spirulina*-fed group (Mustafa et al., 1994). A study on Nile tilapia revealed that feeding 25% of *Chlorella* spp. or 23% of *Scenedesmus*spp. for 90 days lowered carcass protein content, but increased both fat and ash contents (Badwy et al., 2008). On the other hand, the protein and lipid content of the whole body of Nile tilapia fry fed on 10 g live *A. platensis*/kg of feed were significantly higher than the values of the control group (Abdel-Tawwab and Ahmad, 2009). The body protein content was higher, but the lipid content was lower in the *Spirulina*-fed Nile tilapia larvae/juveniles than the values in the control fish that were fed on corn gluten meal-based feed (Hussein et al., 2013). In the current study, the lipid and ash content were lower in the filet of Atlantic salmon fed 10 and 20% of*Desmodesmus* sp., respectively.

Health Status upon Feeding *Desmodesmus* to Atlantic Salmon

Plant ingredients are associated with anti-nutritional factors, which can reduce the bioavailability of some nutrients and negatively affect the growth and health of fish. For instance, there are many anti-nutritional substances in soybean, and these are suspected to induce allergy and alter the growth of fish. It is reported that feeding rainbow trout with high levels (60%) of soya or offering Atlantic salmon with 3.5% soybean agglutinin led to binding of the agglutinin to the intestinal enterocytes, which in turn caused aberrations of the intestinal structure of the carnivorous fishes (Buttle et al., 2001). One of the antinutritional factors, β-conglycinin was found to decrease the antioxidant capacity indices such as SOD, CAT, glutathione-S-transferase, glutathione peroxidase and glutathione reductase activities and glutathione content in the intestine and enterocytes of juvenile Jian carp (*Cyprinus carpio*) (Zhang et al., 2013). Furthermore, feeding Atlantic salmon with genetically modified maize is reported to elevate the SOD and CAT activity in the distal intestine of the fish (Sagstad et al., 2007). However, based on the results of the present study, *Desmodesmus*-feeding did not significantly alter the serum antioxidant indices (TAC, SOD and CAT) and the associated genes (*sod1, nrf2*) in the intestine of alga-fed groups. *sod1* instructs the synthesis of SOD enzyme and nuclear factor erythroid 2–related factor 2 (Nrf2) controls the expression of antioxidant response element–dependent genes (Ma, 2013). GSTA2 (glutathione S-transferase A2) and NQO1 (NADPH: quinone oxidoreductase 1) are two Nrf2-regulated genes (Ma, 2013), and*gstp1* and *nqo1* were altered in zebrafish exposed to toxins (Lam et al., 2008).

Cytokines, namely TGFβ and IL10 are mediators of anti-inflammatory responses, and tumor necrosis factor (TNF) and IL-17A are responsible for pro-inflammatory reactions. IL-17D is important in local immune responses and repair of tissues after structural damage (Starnes et al., 2002). *il17a* and *tfgb* were upregulated in the intestine of Atlantic salmon during the early and/or late inflammatory stages (Marjara et al., 2012), and *il1b*was reduced after recovery from inflammation (Vasanth et al., 2015). Oral intubation of 10 mg of *A. platensis*suspension per carp (*C. carpio*) increased the phagocytic activity, superoxide anion production, *il1b* and *tnfa*, but decreased *il10* in head kidney leukocytes (Watanuki et al., 2006). In the present study, the expression of *il1b, tnfa, il17a* and *il10* were below the detection limit. Marjara et al. (2012) reported the low expression of certain genes including *tnfa* in the distal intestine of Atlantic salmon. In the present study, the mRNA levels of *tgfb* and*il17d* of the algal groups were similar to those of the control fish. The transcripts of *il1b* and *tnfa* were not altered in the intestine of Atlantic salmon fed olive oil, rapeseed oil or soybean oil (Moldal et al., 2014). However, soybean oil feeding was found to lower the *igm* and *igt* levels in the distal intestine of the fish (Moldal et al., 2014). The feeding of the algal biomass in this study did not alter the level of *igt* in the distal intestine of Atlantic salmon. Cathelicidins, a group of antimicrobial peptides, have been implicated in intestinal inflammatory conditions in humans (Schauber et al., 2006). In the present study, the mRNA levels of *cathl1* and *cathl2* were higher than that in the control group, but not significantly different.

We did not observe any inflammatory signals in the micromorphology of the distal intestine. Similarly, feeding longfin yellowtail with *S. limacinum* (Kissinger et al., 2016) or gilthead sea bream with *S. almeriensis* (Vizcaíno et al., 2014) did not alter the intestinal morphology of the fishes.

There are not many studies that have examined the diet-related changes in the intestinal proteins of fish. Distal intestinal proteins (Ahsg, My11, Dld) associated with immune defense, motility of actin filaments and energy metabolism were underexpressed in Atlantic salmon that consumed *Desmodesmus* sp. The functions of AHSG in humans include regulation of protease activity, pathogenesis of colorectal cancer, monocyte recruitment to bone, and bone metabolism (Crawford, 1984; Malone and Richards, 1987; Katsuhito et al., 2011; Fan et al., 2014). Upregulation of AHSG is considered as an indication of squamous cell carcinoma (Tian et al., 2015). Reduction in levels of this glycoprotein in the serum of protein-energy-malnourished children was associated with stunted growth, and compromised defense ability (Abiodun et al., 1985). Although Ahsg was significantly lower in the 4L and 4H groups, the growth of the fish and the expression of immune defense genes in the groups were not significantly different from those of the control group. A myosin type was also altered in the algal groups—myosins are motor proteins that bind to actin, and they convert chemical energy into mechanical work, which enables them to move along actin filaments (Heissler and Sellers, 2016). My11 heavy chain is expressed in the smooth muscle cells of the gastrointestinal tract of mammals, and the expression of the protein was unchanged by hypertrophy (Park et al., 2015), which is associated with inflammation (Blennerhassett et al., 1992). In the present study, My11 was underexpressed in the distal intestine of salmon fed 20% of *Desmodesmus* sp. and we did not observe any inflammation in the distal intestine. DLD is associated with energy metabolism, and this TCA cycle-associated protein was overexpressed in the gut of black tiger shrimp, *Penaeus monodon*, that were orally intubated with white spot syndrome viral antigens (Kulkarni et al., 2014). On the other hand, Dld was underexpressed in the alga-fed groups, significantly in the case of fish that received 10% of the algal biomass. Interestingly, the energy digestibility of feeds that contained the algal biomass was lower compared to that of the control feed, significantly in the case of feeds with 10% of the algal biomass. A decrease in Dld was suggested to help probiotic bacteria such as *Lactobacillus fermentum* to survive in the small intestine (Yang et al., 2007). The physiological relevance of the underexpression of the three proteins in the distal intestine needs to be verified through further studies.

CONCLUSION

Atlantic salmon is able to tolerate the defatted biomass from *Desmodesmus* sp. The results indicate that 20% inclusion level in the feeds of Atlantic salmon does not have any adverse effects on the specific growth rate, condition factor, protein efficiency ratio and whole body proximate composition of fish. Furthermore, the algal inclusion did not impair the health of Atlantic salmon, as evident from the physiological indices, molecular markers of inflammation and intestinal micromorphology. Future studies should consider evaluating the effects of algal inclusion above 20% in long-term feeding studies on salmon, after optimizing the ingredient and feed processing to increase the digestibility of algal nutrients and reduce the feed conversion ratio. Thus, the*Desmodesmus* biomass can effectively replace fishmeal in Atlantic salmon feeds. The defatted microalgal biomass can also be a potential ingredient in the feeds of other aquatic animals.

AUTHOR CONTRIBUTIONS

VK conceived and led this study, and wrote the manuscript. MS was involved in the experimental design and scientific discussions and edited the manuscript. As the leader of the DoE project, MH participated in the project development and revised the manuscript. GV took care of the fish experiments and conducted the gene and protein studies. YG and AP performed the biochemical analysis. DD carried out the histological studies.

Conflict of Interest Statement

The authors declare that the research was conducted in the absence of any commercial or financial relationships that could be construed as a potential conflict of interest.

ACKNOWLEDGMENTS

We thank the Department of Energy (DoE), USA for funding the project "Large-scale production of fuels and feed from marine microalgae" (DE-EE0003371) that was awarded to the Cornell Marine Algae Biofuels Consortium. We thank all the members of the Cornell Marine Algae Biofuels Consortium, especially Charles Green, who led the team. Cellana, San Diego, USA is thanked for providing the algal biomass produced at their facilities in Kailua-Kona, Hawaii. The authors would like to acknowledge the support of the staff at the Research Station of Nord University.

Supplementary Material

The Supplementary Material for this article can be found online at:http://journal.frontiersin.org/article/10.3389/fmars.2016.00067

Supplementary Figure 1. Photomicrographs of the distal intestine of Atlantic salmon from the different study groups. Scale bar: 50 µm.

Supplementary Figure 2. 2-DE gels and the volumes of the differentially expressed protein spots. *Different letters above the bar graphs indicate statistically significant differences. Values are presented as mean ± s.e.m.

REFERENCES

1. Abdel-Tawwab, M., and Ahmad, M. H. (2009). Live Spirulina (*Arthrospira platensis*) as a growth and immunity promoter for Nile tilapia,*Oreochromis niloticus* (L.), challenged with pathogenic *Aeromonas hydrophila*. *Aquacult. Res.* 40, 1037–1046. doi: 10.1111/j.1365-2109.2009.02195.x
2. Abiodun, P. O., Ihongbe, J. C., and Dati, F. (1985). Decreased levels of alpha$_2$ HS-glycoprotein in children with protein-energy-malnutrition. *Eur. J. Pediatr.* 144, 368–369. doi: 10.1007/BF00441779
3. Atack, T. H., Jauncey, K., and Matty, A. J. (1979). The utilization of some single cell proteins by fingerling mirror carp (*Cyprinus carpio*).*Aquaculture* 18, 337–348. doi: 10.1016/0044-8486(79)90037-1
4. Austreng, E. (1978). Digestibility determination in fish using chromic oxide marking and analysis of contents from different segments of the gastrointestinal tract. *Aquaculture* 13, 265–272. doi: 10.1016/0044-8486(78)90008-X
5. Badwy, T. M., Ibrahim, E. M., and Zeinhom, M. M. (2008). "Partial replacement of fish meal with dried microalga (*Chlorella* spp. and*Scenedesmus* spp.) in Nile tilapia (*Oreochromis niloticus*) diets," in *8th International Symposium on Tilapia in Aquaculture* (Cairo), 801–811.
6. Baeverfjord, G., and Krogdahl, A. (1996). Development and regression of soybean meal induced enteritis in Atlantic salmon, *Salmo salar* L., distal intestine: a comparison with the intestines of fasted fish. *J. Fish Dis.* 19, 375–387. doi: 10.1111/j.1365-2761.1996.tb00376.x
7. Bakke, A. (2011). "Pathophysiological and immunological characteristics of soybean meal-induced enteropathy in salmon: contribution of recent molecular investigations," in *Avances en Nutrición Acuícola XI - Memorias del Décimo Primer Simposio Internacional de Nutrición Acuícola*, eds L. E. Cruz-Suárez, D. Ricque-Marie, M. Tapia-Salazar, M. G. Nieto-López,

D. A. Villarreal-Cavazos, J. Gamboa-Delgado, and L. Hernández-Hernández (Monterrey: Universidad Autónoma de Nuevo León), 345–372.

8. Bancroft, J. D., and Gamble, M. (2007). *Theory and Practice of Histological Techniques*. London, UK: Churchill Livingstone.

9. Becker, E. W. (2007). Micro-algae as a source of protein. *Biotechnol. Adv.* 25, 207–210. doi: 10.1016/j.biotechadv.2006.11.002

10. Becker, W. (2004). "Microalgae in human and animal nutrition," in *Microalgal Culture*, ed A. Richmond (Oxford: Blackwell Science Ltd.), 312–351.

11. Blennerhassett, M. G., Vignjevic, P., Vermillion, D. L., and Collins, S. M. (1992). Inflammation causes hyperplasia and hypertrophy in smooth muscle of rat small intestine. *Am. J. Physiol.* 262, G1041–G1046.

12. Burr, G. S., Wolters, W. R., Barrows, F. T., and Hardy, R. W. (2012). Replacing fishmeal with blends of alternative proteins on growth performance of rainbow trout (*Oncorhynchus mykiss*), and early or late stage juvenile Atlantic salmon (*Salmo salar*). *Aquaculture* 334–337, 110–116. doi: 10.1016/j.aquaculture.2011.12.044

13. Buttle, L. G., Burrells, A. C., Good, J. E., Williams, P. D., Southgate, P. J., and Burrells, C. (2001). The binding of soybean agglutinin (SBA) to the intestinal epithelium of Atlantic salmon, *Salmo salar* and rainbow trout, *Oncorhynchus mykiss*, fed high levels of soybean meal. *Vet. Immunol. Immunopathol.* 80, 237–244. doi: 10.1016/S0165-2427(01)00269-0

14. Crawford, S. M. (1984). Alpha-2 HS glycoprotein in the hypercalcaemia of multiple myeloma. *Br. J. Cancer* 49, 813–815. doi: 10.1038/bjc.1984.127

15. Dallaire, V., Lessard, P., Vandenberg, G., and de la Noüe, J. (2007). Effect of algal incorporation on growth, survival and carcass composition of rainbow trout (*Oncorhynchus mykiss*) fry. *Bioresour. Technol.* 98, 1433–1439. doi: 10.1016/j.biortech.2006.05.043

16. Dethloff, G. M., and Schmitt, C. J. (eds.). (2000). "Condition factor and organo-somatic indices," in *Information and Technology Report USGS/BRD-2000–0005: Biomonitoring of Environmental Status and Trends (BEST) Program: Selected Methods for Monitoring Chemical Contaminants and Their Effects in Aquatic Ecosystems* (Columbia, IN: U.S. Geological Survey, Biological Resources Division), 13–18.

17. Energy.Gov: Office of Energy Efficiency Renewable Energy (2016). *Biomass Feedstocks* [Online]. U.S. Department of Energy. Available online at:http://energy.gov/eere/bioenergy/biomass-feedstocks (Accessed March 24, 2016).

18. Enzing, C., Ploeg, M., Barbosa, M., and Sijtsma, L. (2014). "Microalgae-based products for the food and feed sector: an outlook for Europe," eds M. Vigani, C. Parisi, and E. R. Cerezo (Luxembourg: European Commission, Joint Research Centre - Institute for Prospective Technological Studies), 78. Available online at: http://ftp.jrc.es/EURdoc/JRC85709.pdf

19. Fan, N. J., Kang, R., Ge, X. Y., Li, M., Liu, Y., Chen, H. M., et al. (2014). Identification alpha-2-HS-glycoprotein precursor and tubulin beta chain as serology diagnosis biomarker of colorectal cancer. *Diagn. Pathol.* 9, 53. doi: 10.1186/1746-1596-9-53

20. Heissler, S. M., and Sellers, J. R. (2016). "Myosins," in *Encyclopedia of Cell Biology*, eds R. A. Bradshaw and P. D. Stahl (Waltham, MA: Academic Press), 597–607. doi: 10.1016/B978-0-12-394447-4.20059-X

21. Huntley, M. E., Johnson, Z. I., Brown, S. L., Sills, D. L., Gerber, L., Archibald, I., et al. (2015). Demonstrated large-scale production of marine microalgae for fuels and feed. *Algal Res.* 10, 249–265. doi: 10.1016/j.algal.2015.04.016

22. Hussein, E. E.-S., Dabrowski, K., El-Saidy, D. M. S. D., and Lee, B.-J. (2013). Enhancing the growth of Nile tilapia larvae/juveniles by replacing plant (gluten) protein with algae protein. *Aquacult. Res.* 44, 937–949. doi: 10.1111/j.1365-2109.2012.03100.x

23. Katsuhito, M., Masanori, E., and Masaaki, I. (2011). Fetuin-A: a multifunctional protein. *Recent Pat. Endocr. Metab. Immune Drug Discov.* 5, 124–146. doi: 10.2174/187221411799015372

24. Kiron, V., Phromkunthong, W., Huntley, M., Archibald, I., and De Scheemaker, G. (2012). Marine microalgae from biorefinery as a potential feed protein source for Atlantic salmon, common carp and whiteleg shrimp. *Aquacult. Nutr.* 18, 521–531. doi: 10.1111/j.1365-2095.2011.00923.x

25. Kissinger, K. R., García-Ortega, A., and Trushenski, J. T. (2016). Partial fish meal replacement by soy protein concentrate, squid and algal meals in low fish-oil diets containing *Schizochytrium limacinum* for longfin yellowtail *Seriola rivoliana*. *Aquaculture* 452, 37–44. doi: 10.1016/j.aquaculture.2015.10.022

26. Knudsen, D., Urán, P., Arnous, A., Koppe, W., and Frøkiaer, H. (2007). Saponin-containing subfractions of soybean molasses induce enteritis in the distal intestine of Atlantic salmon. *J. Agric. Food Chem.* 55, 2261–2267. doi: 10.1021/jf0626967

27. Kulkarni, A. D., Kiron, V., Rombout, J. H. W. M., Brinchmann, M. F., Fernandes, J. M. O., Sudheer, N. S., et al. (2014). Protein profiling in the gut of *Penaeus monodon* gavaged with oral WSSV-vaccines and live white spot syndrome virus. *Proteomics* 14, 1660–1673. doi: 10.1002/pmic.201300405

28. Lam, S. H., Mathavan, S., Tong, Y., Li, H., Karuturi, R. K. M., Wu, Y., et al. (2008). Zebrafish whole-adult-organism chemogenomics for large-scale predictive and discovery chemical biology. *PLoS Genet.* 4:e1000121. doi: 10.1371/journal.pgen.1000121

29. Ma, Q. (2013). Role of Nrf2 in oxidative stress and toxicity. *Annu. Rev. Pharmacol. Toxicol.* 53, 401–426. doi: 10.1146/annurev-pharmtox-011112-140320

30. Malone, J. D., and Richards, M. (1987). α2HS glycoprotein is chemotactic for mononuclear phagocytes. *J. Cell. Physiol.* 132, 118–124. doi: 10.1002/jcp.1041320116

31. Marjara, I. S., Chikwati, E. M., Valen, E. C., Krogdahl, Å., and Bakke, A. M. (2012). Transcriptional regulation of IL-17A and other inflammatory markers during the development of soybean meal-induced enteropathy

in the distal intestine of Atlantic salmon (*Salmo salar* L.). *Cytokine* 60, 186–196. doi: 10.1016/j.cyto.2012.05.027

32. Matty, A. J., and Smith, P. (1978). Evaluation of a yeast, a bacterium and an alga as a protein source for rainbow trout: I. Effect of protein level on growth, gross conversion efficiency and protein conversion efficiency. *Aquaculture* 14, 235–246. doi: 10.1016/0044-8486(78)90097-2
33. Moldal, T., Løkka, G., Wiik-Nielsen, J., Austbø, L., Torstensen, B. E., Rosenlund, G., et al. (2014). Substitution of dietary fish oil with plant oils is associated with shortened mid intestinal folds in Atlantic salmon (*Salmo salar*). *BMC Vet. Res.* 10:60. doi: 10.1186/1746-6148-10-60
34. Mustafa, M. G., Umino, T., and Nakagawa, H. (1994). The effect of *Spirulina* feeding on muscle protein deposition in red sea bream, *Pagrus major*. *J. Appl. Ichthyol.* 10, 141–145. doi: 10.1111/j.1439-0426.1994.tb00153.x
35. National Research Council of the National Academies (2011). *Nutrient Requirements of Fish and Shrimp*. Washington, DC: The National Academies Press.
36. Norambuena, F., Hermon, K., Skrzypczyk, V., Emery, J. A., Sharon, Y., Beard, A., et al. (2015). Algae in fish feed: Performances and fatty acid metabolism in juvenile Atlantic salmon. *PLoS ONE* 10:e0124042. doi: 10.1371/journal.pone.0124042
37. Norges Standardiseringsforbund (1994). "Norsk Standard 9402E," in *Atlantic Salmon Colour and Fat Measurement* (Oslo).
38. Palmegiano, G. B., Agradi, E., Forneris, G., Gai, F., Gasco, L., Rigamonti, E., et al. (2005). Spirulina as a nutrient source in diets for growing sturgeon (*Acipenser baeri*). *Aquacult. Res.* 36, 188–195. doi: 10.1111/j.1365-2109.2005.01209.x
39. Park, C., Lee, M. Y., Park, P. J., Ha, S. E., Berent, R. M., Fuchs, R., et al. (2015). Serum response factor is essential for prenatal gastrointestinal smooth muscle development and maintenance of differentiated phenotype. *J. Neurogastroenterol. Motil.* 21, 589–602. doi: 10.5056/jnm15063
40. Sagstad, A., Sanden, M., Haugland, Ø., Hansen, A. C., Olsvik, P. A., and Hemre, G. I. (2007). Evaluation of stress- and immune-response biomarkers in Atlantic salmon, *Salmo salar* L., fed different levels of genetically modified maize (Bt maize), compared with its near-isogenic parental line and a commercial suprex maize. *J. Fish Dis.* 30, 201–212. doi: 10.1111/j.1365-2761.2007.00808.x
41. Schauber, J., Rieger, D., Weiler, F., Wehkamp, J., Eck, M., Fellermann, K., et al. (2006). Heterogeneous expression of human cathelicidin hCAP18/LL-37 in inflammatory bowel diseases. *Eur. J. Gastroenterol. Hepatol.* 18, 615–621. doi: 10.1097/00042737-200606000-00007
42. Starnes, T., Broxmeyer, H. E., Robertson, M. J., and Hromas, R. (2002). Cutting edge: IL-17D, a novel member of the IL-17 family, stimulates cytokine production and inhibits hemopoiesis. *J. Immunol.* 169, 642–646. doi: 10.4049/jimmunol.169.2.642
43. Sung-Sam, K., Rahimnejad, S., Kang-Woong, K., and Kyeong-Jun, L. (2013). Partial replacement of fish meal with *Spirulina pacifica* in diets for parrot fish (*Oplegnathus fasciatus*). *Turkish J. Fish. Aquat. Sci.* 13, 197–204. doi: 10.4194/1303-2712-v13_2_01
44. Tian, W.-D., Li, J.-Z., Hu, S.-W., Peng, X.-W., Li, G., Liu, X., et al. (2015). Proteomic identification of alpha-2-HS-glycoprotein as a plasma biomarker of hypopharyngeal squamous cell carcinoma. *Int. J. Clin. Exp. Pathol.* 8, 9021–9031.
45. Tibaldi, E., Chini Zittelli, G., Parisi, G., Bruno, M., Giorgi, G., Tulli, F., et al. (2015). Growth performance and quality traits of European sea bass (*D. labrax*) fed diets including increasing levels of freeze-dried *Isochrysis* sp. (T-ISO) biomass as a source of protein and n-3 long chain PUFA in partial substitution of fish derivatives. *Aquaculture* 440, 60–68. doi: 10.1016/j.aquaculture.2015.02.002
46. Vandesompele, J., De Preter, K., Pattyn, F., Poppe, B., Van Roy, N., De Paepe, A., et al. (2002). Accurate normalization of real-time quantitative RT-PCR data by geometric averaging of multiple internal control genes. *Genome Biol.* 3, research0034.0031–research0034.0011. doi: 10.1186/gb-2002-3-7-research0034
47. Vasanth, G., Kiron, V., Kulkarni, A., Dahle, D., Lokesh, J., and Kitani, Y. (2015). A microbial feed additive abates intestinal inflammation in Atlantic salmon. *Front. Immunol.* 6:409. doi: 10.3389/fimmu.2015.00409
48. Vizcaíno, A. J., López, G., Sáez, M. I., Jiménez, J. A., Barros, A., Hidalgo, L., et al. (2014). Effects of the microalga *Scenedesmus almeriensis* as fishmeal alternative in diets for gilthead sea bream, *Sparus aurata*, juveniles. *Aquaculture* 431, 34–43. doi: 10.1016/j.aquaculture.2014.05.010
49. Watanuki, H., Ota, K., Tassakka, A. C. M. A. R., Kato, T., and Sakai, M. (2006). Immunostimulant effects of dietary *Spirulina platensis* on carp,*Cyprinus carpio*. *Aquaculture* 258, 157–163. doi: 10.1016/j.aquaculture.2006.05.003
50. Yang, F., Wang, J., Li, X., Ying, T., Qiao, S., Li, D., et al. (2007). 2-DE and MS analysis of interactions between *Lactobacillus fermentum* I5007 and intestinal epithelial cells. *Electrophoresis* 28, 4330–4339. doi: 10.1002/elps.200700166
51. Ytrestøyl, T., Aas, T. S., and Åsgård, T. (2015). Utilisation of feed resources in production of Atlantic salmon (*Salmo salar*) in Norway.*Aquaculture* 448, 365–374. doi: 10.1016/j.aquaculture.2015.06.023
52. Zhang, C. (2013). *Determination of the Digestibility of a Whole-Cell DHA-Rich Algal Product and Its Effect on the Lipid Composition of Rainbow Trout and Atlantic salmon*. Saskatoon: University of Saskatchewan.
53. Zhang, J.-X., Guo, L.-Y., Feng, L., Jiang, W.-D., Kuang, S.-Y., Liu, Y., et al. (2013). Soybean β-conglycinin induces inflammation and oxidation and causes dysfunction of intestinal digestion and absorption in fish. *PLoS ONE* 8:e58115. doi: 10.1371/journal.pone.0058115

CHAPTER 6 SUPPLEMENTATION WITH SODIUM SELENITE AND SELENIUM-ENRICHED MICROALGAE BIOMASS

Antonín Kouba, Josef Velíšek, Alžběta Stará, Jiří Masojídek, and Pavel Kozák

INTRODUCTION

Selenium (Se) is a trace element essential for all living organisms, where it is a constituent of selenoproteins [1,2]. In aquaculture, it is used as a supplement to live feed, improving their nutritional profile to fulfil requirements of cultured fish [3–5]. Elevation of dietary Se helps fish to withstand stress [6, 7] and moderates toxicity of heavy metals such as mercury and cadmium [8, 9]. Selenium contributes to stabilizing meat quality of farmed animals [10, 11]. Studies on humans show its anticancer properties [12, 13] and the positive effects on reproduction are well known [14, 15]. However, selenium-related studies also confirm that the concentration range in which Se is considered essential or toxic is quite narrow (see further). Various forms of Se supplementation (selenite, selenate, selenomethionine (Se-Met), selenocysteine (Se-Cys), and Se-enriched yeasts) have been studied in fish species including rainbow trout Oncorhynchus mykiss [16], Atlantic salmonSalmo salar [17, 18], channel catfish Ictalurus punctatus [19], hybrid striped bass Morone chrysops × M. saxatilis[20], common carp Cyprinus carpio [21], and crucian carp Carassius auratus gibelio [22]. Although Zhou et al. [23] confirmed that Se nanoparticles are more effective in increasing muscle Se content compared to Se-Met in the crucian carp, studies mentioned above confirmed that organic forms are more digestible, better accumulated, and more biologically active, while being less toxic than inorganic forms. Recently, supplementation of Se-enriched garlic, primarily containing selenium as –γ glutamyl selenomethylselenocysteine and selenomethylselenocysteine, has been proposed as an alternative in producing Se-enriched meat of African catfish [24, 25]. However, Se-enriched yeasts are currently the supplement of choice.

Feeding microalgae to fish has been used successfully to provide an important natural source of bioactive compounds, such as carotenoids, fatty acids, polysaccharides, amino acids, and vitamins [26]. Progress in microalgae cultivation has included development of fed-batch phototrophic cultivation of the Se-enriched microalgae (Masojídek et al., unpublished data). This source of Se has not been evaluated as a fish dietary supplement, so we compared Se-enriched microalgae biomass and sodium selenite as dietary supplements for indoor culture of common barbel (Barbus barbus L.) with respect to effects on fish growth, Se accumulation, and the activity of selected enzymes in certain tissues.

Citation: Antonín Kouba, Josef Velíšek, Alžběta Stará, Jiří Masojídek, and Pavel Kozák, "Supplementation with Sodium Selenite and Selenium-Enriched Microalgae Biomass," BioMed Research International, vol. 2014, Article ID 408270, 8 pages, 2014. doi:10.1155/2014/408270

MATERIALS AND METHODS

Diets

Experimental purified casein-based diets followed formulation of Wang et al. [22] with some modifications including the use of rapeseed oil and the commercially available vitamin/mineral premix (Roboran H, Univit Ltd.) with 8.39 g kg^{-1} replaced with microalgae Chlorella spp. (cf. C. vulgaris Beijerinck) biomass (Table 1). Experimental diets were supplemented with sodium selenite or Se-enriched microalgae to obtain Se concentrations of 0.3 mg kg^{-1} dry weight (dw) from sodium selenite (Se 0.3) and 0.3 and 1.0 mg kg^{-1} from Se-enriched microalgae biomass, (Algadiet 0.3 and Algadiet 1.0) (Table 2). A control diet contained Se-free microalgae biomass. Chemical analysis showed Se concentrations of 0.06, 0.29, 0.31, and 1.04 mg kg^{-1} dw in control, Se 0.3, Algadiet 0.3, and Algadiet 1.0 diet, respectively. An Se level near 0.3 mg kg^{-1} in the diet has been stated as a requirement for various fish species [1, 27], while a concentration of 1.0 mg kg^{-1} may be associated with the onset of oxidative damage leading to impaired antioxidant status [28, 29].

Table 1: Formulation, mean approximate composition, and gross energy of experimental diets.

Ingredients (g kg^{-1})	
Casein	320.0
Gelatine	80.0
Dextrine	280.0
Cellulose	190.0
Rapeseed oil	60.0
Carboxymethyl cellulose	20.0
Vitamin and mineral premix[a]	41.61
Microalgae biomass[b]	8.39
Proximate composition (g kg^{-1} dw)	
Dry matter	888.1
Crude protein	336.4
NFE[c]	474.6
Crude lipid	21.4
Crude fibre	139.6
Ash	28.0
Gross energy (MJ kg^{-1})[d]	16.9

Table 2: Sodium selenite, Se-free and/or Se-enriched microalgae supplement, and calculated concentration of selenium (mg kg^{-1} dw) in experimental diets.

Se source/treatment	Control	Se 0.3	Algadiet 0.3	Algadiet 1.0
Sodium selenite	–	+	–	–
Se-free[a]/Se-enriched[b] microalgae biomass ratio (%)	100/0	100/0	30/70	0/100
Calculated additive concentration	0.0	0.3	0.3	1.0

Components of artificial diets were sourced as follows: casein, gelatine, and dextrin were purchased from VWR International (Radnor, PA, USA). Cellulose and carboxymethyl cellulose were obtained from Alfa Aesar GmbH & Co KG (Karlsruhe, Germany). Microalgae were cultivated at the Institute of Microbiology in Třeboň, Czech Republic. Pure sodium selenite was supplied by Sigma-Aldrich (St. Louis, MO, USA). Ingredients were mixed in a commercial food mixer. Oil was gradually added during mixing. Water was slowly blended into the mix to obtain dough of suitable texture for processing through a cold extruder. Obtained spaghetti-like strips were manually cut, dried in a household fruit dryer at 60°C for 24 h, and frozen at −20°C until feeding. The composition of the diets (Table 1) was analyzed in the accredited laboratory of the State Veterinary Institute in Prague, Czech Republic.

Animals and Husbandry

Common barbel (Barbus barbus L.) yearlings weighing 48.4 ± 6.1 g (range 40.0–70.7 g) were stocked in outdoor experimental feeding cages (60 × 40 × 30 cm) supplied with microfiltrated (109 μm pore size) water from the Blanice River, South Bohemia. Initial standard body length and weight of stocked fish (Table 3) were not significantly different among groups ($F = 1.03$, $P = 0.381$, and $F = 0.16$, $P = 0.922$, resp.). Groups of 10 fish were acclimatized and fed the control diet for 10 days prior to the experimental period of 6 weeks. Fish were provided four equal feedings per day (08:00, 12:00, 15:00, and 18:00 h), initially at a rate of 1.5% of live weight to ensure that all feed was rapidly consumed. Feeding rate was adjusted according to batch weighing of fish every two weeks. Fish were not fed on the final day before sampling at the end of the experiment. Water temperature (19.2 ± 1.5°C) was registered hourly using Minikin loggers (Environmental Measuring Systems, Brno, Czech Republic). Dissolved oxygen concentration ((8.5 ± 0.8 mg L^{-1}) and pH (7.5 ± 0.1) were measured daily by means of oximeter Oxi 330i and pH meter pH/Cond 340i (WTW GmbH, Weilheim, Germany), respectively. The trial was conducted in triplicate.

Table 3: Initial and final size indices of common barbel (Barbus barbus L.) yearlings fed on control diet containing Se-free microalgae biomass and diets supplemented by either sodium selenite or Se-enriched microalgae (replacing Se-free microalgae) to obtain Se concentrations of 0.3 mg kg^{-1} from sodium selenite (Se 0.3) or 0.3 and 1.0 mg kg^{-1} from Se-enriched microalgae biomass, Algadiet 0.3 and Algadiet 1.0, respectively.

Growth indices/treatment	Control	Se 0.3	Algadiet 0.3	Algadiet 1.0
Initial standard body length (mm)	14.2 ± 0.7	14.3 ± 0.8	14.4 ± 0.6	14.4 ± 0.6
Initial weight (g)	47.8 ± 6.3	48.5 ± 6.9	48.9 ± 5.9	48.3 ± 5.5
Final standard body length (mm)	16.1 ± 0.9	16.3 ± 0.7	16.3 ± 0.8	16.5 ± 0.7
Final weight (g)	54.2 ± 6.9	52.1 ± 6.6	52.4 ± 5.5	54.6 ± 7.5

Sampling Protocol

At the end of the trial, fish were measured to the nearest 0.1 mm (standard length measured from the tip of the head to the base hypural plate at caudal flexion) and weighed to the nearest 1.0 g. Eight randomly selected fish from each replicate were sacrificed (stunned by a blow with a blunt object over the head) and sampled for biochemical analysis. The study was conducted according to the principles of the Ethical Committee for the Protection of Animals in Research of the University of South Bohemia, Faculty of Fisheries and Protection of Waters, Research Institute of Fish Culture and Hydrobiology, Vodňany, based on the EU harmonized animal welfare act of Czech Republic. The principles of laboratory animal care and the national laws 246/1992 and regulations on animal welfare were followed (Ref. number 22761/2009-17210).

Biochemical Blood Plasma Indices

Blood was drawn from the vena caudalis and samples stabilized with 40 IU sodium heparin per 1 mL. Blood was separated by centrifugation at 12 000 ×g for 10 min at 4°C, and plasma samples were held at −80°C until analysis. The following biochemical indices were evaluated: aspartate aminotransferase (AST), alanine aminotransferase (ALT), lactate dehydrogenase (LDH), and creatine kinase (CK). For the biochemical analysis of plasma, the VETTEST 8008 analyzer (IDEXX Laboratories Inc., USA) was used.

Tissue Samples and Preparation of Postmitochondrial Supernatant

After blood sampling, the muscle, liver, and intestine were quickly dissected, immediately frozen, and stored at −80°C. For analysis, frozen tissue samples were weighed and homogenized (1 : 10, w/v) with an Ultra Turrax homogenizer (Ika, Germany) using 50 mM potassium phosphate buffer, pH 7.0, containing 0.5 mM EDTA. The homogenate was divided into two portions, one for measuring thiobarbituric acid reactive substances (TBARS) and a second was centrifuged at 12 000 ×g for 30 min at 4°C to obtain the postmitochondrial supernatant for other antioxidant parameter analyses.

Indices of Oxidative Stress

The TBARS method was used to evaluate lipid peroxidation [30]. The concentration of TBARS was calculated by the absorption at 535 nm and a molar extinction coefficient of 156 mM cm^{-1}. The value was expressed as nanomoles of TBARS per mg of wet tissue.

Antioxidant Parameters

Total superoxide dismutase (SOD; EC 1.15.1.1) activity was determined by the method of S. Marklund and G. Marklund [31]. This assay depends on the autoxidation of pyrogallol. Superoxide dismutase activity was assessed spectrophotometrically at 420 nm and expressed as the amount of enzyme per milligram of protein. The catalase (CAT; EC 1.11.1.6) activity was assayed spectrophotometrically as H_2O_2 breakdown at 240 nm [32]. Glutathione peroxidase (GPx; EC 1.11.1.9) activity was calculated from the rate of NADPH oxidation at 340 nm in the reaction with glutathione reductase (GR; EC 1.6.4.2). The specific activity was determined using the extinction coefficient 6.22 mM cm^{-1} [33]. Glutathione reductase activity was determined spectrophotometrically, measuring NADPH oxidation at 340 nm [34]. One unit of

CAT, GPx, or GR activity is defined as the amount of enzyme that consumes 1 mol L^{-1} of substrate or generates 1 mol L^{-1} of product per min expressed in international units per mg of protein.

Protein Estimation

Protein levels were estimated spectrophotometrically (Tecan Infinite M200, Schoeller Instruments, Prague, Czech Republic) using bovine serum albumin as a standard [35].

Selenium Concentration

Concentration of selenium was determined in the microalgae biomass, experimental diets, and fish tissues by means of ICP-MS analysis. Muscle of four randomly selected individuals per replication was analysed. Insufficient liver tissue for analysis required that we pair target tissue from two randomly chosen fish (final $n = 4$ per replication). The detection limit of Se was 0.05 mg kg^{-1} dw in all matrices.

Statistical Analysis

Statistical analysis was carried out using Statistica software 9.0 for Windows (StatSoft, Czech Republic). Data were first tested for normality and homoscedasticity with Kolmogorov-Smirnov and Levene's tests, respectively. If those conditions were satisfied, one-way analysis of variance (ANOVA) was employed to reveal significant differences in measured variables among control and experimental groups. When a difference was detected ($P < 0.05$), Tukey's multiple comparison test was applied to identify which treatments were significantly different. If the conditions for ANOVA were not satisfied, the nonparametric Kruskal-Wallis test followed by multiple comparisons of mean ranks for all groups was used [36]. All data are expressed as means ± s.d.

RESULTS

No mortality was observed during the trial. Mean standard body length was 16.1 to 16.5 mm and weight was 52.1 to 54.6 g, at the end of the experiment (Table 3). Values were not significantly different ($F = 1.47$, $P = 0.228$, and $F = 0.93$, $P = 0.428$, resp.), and no clear trend of treatments was observed.

The final Se concentration in muscle of fish from the control group was 0.71 ± 0.08 mg kg^{-1} dw. A concentration of 0.80 ± 0.10 and 0.88 ± 0.09 mg kg^{-1} dw was found in the muscle of fish fed Se 0.3 and Algadiet 0.3, respectively, with the latter being significantly higher than that of the control ($P < 0.002$). The Algadiet 1.0 group showed Se concentration of 1.18 ± 0.12 mg kg^{-1} dw, significantly higher than other groups ($P < 10^{-3}$ in all cases). Similar trends with greater mean concentrations of Se were found in liver, 2.22, 2.92, 3.18, and 3.71 mg kg^{-1} dw in the control, Se 0.3, Algadiet 0.3, and 1.0 groups, respectively.

Activity of ALT and CK was significantly ($P < 0.01$) greater in the Se 0.3 and Algadiet 1.0 groups as compared to the control group. The AST and LDH activity was similar in all groups (Table 4).

Table 4: Blood enzymes activity in common barbel (Barbus barbus L.) yearlings fed on control diet containing Se-free microalgae biomass and diets supplemented by either sodium selenite or Se-enriched microalgae (replacing Se-free microalgae) to obtain Se concentrations of 0.3 mg kg^{-1} from sodium selenite (Se 0.3) or 0.3 and 1.0 mg kg^{-1} from Se-enriched microalgae biomass Algadiet 0.3 and Algadiet 1.0, respectively.

Indices	Test groups			
	Control	Se 0.3	Algadiet 0.3	Algadiet 1.0
AST (μkat L^{-1})	22.2 ± 5.6^{a}	23.6 ± 5.0^{a}	20.9 ± 5.1^{a}	19.7 ± 4.6^{a}
ALT (μkat L^{-1})	1.9 ± 0.7^{a}	2.6 ± 0.9^{b}	2.2 ± 0.8^{ab}	3.3 ± 0.7^{c}
LDH (μkat L^{-1})	32.0 ± 7.6^{a}	34.6 ± 7.0^{a}	31.4 ± 8.3^{a}	29.0 ± 6.5^{a}
CK (μkat L^{-1})	35.2 ± 8.6^{a}	42.5 ± 7.6^{b}	30.6 ± 6.4^{a}	42.2 ± 6.7^{b}

Among the test groups, none showed significant differences from the control in TBARS level, GPx, or SOD activity in muscle, liver, or intestine (Table 5).

Table 5: Oxidative stress and antioxidant responses in tissue of common barbel (Barbus barbus L.) yearlings fed on control diet containing Se-free microalgae biomass and on diets supplemented by either sodium selenite or Se-enriched microalgae (replacing Se-free microalgae) to obtain Se concentrations of 0.3 mg kg^{-1} from sodium selenite (Se 0.3) or 0.3 and 1.0 mg kg^{-1} from Se-enriched microalgae biomass Algadiet 0.3 and Algadiet 1.0, respectively.

Indices	Tissues	Test groups			
		Control	Se 0.3	Algadiet 0.3	Algadiet 1.0
TBARS (nmol g ww^{-1})	Muscle	8.4 ± 3.8^{a}	7.5 ± 4.4^{a}	7.1 ± 3.6^{a}	6.5 ± 3.6^{a}
	Liver	32.8 ± 9.1^{a}	33.3 ± 8.3^{a}	32.3 ± 9.8^{a}	31.4 ± 8.1^{a}
	Intestine	3.4 ± 1.1^{a}	3.1 ± 0.9^{a}	2.9 ± 0.8^{a}	2.8 ± 0.9^{a}
SOD (U mg^{-1} protein)	Muscle	21.9 ± 4.5^{a}	19.3 ± 5.4^{a}	20.5 ± 5.4^{a}	21.4 ± 4.2^{a}
	Liver	141.9 ± 21.8^{a}	127.1 ± 21.2^{a}	137.0 ± 15.3^{a}	164.0 ± 26.3^{a}
	Intestine	5.1 ± 1.6^{a}	4.8 ± 1.1^{a}	5.0 ± 1.3^{a}	4.3 ± 1.2^{a}
CAT (mU mg^{-1} protein)	Muscle	2.4 ± 0.8^{a}	3.7 ± 0.8^{b}	2.7 ± 1.0^{a}	4.6 ± 1.1^{c}
	Liver	26.9 ± 8.1^{a}	41.7 ± 8.6^{bc}	35.4 ± 7.7^{b}	47.1 ± 9.6^{c}
	Intestine	0.8 ± 0.3^{a}	0.9 ± 0.3^{a}	0.8 ± 0.2^{a}	0.8 ± 0.2^{a}
GPx (U mg^{-1} protein)	Muscle	66.4 ± 14.9^{a}	63.0 ± 10.7^{a}	64.8 ± 9.4^{a}	62.5 ± 8.5^{a}
	Liver	194.4 ± 45.1^{a}	185.6 ± 35.4^{a}	177.4 ± 32.9^{a}	198.2 ± 26.0^{a}
	Intestine	29.6 ± 6.6^{a}	29.5 ± 7.6^{a}	27.9 ± 6.4^{a}	25.0 ± 6.3^{a}
GR (U mg^{-1} protein)	Muscle	17.4 ± 7.3^{b}	7.3 ± 3.2^{a}	10.0 ± 3.6^{a}	6.8 ± 2.8^{a}
	Liver	6.0 ± 2.3^{a}	6.9 ± 2.0^{a}	6.2 ± 2.3^{a}	6.0 ± 2.6^{a}
	Intestine	8.1 ± 2.0^{a}	7.7 ± 2.7^{a}	7.1 ± 2.6^{a}	6.7 ± 2.5^{a}

The activity of CAT in liver was significantly () higher than the in the control in all Se-supplemented groups. In case of muscle, higher CAT activities were seen in the Se 0.3 and Algadiet 1.0 groups (Table 5) compared to the control. No differences were found in CAT activity in intestine. Significantly lower () muscle GR activity was observed in all treatment groups compared to the control. GR activity observed in liver or intestine did not differ from the enzyme activity in the control group.

DISCUSSION

Selenium is an important micronutrient in animals as well as in humans [37, 38]. Its deficiency has various negative impacts [8, 39, 40]; however, the line between Se nutritive requirements and toxicity is narrow [41–44]. Research has focussed on comparing doses and forms of Se supplementation. Studies on fish have confirmed that organic forms of Se are more digestible, better accumulated in tissue, and more biologically active than inorganic [6, 19, 21, 22]. In our experiment using Se-enriched microalgae biomass as feed supplement for common barbel, no significant differences in final fish size were found. However, Se concentration in muscle and liver of Se 0.3 and Algadiet 0.3 supplemented groups suggested that Se from Se-enriched microalgae is more readily accumulated in tissue.

Recently, Se-enriched yeasts containing 54–74% total Se in the form of selenomethionine (Se-Met) [45] have been promoted. This form of selenium is considered to facilitate adequate digestibility and consequent biological activity, since Se-Met is directly incorporated into protein in place of methionine (Met), as does not discriminate between Met and Se-Met [46]. Reports of content of Se-Met in Chlorella biomass are conflicting. De Alcantara et al. [47] assumed intracellularly fixed selenium in the Chlorella biomass to be Se-Met. More than 70% of protein-bound selenium has been found to be Se-Met, although this constituted only a minor fraction (0.7%) of the total Se in the Chlorella biomass according to Fan et al. [48]. Dimethylselenonium propionate has been reported as the principal Se-containing compound in Chlorella [49]. Neumann et al. [50] found 24–39% of total selenium to be in the form of Se-Met when incubating Chlorella in differing mineral-nutrient solutions. However, content of selenocysteine (Se-Cys) was higher, at 48 to 76%. Differing Se compounds in Se-enriched yeasts and Se-enriched microalgae might suggest using the former, as Se-Met is probably the most active compound. Se-enriched microalgae may, however, benefit from the presence of specific bioactive compounds such as antioxidants, pigments, fatty acids, polysaccharides, and immunoactive substances.

Blood enzyme levels provide vital information to aid in fish health assessment [51, 52]. Biochemical alterations are usually the first detectable and quantifiable responses to environmental change. Alteration of blood biochemical composition might be indicative of unsuitable environmental conditions or the presence of stress factors such as toxic chemicals, overcrowding, and some common aquaculture procedures [53, 54]. We analysed LDH, CK, and the transaminases ALT and AST on the basis that alterations in the activity of these enzymes might be stress induced and correspond to tissue damage. Increased ALT and CK activity indicates the induction of transamination processes that occur as amino acid input into the tricarboxylic acid cycle. In this way, fish may have coped with the energy deficit during stress associated with the inorganic Se 0.3 and organic Algadiet 1.0 treatments. Differences in activity of ALT and CK in groups with comparable selenium concentrations (Se 0.3 and Algadiet 0.3) suggest higher toxicity of the inorganic form. Reports concerning effects of selenium on fish blood biochemical profile are scarce. Increased activity of AST and ALT was found in African catfish fed 0.5 g kg^{-1} Se-enriched yeasts, suggesting that high Se levels in the diet exhibit a toxic effect related to liver and kidney dysfunction [55].

Selenium is essential for the proper functioning of the antioxidant enzymes, which protect against oxidative stress

[56, 57]. Maintaining cellular oxidative homeostasis involves the antioxidant defence system, including enzymes such as CAT, GR, SOD, and GPx, which can be employed as biomarkers of oxidative stress [29, 40,58]. In this study, the CAT activity in liver was significantly higher in all Se-supplemented groups, but in muscle was higher than the control group only in the Se 0.3 and Algadiet 1.0 groups. CAT activity prevents adverse effects of oxidative stress in cells. This enzyme is mainly located in the peroxisomes and is responsible for the reduction of hydrogen peroxide produced in the metabolic pathways of long-chain fatty acids [59]. Increased CAT activity may have indicated that its antioxidant defence was exhausted by the level of hydroperoxide products, reflecting a possible failure of the antioxidant system caused by the Se 0.3 and the higher dose of organic Se in the Algadiet 1.0 group. Increased CAT activity associated with elevated dietary selenium has been reported in various fish [21, 28, 29] as well as crustaceans [60–62].

Glutathione reductase activity maintains the cytosolic concentration of reduced glutathione [63, 64]. Significantly ($P < 0.01$) lower muscle GR activity was observed in all Se-supplemented groups compared to the control. Decreased GR activity may lead to the depletion of glutathione if its loss cannot be compensated for by the synthesis of new glutathione molecules [65]. Lower muscle GR activity has also been observed in coho salmon [40], largemouth bass [66], tench [29], and red swamp crayfish Procambarus clarkii [62, 67].

Results of the present study support the hypothesis that alteration in activity of CAT and GR enzymes in muscle and liver can counteract the prooxidant effect induced by elevated dietary Se. Oxidative stress indicators in muscle and liver of inorganic Se and higher concentration of organic Se-treated common barbel suggest that these tissues may possess differing abilities to counteract Se-prooxidant effects, with muscle showing a higher level of oxidative stress. Se supplementation at levels equal to or higher than those employed in this study (Se 0.3 and Algadiet 1.0) may accelerate the antioxidant condition of barbell, and, therefore, the optimal Se dietary concentration may be lower in this species.

CONCLUSION

Selenium concentrations and enzyme activity in tissues of common barbel suggest that supplementation with Se-enriched Chlorella biomass is more effective than inorganic sodium selenite. Direct comparison of Se-enriched microalgae with organic forms (especially Se-Met) and Se-enriched yeasts is needed. Efficacy of Se-enriched microalgae when compared with Se-enriched yeasts may be lower in commonly investigated indices. Use of Se-enriched microalgae biomass in aquaculture feedstuffs could be supported by the presence of specific bioactive compounds like polyunsaturated fatty acids, pigments, and vitamins.

CONFLICT OF INTERESTS

The authors declare that there is no conflict of interests regarding the publication of this paper.

ACKNOWLEDGMENTS

The authors thank Ms. Magda Sergejevová, Mr. Petr Novotný, and Ms. Soňa Pekařová for technical assistance in production of Se-enriched microalgae biomass. The Czech Science Foundation supported this work through Project 521/09/0656. Partial funding was provided by the Projects CENAKVA CZ.1.05/2.1.00/01.0024, CENAKVA II (the results of the Project LO1205 were obtained with a financial support from the MEYS of the CR under the NPU I program), Algatech CZ.1.05/2.1.00/03.0110, and by the Grant Agency of the University of South Bohemia (087/2013/Z). The authors also thank Alan Pike and Kathleen Hills for language correction.

REFERENCES

1. S. P. Lall, "The minerals," in Fish Nutrition, J. E. Halver and R. W. Hardy, Eds., pp. 259–308, Academic Press, San Diego, Calif, USA, 2002.
2. J. W. Sweetman, S. Torrecillas, A. Dimitroglou, S. Rider, S. J. Davies, and M. S. Izquierdo, "Enhancing the natural defences and barrier protection of aquaculture species," Aquaculture Research, vol. 41, no. 3, pp. 345–355, 2010.
3. K. Hamre, T. A. Mollan, Ø. Sæle, and B. Erstad, "Rotifers enriched with iodine and selenium increase survival in Atlantic cod (Gadus morhua) larvae," Aquaculture, vol. 284, no. 1–4, pp. 190–195, 2008.
4. A. R. A. Ribeiro, L. Ribeiro, M. T. Dinis, and M. Moren, "Protocol to enrich rotifers (Brachionus plicatilis) with iodine and selenium," Aquaculture Research, vol. 42, no. 11, pp. 1737–1740, 2011. ·
5. S. Penglase, K. Hamre, J. W. Sweetman, and A. Nordgreen, "A new method to increase and maintain the concentration of selenium in rotifers (Brachionus spp.)," Aquaculture, vol. 315, no. 1-2, pp. 144–153, 2011.
6. F. Z. Küçükbay, H. Yazlak, I. Karaca et al., "The effects of dietary organic or inorganic selenium in rainbow trout (Oncorhynchus mykiss) under crowding conditions," Aquaculture Nutrition, vol. 15, no. 6, pp. 569–576, 2009.
7. S. A. Rider, S. J. Davies, A. N. Jha, A. A. Fisher, J. Knight, and J. W. Sweetman, "Supra-nutritional dietary intake of selenite and selenium yeast in normal and stressed rainbow trout (Oncorhynchus mykiss): implications on selenium status and health responses," Aquaculture, vol. 295, no. 3-4, pp. 282–291, 2009.
8. T. Watanabe, V. Kiron, and S. Satoh, "Trace minerals in fish nutrition," Aquaculture, vol. 151, no. 1–4, pp. 185–207, 1997.
9. L. J. Raymond and N. V. C. Ralston, "Selenium›s importance in regulatory issues regarding mercury," Fuel Processing Technology, vol. 90, no. 11, pp. 1333–1338, 2009.
10. D. Mikulski, J. Jankowski, Z. Zduńczyk, M. Wróblewska, K. Sartowska, and T. Majewska, "The effect of selenium source on performance, carcass traits, oxidative status of the organism, and meat quality of turkeys," Journal of Animal and Feed Sciences, vol. 18, no. 3, pp. 518–530, 2009.
11. G. Ripoll, M. Joy, and F. Muñoz, "Use of dietary vitamin E and selenium (Se) to increase the shelf life of modified

atmosphere packaged light lamb meat," Meat Science, vol. 87, no. 1, pp. 88–93, 2011.

12. M. P. Rayman, "Selenium in cancer prevention: a review of the evidence and mechanism of action,"Proceedings of the Nutrition Society, vol. 64, no. 4, pp. 527–542, 2005.

13. H. Fritz, D. Kennedy, D. Fergusson et al., "Selenium and lung cancer: a systematic review and meta analysis," PLoS One, vol. 6, no. 11, Article ID e26259, 2011. ·

14. H. D. Mistry, F. B. Pipkin, C. W. G. Redman, and L. Poston, "Selenium in reproductive health," American Journal of Obstetrics and Gynecology, vol. 206, no. 1, pp. 21–30, 2012.

15. A. A. Turanov, M. Malinouski, and V. N. Gladyshev, "Selenium and male reproduction," in Selenium, D. L. Hatfield, M. J. Berry, and V. N. Gladyshev, Eds., pp. 409–417, Springer, New York, NY, USA, 2012.

16. S. A. Rider, S. J. Davies, A. N. Jha, R. Clough, and J. W. Sweetman, "Bioavailability of co-supplemented organic and inorganic zinc and selenium sources in a white fishmeal-based rainbow trout (Oncorhynchus mykiss) diet," Journal of Animal Physiology and Animal Nutrition, vol. 94, no. 1, pp. 99–110, 2010.

17. J. G. Bell and C. B. Cowey, "Digestibility and bioavailability of dietary selenium from fishmeal, selenite, selenomethionine and selenocystine in Atlantic salmon (Salmo salar)," Aquaculture, vol. 81, no. 1, pp. 61–68, 1989.

18. M. Lorentzen, A. Maage, and K. Julshamn, "Effects of dietary selenite or selenomethionine on tissue selenium levels of Atlantic salmon (Salmo salar)," Aquaculture, vol. 121, no. 4, pp. 359–367, 1994.

19. C. Wang and R. T. Lovell, "Organic selenium sources, selenomethionine and selenoyeast, have higher bioavailability than an inorganic selenium source, sodium selenite, in diets for channel catfish (Ictalurus punctatus)," Aquaculture, vol. 152, no. 1–4, pp. 223–234, 1997.

20. P. A. Cotter, S. R. Craig, and E. Mclean, "Hyperaccumulation of selenium in hybrid striped bass: a functional food for aquaculture?" Aquaculture Nutrition, vol. 14, no. 3, pp. 215–222, 2008.

21. A. Jovanovic, G. Grubor-Lajsic, N. Djukic, G. Gardinovacki, A. Matic, and M. Spasic, "The effect of selenium on antioxidant system in erythrocytes and liver of the carp (Cyprinus carpio L.)," Critical Reviews in Food Science and Nutrition, vol. 37, no. 5, pp. 443–448, 1997.

22. Y. Wang, J. Han, W. Li, and Z. Xu, "Effect of different selenium source on growth performances, glutathione peroxidase activities, muscle composition and selenium concentration of allogynogenetic crucian carp (Carassius auratus gibelio)," Animal Feed Science and Technology, vol. 134, no. 3-4, pp. 243–251, 2007.

23. X. Zhou, Y. Wang, Q. Gu, and W. Li, "Effects of different dietary selenium sources (selenium nanoparticle and selenomethionine) on growth performance, muscle composition and glutathione peroxidase enzyme activity of crucian carp (Carassius auratus gibelio)," Aquaculture, vol. 291, no. 1-2, pp. 78–81, 2009.

24. E. Schram, Z. Pedrero, C. Cámara, J. W. Van Der Heul, and J. B. Luten, "Enrichment of African catfish with functional selenium originating from garlic," Aquaculture Research, vol. 39, no. 8, pp. 850–860, 2008.

25. E. Schram, R. A. A. M. Schelvis-Smit, J. W. van der Heul, and J. B. Luten, "Enrichment of the African catfish Clarias gariepinus (Burchell) with functional selenium originating from garlic: effect of enrichment period and depuration on total selenium level and sensory properties," Aquaculture Research, vol. 41, no. 6, pp. 793–803, 2010.

26. O. Zmora and A. Richmond, "Microalgae for aquaculture: microalgae production for aquaculture," inHandbook of Microalgal Cultures, A. Richmond, Ed., pp. 365–379, Blackwell Science, Oxford, UK, 2004.

27. S. J. Kaushik, "Mineral nutrition," in Nutrition and Feeding of Fish and Crustaceans, J. Guillaume, S. J. Kaushik, P. Bergot, and R. Mtailler, Eds., pp. 169–181, Praxis, Chichester, UK, 2001.

28. A. C. Elia, M. Prearo, N. Pacini, A. J. M. Dörr, and M. C. Abete, "Effects of selenium diets on growth, accumulation and antioxidant response in juvenile carp," Ecotoxicology and Environmental Safety, vol. 74, no. 2, pp. 166–173, 2011.

29. N. Pacini, M. C. Abete, A. J. M. Dörr, M. Prearo, M. Natali, and A. C. Elia, "Detoxifying response in juvenile tench fed by selenium diet," Environmental Toxicology and Pharmacology, vol. 33, no. 1, pp. 46–52, 2012.

30. V. I. Lushchak, T. V. Bagnyukova, V. V. Husak, L. I. Luzhna, O. V. Lushchak, and K. B. Storey, "Hyperoxia results in transient oxidative stress and an adaptive response by antioxidant enzymes in goldfish tissues,"International Journal of Biochemistry and Cell Biology, vol. 37, no. 8, pp. 1670–1680, 2005.

31. S. Marklund and G. Marklund, "Involvement of the superoxide anion radical in the autoxidation of pyrogallol and a convenient assay for superoxide dismutase," European Journal of Biochemistry, vol. 47, no. 3, pp. 469–474, 1974.

32. R. F. Beers Jr. and I. W. Sizer, "A spectrophotometric method for measuring the breakdown of hydrogen peroxide by catalase," The Journal of Biological Chemistry, vol. 195, no. 1, pp. 133–140, 1952.

33. R. A. Lawrence and R. F. Burk, "Glutathione peroxidase activity in selenium deficient rat liver,"Biochemical and Biophysical Research Communications, vol. 71, no. 4, pp. 952–958, 1976.

34. I. Carlberg and B. Mannervik, "Purification and characterization of the flavoenzyme glutathione reductase from rat liver," Journal of Biological Chemistry, vol. 250, no. 14, pp. 5475–5480, 1975.

35. M. M. Bradford, "A rapid and sensitive method for the quantitation of microgram quantities of protein utilizing the principle of protein dye binding," Analytical Biochemistry, vol. 72, no. 1-2, pp. 248–254, 1976.

36. J. H. Zar, Biostatistical Analysis, Prentice-Hall, Upper Saddle River, NJ, USA, 3rd edition, 1996.

37. M. P. Rayman, "The importance of selenium to human health," The Lancet, vol. 356, no. 9225, pp. 233–241, 2000.

38. S. J. Hamilton, "Review of selenium toxicity in the aquatic food chain," Science of the Total Environment, vol. 326, no. 1–3, pp. 1–31, 2004.

39. L. D. Koller and J. H. Exon, "The two faces of selenium—deficiency and toxicity—are similar in animals and man," Canadian Journal of Veterinary Research, vol. 50, no. 3, pp. 297–306, 1986.

40. S. P. Felton, M. L. Landolt, R. Grace, and A. Palmisano, "Effects of selenium dietary enhancement on hatchery-reared coho salmon, Oncorhynchus kisutch (Walbaum), when compared with wild coho: hepatic enzymes and seawater adaptation evaluated," Aquaculture Research, vol. 27, no. 2, pp. 135–142, 1996.

41. J. W. Hilton and P. V. Hodson, "Effect of increased dietary carbohydrate on selenium metabolism and toxicity in rainbow trout (Salmo gairdneri)," Journal of Nutrition, vol. 113, no. 6, pp. 1241–1248, 1983.

42. B. D. Hicks, J. W. Hilton, and H. W. Ferguson, "Influence of dietary selenium on the occurrence of nephrocalcinosis in the rainbow trout, Salmo gairdneri Richardson," Journal of Fish Diseases, vol. 7, no. 5, pp. 379–389, 1984.

43. A. D. Lemly, "Teratogenic effects of selenium in natural populations of freshwater fish," Ecotoxicology and Environmental Safety, vol. 26, no. 2, pp. 181–204, 1993.

44. A. D. Lemly, Selenium Assessment in Aquatic Ecosystems: A Guide for Hazard Evaluation and Water Quality Criteria,, Springer, New York, NY, USA, 2002.

45. M. P. Rayman, "The use of high-selenium yeast to raise selenium status: how does it measure up?" British Journal of Nutrition, vol. 92, no. 4, pp. 557–573, 2004.

46. G. N. Schrauzer, "Selenomethionine: a review of its nutritional significance, metabolism and toxicity,"Journal of Nutrition, vol. 130, no. 7, pp. 1653–1656, 2000.

47. S. De Alcantara, C. Cerqueira Lopes, and K. Wagener, "Controlled introduction of selenium intoChlorella cells," Indian Journal of Experimental Biology, vol. 36, no. 12, pp. 1286–1288, 1998.

48. T. W.-M. Fan, A. N. Lane, D. Martens, and R. M. Higashi, "Synthesis and structure characterization of selenium metabolites," Analyst, vol. 123, no. 5, pp. 875–884, 1998.

49. E. H. Larsen, M. Hansen, T. Fan, and M. Vahl, "Speciation of selenoamino acids, selenonium ions and inorganic selenium by ion exchange HPLC with mass spectrometric detection and its application to yeast and algae," Journal of Analytical Atomic Spectrometry, vol. 16, no. 12, pp. 1403–1408, 2001.

50. P. M. Neumann, M. P. De Souza, I. J. Pickering, and N. Terry, "Rapid microalgal metabolism of selenate to volatile dimethylselenide," Plant, Cell and Environment, vol. 26, no. 6, pp. 897–905, 2003.

51. M. R. Pincus, "Interpreting laboratory results: reference values and decision making," in ClInical Diagnosis and Management by Laboratory Methods, J. B. Henry, Ed., pp. 74–91, W. B. Saunders, Philadelphia, Pa, USA, 1996.

52. A. Cnaani, S. Tinman, Y. Avidar, M. Ron, and G. Hulata, "Comparativestudyofbiochemicalparametersinresponse to stress in Oreochromis aureus, O. mossambicus and two strains of O. niloticus," Aquaculture Research, vol. 35, no. 15, pp. 1434–1440, 2004.

53. J. J. Stegeman, M. Brouwer, R. T. Di Giulio et al., "Molecular responses to environmental contamination: enzyme and protein systems as indicators of chemical exposure and effect," in Biomarkers—Biochemical, Physiological, and Histological Markers of Anthropogenic Stress, R. J. Huggett, R. A. Kimerl, P. M. Mehrl, and H. L. Bergman, Eds., pp. 235–335, Lewis, Chelsea, Mich, USA, 1992.

54. L. J. G. Barcellos, L. C. Kreutz, C. De Souza et al., "Hematological changes in jundiá (Rhamdia quelenQuoy and Gaimard Pimelodidae) after acute and chronic stress caused by usual aquacultural management, with emphasis on immunosuppressive effects," Aquaculture, vol. 237, no. 1–4, pp. 229–236, 2004.

55. M. Abdel-Tawwab, M. A. A. Mousa, and F. E. Abbass, "Growth performance and physiological response of African catfish, Clarias gariepinus (B.) fed organic selenium prior to the exposure to environmental copper toxicity," Aquaculture, vol. 272, no. 1–4, pp. 335–345, 2007.

56. L. L. Miller, F. Wang, V. P. Palace, and A. Hontela, "Effects of acute and subchronic exposures to waterborne selenite on the physiological stress response and oxidative stress indicators in juvenile rainbow trout," Aquatic Toxicology, vol. 83, no. 4, pp. 263–271, 2007.

57. Z. Pedrero and Y. Madrid, "Novel approaches for selenium speciation in foodstuffs and biological specimens: a review," Analytica Chimica Acta, vol. 634, no. 2, pp. 135–152, 2009.

58. L. Atencio, I. Moreno, Á. Jos et al., "Effects of dietary selenium on the oxidative stress and pathological changes in tilapia (Oreochromis niloticus) exposed to a microcystin-producing cyanobacterial water bloom," Toxicon, vol. 53, no. 2, pp. 269–282, 2009.

59. B. Halliwell and J. M. C. Gutteridge, Free Radicals in Biology and Medicine, Oxford University Press, Oxford, UK, 3rd edition, 1999.

60. H.-W. Wang, H.-M. Xu, G.-H. Xiao et al., "Effects of selenium on the antioxidant enzymes response ofNeocaridina heteropoda exposed to ambient nitrite," Bulletin of Environmental Contamination and Toxicology, vol. 84, no. 1, pp. 112–117, 2010.

61. W.-N. Wang, A.-L. Wang, and Y.-J. Zhang, "Effect of dietary higher level of selenium and nitrite concentration on the cellular defense response of Penaeus vannamei," Aquaculture, vol. 256, no. 1–4, pp. 558–563, 2006.

62. A. J. M. Dörr, N. Pacini, M. C. Abete, M. Prearo, and A. C. Elia, "Effects of a selenium-enriched diet on

antioxidant response in adult crayfish (Procambarus clarkii)," Chemosphere, vol. 73, no. 7, pp. 1090–1095, 2008.

63. R. H. Schirmer, R. L. Krauth-Siegel, and G. E. Schulz, "Glutathione reductase," in Coenzymes and Cofactors: Glutathione, D. Dolphi, R. Poulso, and O. Avaramovi, Eds., pp. 553–596, John Wiley & Sons, New York, NY, USA, 1989.

64. J. Cazenave, M. D. L. A. Bistoni, S. F. Pesce, and D. A. Wunderlin, "Differential detoxification and antioxidant response in diverse organs of Corydoras paleatus experimentally exposed to microcystin-RR," Aquatic Toxicology, vol. 76, no. 1, pp. 1–12, 2006.

65. J. Zhang, H. Shen, X. Wang, J. Wu, and Y. Xue, "Effects of chronic exposure of 2,4-dichlorophenol on the antioxidant system in liver of freshwater fish Carassius auratus," Chemosphere, vol. 55, no. 2, pp. 167–174, 2004.

66. Y. Zhu, Y. Chen, Y. Liu, H. Yang, G. Liang, and L. Tian, "Effect of dietary selenium level on growth performance, body composition and hepatic glutathione peroxidase activities of largemouth bassMicropterus salmoide," Aquaculture Research, vol. 43, no. 11, pp. 1660–1668, 2011.

67. A. C. Elia, A. J. M. Dörr, M. Prearo, M. I. Taticchi, and M. C. Abete, "Detoxification enzymes of freshwater crayfish Procambarus clarkii fed a diet enriched in selenium: preliminary results," Marine and Freshwater Behaviour and Physiology, vol. 40, no. 3, pp. 195–199, 2007.

68. National Research Council (NRC), Nutrient Requirements of Fish, National Academy Press, Washington, DC, USA, 1993.

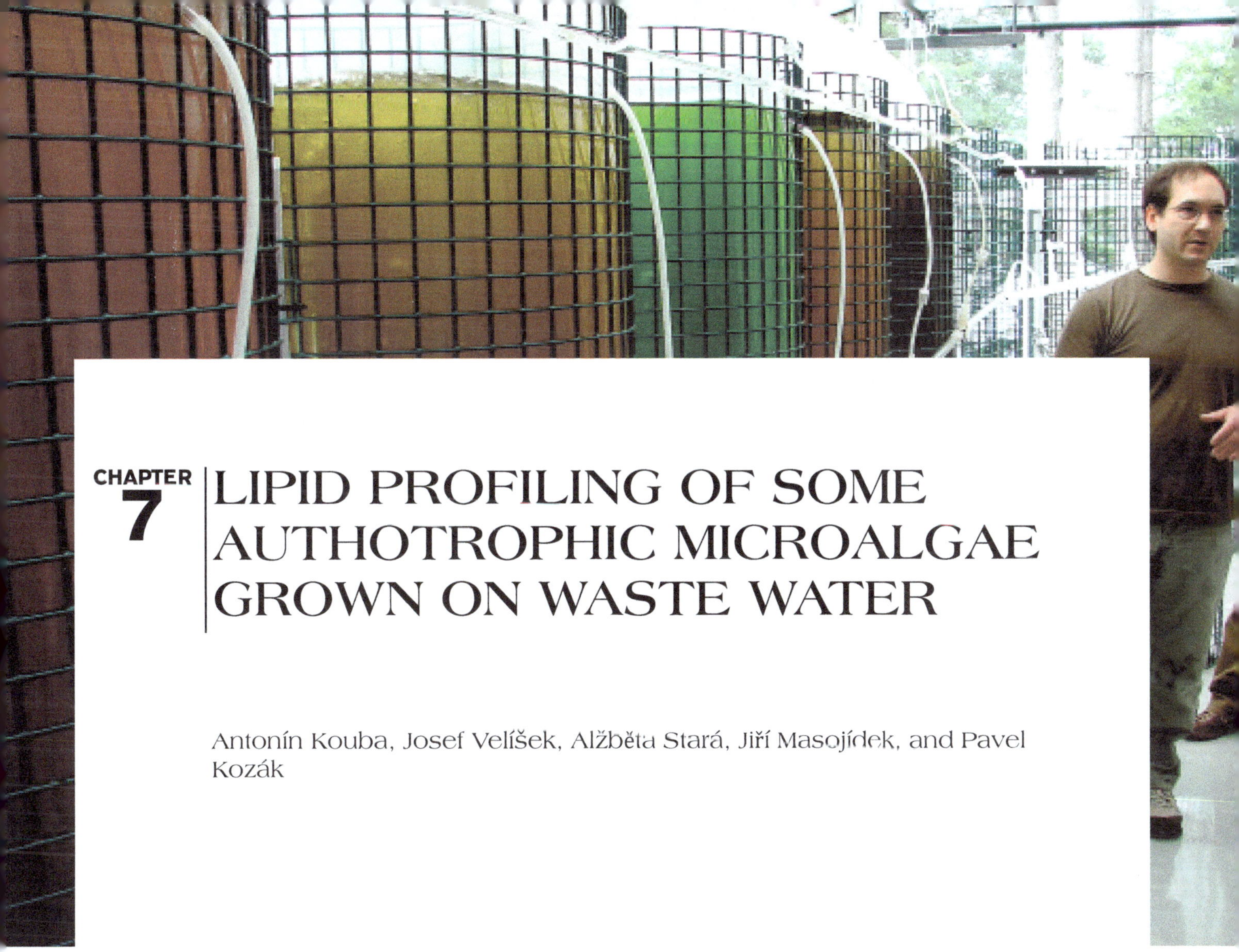

CHAPTER 7 LIPID PROFILING OF SOME AUTHOTROPHIC MICROALGAE GROWN ON WASTE WATER

Antonín Kouba, Josef Velíšek, Alžběta Stará, Jiří Masojídek, and Pavel Kozák

Microalgae can be a new source of lipids for the aquaculture industry. Moreover, their potential as natural sources of antioxidants has gained recent attention. About 40 species of microalgae are used in aquaculture worldwide. A full characterization of lipid components is critical for selecting the most suitable microalgae and downstream processing for food and feed production. The present study is part of a big project funded by GUDP (green development and demonstration program of ministry of agriculture and fisheries of Denmark) which aims at developing new processing technologies, so that microalgae-biomass can be used as an alternative valuable resource in fish feed. In this work, 10 fresh water and marine microalgae from Chlorella, Scenedesmus, Haematococcus, Nannochloropsis, Nannochloropsis and Dunialiella species grown in waste water in Kalundborg micro algal facility were harvested by membrane microfiltration and analyzed for fatty acid (GC), triacylglycerol (HPLC), sterol (GC) and tochol (HPLC) composition and also for amounts of phospholipids . Lipid composition in micro algae varied strongly between species.

Citation: Safafar, Hamed; Jacobsen, Charlotte; Møller, Per / Lipid profiling of some authotrophic microalgae grown on waste water. 2014. Abstract from 2nd International Conference on Algal Biorefinery, Lyngby, Denmark.

CHAPTER 8

EFFECT OF DIETARY (N-3) HIGHLY UNSATURATED FATTY ACIDS ON GROWTH AND SURVIVAL OF FAT SNOOK (CENTROPOMUS PARALLELUS, PISCES: CENTROPOMIDAE) LARVAE DURING FIRST FEEDING

M.E.B. Seiffert, V.R. Cerqueira and L.A.S. Madureira

Key words: Fish, larvae, Centropomus parallelus, feeding, highly unsaturated fatty acid

INTRODUCTION

Fat snook (robalo or camorim), *Centropomus parallelus* Poey, 1860, is a highly valued sport fish which inhabits the tropical and subtropical western Atlantic (1). Relatively little is known about the biological requirements of this species as a basis for commercial cultivation or restocking programs.

One of the greatest obstacles to the mass production of snook fingerlings is the high mortality that occurs in the first two weeks. This mortality may be associated with many factors but, according to previous work (2), priority should be given to the investigation of the influence of light intensity on the feeding habits of larvae and the nutritional quality of live feed.

The rotifer *Brachionus* spp is widely used in aquaculture because they are an ideal link in the food chain for different stages of fish and shrimp larvae. Its nutritional value is strictly dependent on the nutritional value of its food source, particularly the content of highly unsaturated fatty acids ((n-3) HUFA), like docosahexaenoic acid (DHA, 22:6(n-3)) and eicosapentaenoic acid (EPA, 20:5(n-3)). Low dietary HUFA levels can lead to high mortality in fish larviculture (3-5). Koven et al. (6) suggested that such acids function as essential components of biomembranes, and their levels in the tissue phospholipid fraction are associated with larval growth. Rainuzzo et al. (7) emphasized the relevance of DHA to the development of neural tissues such as brain and retina, considering that the larval head constitutes a significant part of the body mass, and that predatory fish larvae rely on vision to capture their food.

Sorgeloos et al. (8) reported a strong correlation between the dietary EPA content and survival, and between DHA and growth of the Asian sea bass larvae. Watanabe (9) concluded that DHA and EPA increased survival and growth of several marine fish larvae. Kanazawa (10) observed that high DHA levels increased the tolerance of red sea bream larvae to various stressful conditions. Barreto and Cavalcanti (11) reviewed live food enrichment techniques with HUFA and its importance to the larvae of marine organisms. All of these findings are relevant to the production of high quality larval fish; however,

Citation: M.E.B. Seiffert, V.R. Cerqueira and L.A.S. Madureira, Effect of dietary (n-3) highly unsaturated fatty acids on growth and survival of fat snook (Centropomus parallelus, Pisces: Centropomidae) larvae during first feeding, doi.org/10.1590/S0100-879X2001000500013

there are no data concerning *Centropomus* species or other indigenous Brazilian marine finfish.

In the present study we compared survival, growth, and swim bladder inflation of fat snook larvae fed rotifers reared on different diets, resulting in different fatty acid compositions.

MATERIAL AND METHODS

Origin of the larvae

Wild male and female snook captured by local fishermen in coastal waters near Lagoa da Conceição (Florianópolis, SC, Brazil) were induced to spawn within three days with an intramuscular injection of human chorionic gonadotropin (500 and 1000 IU/kg body weight, respectively). Gametes were obtained by hand-stripping followed by artificial fertilization. In the incubation process, fiberglass cylindric-conical tanks of 36-liter capacity were placed inside a rectangular tank of 1000-liter capacity, with a stocking density of approximately 2000 eggs/liter. Salinity (35 g/l) and temperature (26°C) were kept constant, and water renewal rate was 2 l/min.

Live feed culture

Microalgae *Nannochloropsis oculata* and rotifers *Brachionus rotundiformis* (strain obtained from Dr. A. Hagiwara, Nagasaki University, Japan) were cultured in a greenhouse as a semi-continuous system (harvesting 20-30% daily), using fiberglass tanks of 2500- and 500-liter capacity, respectively. A mixture of agricultural-grade fertilizers was the culture medium used for microalgae (1 liter of freshwater solution, containing 150 g ammonium sulfate, 25 g super phosphate, and 7.5 g urea, for 1000 liters of seawater added to the culture). Temperature range was 22-26°C. One of the rotifer cultures received 100 x 10^3 microalga cells per individual/day, while the other received 50 x 10^3 microalga cells per individual/day and 0.5 µg of baker›s yeast per individual/day, as recommended in the literature (12). Rotifer enrichment was done with a commercial emulsion (Selco® from INVE Aquaculture NV, Baasrode, Belgium), in 40-liter cylindric-conical tanks at a density of 400 x 10^3 individuals/liter over 15 h with 0.2 g/l of the product, following manufacturer instructions.

Rotifers were fed three different diets which represented the three treatments, with three replicates: treatment A, rotifers fed only microalgae; treatment B, rotifers fed microalgae and baker›s yeast (1:1); treatment C, rotifers fed microalgae and baker›s yeast (1:1) enriched with commercial emulsion.

Larviculture

Approximately 30 h after hatching, larvae were placed in 9 black cylindric-conical fiberglass tanks (36 liters) at a density of 30 individuals/liter. Each tank was equipped with an airlifting system. Daylight fluorescent lights provided a light intensity of 1000 lux at the water surface, with a photoperiod of 14 h, from 9:00 to 23:00 h. Larvae were fed rotifers from day 3 after hatching onwards, at a density of 10/ml. Microalgae were added to the tanks at 200 x 10^3 cells/ml daily.

Water exchange was approximately 15% per day. Temperature was monitored at 8:00 h daily, while salinity, ammonia, pH and dissolved oxygen were measured every six days, in one tank of each treatment. Temperature (21.3-26.0°C), non-ionized ammonia (0.07-0.10 mg/l), dissolved oxygen (4.0-5.5 mg/l), salinity (35 g/l) and pH (8.0) did not differ significantly among treatments.

On the 14th day after hatching all surviving larvae were counted and 30 larvae per treatment were measured (standard length) using a dissecting microscope equipped with an ocular micrometer. Growth was considered as the difference between final and initial standard length. Individuals with a functional swim bladder in each treatment were counted.

Fatty acid analysis

Every three days, approximately one million rotifers were filtered from the different cultures through 65-µm mesh screens, and then frozen at -20°C. After lyophilization of the samples and addition of internal standards (nonadecanol and colestane dissolved in chloroform) the fatty acids were extracted with 1:3 dichloromethane-methanol (3x) by ultrasound, followed by centrifugation.

The total combined extracts were evaporated and the fatty acids methylated under N_2 using anhydrous 5% methanolic HCl in toluene at 50°C for 12 h (13). After esterification, 1 ml aqueous NaCl (5%) was added and the methylated products were extracted into dichloromethane. The samples were then analyzed using a Shimadzu Gas Chromatographer (GC) model 14A equipped with a split/splitless injector and flame ionization detector, using a 25 m x 0.25 mm ID CBP1 column (film thickness 0.25 µm). The temperature program was 40-300°C at 10°C/min. Nitrogen was used as a carrier gas. GC-mass spectrometry (MS) analyses (EI mode, 70 eV) were carried out on a QP2000A Shimadzu GC-MS. Conditions for GC-MS analyses were as for GC except that helium was used as the carrier gas (14).

There were no chemical reactions to identify the position of the first unsaturation in the fatty acids (series (n-3), (n-6), etc.). The 20:5(n-3) and 22:6(n-3) fatty acids were compared to a standard product (Super Selco® from INVE Aquaculture NV).

Statistical analysis

Survival and swim bladder inflation data were compared among treatments by the nonparametric chi-square test. Final standard length (N = 30 for each treatment) was compared by one-way analysis of variance (ANOVA). A significance level of 5% was used for both tests (15).

RESULTS AND DISCUSSION

Rotifers fed microalgae showed a higher content of most fatty acids (Table 1), especially 16:0, 16:1, and 20:5(n-3) compared to rotifers fed microalgae and baker's yeast. Replacement of 50% of the microalgae with yeast yielded a reduction in total fatty acids corresponding to 162% for saturated fatty acids, 61% for monounsaturated fatty acids, and 117% for polyunsaturated fatty acids. However, there was no decrease in the amount of some fatty acids (18:3(n-3) and 22:6(n-3), for example). In the oil-enriched rotifers there was an increase in the level of almost all fatty acids. The most notable were 20:5(n-3), 22:6(n-3), 22:5(n-3), 16:1, 16:0, and 14:0. The enrichment process increased the total saturated fatty acid content by 380%, the monounsaturated fatty acid content by 328%, and the polyunsaturated fatty acid content by 744% in relation to the rotifers fed microalgae and baker's yeast.

The levels of HUFA tested in treatment C of the present study were within the range recommended in the literature for other marine species. Lavens and Sorgeloos (16) stated that rotifers grown on a formulated feed had an excellent HUFA composition, containing 5.4, 4.4 and 15.6 mg/g dry weight of EPA, DHA and (n-3) HUFA, respectively. Rodríguez et al. (17) found that the minimum requirement for larval gilthead sea bream (0-17 days) fed rotifers was 6 mg/g of EPA, 8 mg/g of DHA and 15 mg/g dry weight of (n-3) HUFA.

There was no significant effect of treatment on survival ($c^2_o = 5.8 < c^2_c = 7.8$) or swim bladder inflation ($c^2_o = 3.7 < c^2_c = 7.8$). Survival rates were particularly low, the highest being 2.7% in treatment A. Swim bladder inflation rates were 11.9-34.7% (Table 2).

One important observation about survival, which was not quantified, is that most losses occurred from 6 to 9 days after hatching for all treatments. By this time, with the yolk sac exhausted, most larvae had not fed, were motionless, apparently weak and starving in comparison to a few healthy individuals.

About 30 h after hatching larvae presented a mean standard length of 2.45 ± 0.038 mm (N = 30). After 14 days of rearing, there were no significant differences among treatments in final length (ANOVA: F = 1.35, d.f. = 2, P = 0.263) or growth (Table 2).

Despite the differences in the composition of essential fatty acids in each rotifer treatment, there was no influence on the larviculture performance of fat snook. Similarly, Tamaru et al. (18), by replacing 50% of the microalgae with baker›s yeast in the rotifer diet, obtained small differences in survival and growth of milkfish and striped mullet larvae. While Koven et al. (6) observed that enriching rotifers with different oils enhanced the growth rate of gilthead sea bream larvae, but found no influence on survival and swim bladder inflation. Mourente et al. (19) and Rodríguez et al. (17), working with the same species, also observed a positive effect of (n-3) HUFA only on the growth of first-feeding larvae (0-17 days). Dhert et al. (20), rearing Asian sea bass larvae (0-21 days), obtained no improvement in growth or survival of larvae fed (n-3) HUFA-enriched food; however, an increase in larval resistance to the salinity stress test was observed.

Some authors have concluded that DHA is superior to EPA as an essential fatty acid for larval fish. Watanabe (9) studied the larviculture of five marine fish (red sea bream, yellowtail, striped knifejaw, Japanese flounder, and turbot) and found that DHA was necessary in greater quantities than EPA, and that its presence in adequate quantities increased survival and growth. Similarly, Rodríguez et al. (21), rearing gilthead sea bream larvae, observed that the best growth rate was achieved with a DHA:EPA ratio of about 1:3. Ibeas et al. (22) obtained the best growth of gilthead sea bream larvae with a ratio of 2, while in the juvenile stage the best ratio was found to be 0.5. The highest DHA:EPA ratio found in the present study was 0.46 (treatment C with oil-enriched rotifers).

Some investigators have proposed that the presence of (n-3) HUFA may not be important for young (newly hatched) larvae. Kraul (23) reported that the first stage survival (0-9 days) of mahimahi larvae was not affected by the content of fatty acids in the feed, and suggested that this is probably due to high levels of endogenous EPA and DHA stored in the larval yolk sac. Watanabe (9) reported that DHA is high in the eggs of marine fish and rapidly decreases during the first ten days after hatching. Mourente et al. (19) suggested that the presence of DHA in the diet of larval gilthead sea bream may not be necessary, because they could retain DHA from the yolk. Thus, if the first-feeding fat snook does not depend on the (n-3) HUFA content of the exogenous feed, as observed with other species, rotifer enrichment at this stage may not be necessary, thus simplifying and lowering the cost of larviculture.

The (n-3) HUFA requirements of marine larval fish are highly variable. Such differences could be explained by species-specific nutritional requirements, culture period, or quality of larvae. The latter can be influenced by both the nutritional conditions during gonad formation (24) and the process used for spawning induction of broodstock (25). Tamaru et al. (26) also correlated differences in growth and survival of striped mullet larvae to the quality of broodstock nutrition. Abi-Ayad et al. (27) found a correlation with DHA and EPA levels in the broodstock diet, and stress resistance and survival of perch larvae.

The low survival rates (under 2.7%) obtained in the present study can also be correlated with factors other than the nutritional value of the rotifers. For example, the use of wild spawners, probably not well acclimated, and artificial reproduction methods could have resulted in low quality larvae. Wallace et al. (28) observed in common snook, *Centropomus undecimalis*, that adult fish appeared to be unusually traumatized by the stress of capture and translocation, and that 48 h after capture ovaries began to degenerate. The swim bladder inflation rates observed (lower than 34.7%) are indicative of a state of weakness. Brügger (29), working with fat snook larvae of the same age under conditions very similar to those of the present study, obtained highly variable rates (37 to 100%) of swim bladder inflation. It was observed in other species that lack of an inflated swim bladder affects mainly larval growth, but affects survival only later during weaning (30). With snook, inflation occurs very early, around 5-7 days after hatching, and larvae are able to feed even without a normally inflated swim bladder (2,29). This phenomenon has already been described for several marine species (30,31). Concerning larval mortality, other studies (2,29, and Cerqueira VR, unpublished results) achieved survival rates up to 20% and occasionally higher with fat snook larvae fed non-enriched rotifers (low HUFA levels).

Another sign of low quality larvae was a period between 6 and 9 days after hatching of high mortality rates coincident with yolk sac depletion. Larvae might have had difficulty capturing prey if there was some kind of weakness or malformation (induced, for example, by a biochemical deficiency in the eggs), and might have starved to death. The transition from endogenous to exogenous feeding is normally a vulnerable period for many species, when endogenous factors may acquire some importance (32). Rodríguez et al. (17) observed a notable mortality of gilthead sea bream larvae during the first 2-3 days of external feeding. Brügger (29) also noticed the same problem in some experiments with fat snook.

Finally, the improvement of the dietary value of rotifers was not sufficient to solve the problem of larval rearing. The high mortality obtained just after yolk absorption was probably due to some other factor. In order to improve survival of first-feeding fat snook larvae, factors other than the dietary HUFA level should also be considered in the future.

REFERENCES

1. Rivas LR (1986). Systematic review of the perciform fishes of the genus *Centropomus*. *Copeia*, 3: 579-611.
2. Cerqueira VR, MacChiavello JAG & Brügger AM (1995). Produção de alevinos de robalo, *Centropomus parallelus* Poey, 1860, através de larvicultura intensiva em laboratório. In: Castagnolli N (Editor), *Anais do Encontro Nacional de Aqüicultura, VII SIMBRAq, II ENBRAPOA*. Peruíbe, SP, Brazil. Publicação ACIESP, 191-197.
3. Scott AP & Middleton C (1979). Unicellular algae as a food for turbot (*Scophthalmus maximus* L.) larvae - The importance of dietary long-chain polyunsaturated fatty acids. *Aquaculture*, 18: 227-240.
4. Watanabe T, Kitajima C & Fujita S (1983). Nutritional values of live organisms used in Japan for mass propagation of fish. *Aquaculture*, 34: 115-143.
5. Caric M, Sanko-Njire J & Skaramuca B (1993). Dietary effects of different feeds on the biochemical composition of the rotifer (*Brachionus plicatilis* Müller). *Aquaculture*, 110: 141-150.
6. Koven WM, Tandler A, Kissil GW, Skland D, Friezlander O & Harel M (1990). The effect of dietary (n-3) polyunsaturated fatty acids on growth, survival and swim bladder development in *Sparus aurata* larvae. *Aquaculture*, 91: 131-141.
7. Rainuzzo JR, Reitan KI & Olsen Y (1997). The significance of lipids at early stages of marine fish: a review.*Aquaculture*, 155: 103-115.
8. Sorgeloos P, Léger PH & Lavens P (1988). Improved larval rearing of European and Asian seabass, seabream, mahimahi, siganid and milkfish using enrichment diets for *Brachionus* and *Artemia*. *World Aquaculture*, 19: 78-79.
9. Watanabe T (1993). Importance of docosahexaenoic acid in marine larval fish. *Journal of the World Aquaculture Society*, 24: 153-161.
10. Kanazawa A (1997). Effects of docosahexaenoic acid and phopholipids on stress tolerance of fish.*Aquaculture*, 155: 129-134.
11. Barreto O & Cavalcanti D (1997). Enriquecimento de alimentos vivos para alimentação de larvas de organismos marinhos: uma breve revisão. *Boletim do Instituto de Pesca*, 24: 139-159.
12. Fulks W & Main K (1991). Rotifer (*Brachionus plicatilis*) production systems. In: Fulks W & Main K (Editors),*Rotifer and Microalgae Culture Systems. Proceedings of a US-Asia Workshop*. Oceanic Institute, Honolulu, 23-46.
13. Christie WW (1982). *Lipid Analysis*. Pergamon Press, New York.
14. Madureira LAS, Conte MH & Eglinton G (1995). Early diagenesis of lipid biomarker compounds in North Atlantic sediments. *Paleoceanography*, 10: 627-642.
15. Sokal RR & Rohlf FJ (1981). *Biometry. The Principles and Practice of Statistics in Biological Research*. 2nd edn. WH Freeman, New York.
16. Lavens P & Sorgeloos P (1996). *Manual on the Production and Use of Live Food for Aquaculture*. FAO Fisheries Technical Paper No. 361. FAO, Rome.
17. Rodríguez C, Pérez JA, Badía P, Izquierdo MS, Fernández-Palacios H & Lorenzo Hernández A (1998). The n-3 highly unsaturated fatty acid requirements of gilthead seabream (*Sparus aurata* L.) larvae when using an appropriate DHA/EPA ratio in the diet. *Aquaculture*, 169: 9-23.
18. Tamaru CS, Lee C & Ako H (1991). Improving the larval rearing of striped mullet (*Mugil cephalus*) by manipulating quantity and quality of the rotifer, *Brachionus plicatilis*. In: Fulks W & Main K (Editors), *Rotifer and Microalgae Culture Systems. Proceedings of a US-Asia Workshop*. Oceanic Institute, Honolulu, 89-103.
19. Mourente G, Rodriguez A, Tocher DR & Sargent JR (1993). Effects of dietary docosahexaenoic acid (DHA; 22-6w3) on lipid and fatty acid compositions and growth in gilthead sea bream (*Sparus aurata* L.) larvae during first feeding. *Aquaculture*, 112: 79-98.
20. Dhert P, Lavens P, Duray M & Sorgeloos P (1990). Improved larval survival at metamorphosis of Asian seabass (*Lates calcarifer*) using (n-3)-HUFA enriched live food. *Aquaculture*, 90: 63-74.
21. Rodríguez C, Pérez JA, Días M, Izquierdo MS, Fernández-Palacios H & Lorenzo A (1997). Influence of EPA/DHA ratio in rotifers on gilthead seabream (*Sparus aurata*) larval development. *Aquaculture*, 150: 77-89.
22. Ibeas C, Cejas JR, Flores R, Badía P, Gómez T & Lorenzo Hernández A (1997). Influence of eicosapentaenoic to docosahexaenoic acid ratio (EPA/DHA) of dietary lipids on growth and fatty acid composition of gilthead seabream (*Sparus aurata*) juveniles. *Aquaculture*, 150: 91-102.
23. Kraul S (1993). Larviculture of the mahimahi *Coryphaena hippurus* in Hawaii, USA. *Journal of the World Aquaculture Society*, 24: 410-421.
24. Lavens P & Sorgeloos P (1991). Variation in egg and larval quality in various fish and crustacean species. In: Lavens P, Sorgeloos P, Jaspers E & Ollevier F (Editors), *Larvi'91 - Fish and Crustacean Larviculture Symposium*. Vol. 15. European Aquaculture Society, Special Publication, 221-222.
25. Lam TJ (1991). Hormones and egg/larval quality in fish. In: Lavens P, Sorgeloos P, Jaspers E & Ollevier F (Editors), *Larvi'91 - Fish and Crustacean Larviculture Symposium*. Vol. 15. European Aquaculture Society, Special Publication, 197.
26. Tamaru CS, Murashige R & Lee C (1994). The paradox of using background phytoplankton during the larval culture of striped mullet, *Mugil cephalus* L. *Aquaculture*, 119: 167-174.

27. Abi-Ayad S-M E-A, Melard C & Kestemont P (1997). Effects of (n-3) fatty acids in Eurasian perch broodstock diet on egg fatty acid composition and larvae stress resistance. *Aquaculture International*, 5: 161-168.

28. Wallace RA, Boyle SM, Grier HJ, Selman K & Petrino TR (1993). Preliminary observations on oocyte maturation and other aspects of reproductive biology in captive female snook, *Centropomus undecimalis*. *Aquaculture*, 116: 257-273.

29. Brügger AM (1995). Efeito da intensidade luminosa na sobrevivência e na insuflação da bexiga gasosa de larvas de robalo *Centropomus parallelus* (Poey, 1860). Master›s thesis, Universidade Federal de Santa Catarina, Florianópolis, SC, Brazil.

30. Chatain B (1997). Development and achievements of marine fish-rearing technology in France over the last 15 years. *Hydrobiologia*, 358: 7-11.

31. Planas M & Cunha I (1999). Larviculture of marine fish: problems and perspectives. *Aquaculture*, 177: 171-190.

32. Kamler E (1992). Mixed feeding period. In: *Early Life History of Fish*. Chapman & Hall, London, 176-181.

CHAPTER 9 MONTMORILLONITE CLAY MINERALS WITH OR WITHOUT MICROALGAEB AS A FEED ADDITIVE IN LARVAL WHITE LEG SHRIMP (LITOPENAEUS VANNAMEI)

Harry W. Palm, Hendrik Sörensen and Ulrich Knaus

Keywords: Cannibalism; Litopenaeus vannamei; Mixed layer clay minerals; Montmorillonite-illite/muscovite; Survival; Size distribution; Microalgae

INTRODUCTION

Brackish water shrimps belonging to the family Penaeidae are a highly valuable food commodity. Although shrimps contributed only 6.5% to the total world aquaculture production by quantity, they represented 14% of the value of aquaculture products in 2012. An increasing demand, limited production volumes and rising shrimp prices are responsible for this trend [1]. One of the main obstacles in shrimp production is disease problems, which regularly lead to higher mortality rates and production losses. Measures of disease control, stabilization of the immune system and improving the health status of the shrimps have been focused in former investigations [2-6]. The reduction of antibiotics and other chemicals used in shrimp production is important for an environmentally sustainable development of the global aquaculture [4,7-11]. Different feed additives, bioactive components and probiotics have been reported to support shrimp health and growth [12- 15] and have been discussed as an alternative for antibiotic use [3]. Feed additives like vitamins, minerals or plant compounds are well known for their positive effects and commonly used in shrimp aquaculture [15-17]. Especially algae and their compounds as feed ingredients have been suggested to support shrimp production [18-23]. Clay minerals have been demonstrated to have positive effects onto animal and even human health [24-27]. As an animal feed additive they bind and eliminate mycotoxins [28], improve weight gain, feed efficiency and egg production in poultry farming [29-31], and increase the average daily weight gain and performance in pig [32-35] and sheep farming [36]. Earlier studies suggest that the use of clay minerals might have beneficial effects during cultivation of aquatic organisms [37,38]. Clay was used to increase turbidity and to reduce the bacteria level in the production water of cod larvae with indirect positive effects onto growth and survival [39]. Direct positive effects of different clay minerals as a feed additive onto growth and feed conversion in salmonid fish aquaculture were documented [40]. Friedland montmorillonite–illite/muscovite alternating sequence clay minerals have been reported being a highly efficient absorbent of mycotoxin from animal feed. 100% of Aflatoxin B1 and

Citation: Harry W. Palm, Hendrik Sörensen and Ulrich Knaus, Montmorillonite Clay Minerals With or Without Microalgae as a Feed Additive in Larval White Leg Shrimp (Litopenaeus vannamei), https://www.jscimedcentral.com/Aquaculture/aquaculture-2-1008.php

Fumonisin, 80% of Ochratoxin and Zearalenone and 60% of T2-Toxin were absorbed during experimental investigations [41]. A significant reduction of inflammatory bowel diseases was reported by [26] through in situ perfusion of an extract of this clay mineral with rats. Friedland clay minerals are a highly efficient phosphate binder and assumed to have positive effects onto human kidney diseases [27]. As a feed additive, this specific clay mineral increased the body weight and feed intake in pig farming, milk production and feed intake in cow farming, and induced a significant increase in body weight gain and feed intake in chicken farming [42-44]. So far, clay minerals have not been applied in shrimp aquaculture, though these animals are highly intolerant to mycotoxins [45,46]. L. vannamei is surrounded by different clay minerals in the natural habitat, with a possible effect onto infectious shrimp disease agents. The purpose of the present study is to apply Friedland clay minerals as a feed additive for shrimp culture, with and without addition of the green algae Chlorella vulgaris. Possible reasons for the reported effects onto shrimp mortality, total biomass, growth performance and size range are discussed.

MATERIALS AND METHODS

Experimental design

The experiments were conducted in four identical recirculation systems, each with a water volume of 750 L and a flow through of 250% per hour. Each system consisted of 4 glass aquaria (100 cm x 50 cm x 50 cm) with a volume of approx. 165 L each combined with a filter unit including a biofilter, protein skimmer, sump and pipes (90 L). For each experiment, early stage larval White-Leg-Shrimps (Litopenaeus vannamei) were obtained as PL 1postlarvae (12 days post hatch) from Shrimp Improvement Systems LLC (SIS), 88081 Overseas Highway, Islamorada, FL 33036, USA. Before the experiment the shrimp larvae were acclimatized to the experimental conditions in a special 500 L acclimatization tank, and fed first with living Artemia for 2 days, and weaned with Artemia combined with the reference feed for 7-10 days. During the feeding experiments all systems were filled with purified water, salted to a salinity of 13-15‰, and all tanks were randomly stocked with 100-150 PL 1 White-Leg-Shrimps (Litopenaeus vannamei). The water quality was monitored using Hach-Lange spectral photometer DR 3900 and Hach HQ40d portable meter (Hach Lange GmbH, Germany), and maintained in the optimum range (temperature 29-31 ° C, salinity 13-16‰, pH 7.7-8.0, dissolved oxygen 6.7-8.0 mg L^{-1}, total NH3 0.01-0.07 mg L^{-1}, nitrite 0.01-0.11 mg L^{-1}, and nitrate 2.0-16.6 mg L^{-1}) during the run of the experiments. The feed amount for each aquarium was calculated daily according to the biomass of the shrimp (5- 15% measured) during Experiment I, and kept unchanged during a prestudy and Experiments II and III. The reference diets were fed by automatic feeders 6-8 times a day.

Experiments (prestudy, I-III)

A prestudy with stocking densities of 50, 100, 150 and 200 PL 1 in each system with the use of the commercial feed Le Goussant was carried out for 60 days, monitoring survival rates and total biomass in four systems. During Experiment I the aquaria in each of the four systems were stocked with 150 post larvae (equals 300 PL m-2). As a control standard the commercial feed from the prestudy was used. The same feed was mixed with the feed additive of 1% clay, 2% clay and a mixture of 1.5% clay and 0.5% algae paste (Chlorella vulgaris) originally obtained as a dilution from an algae bioreactor. This trial was carried out for 60 days, and the feeding was adapted to the weekly-calculated biomass in each tank. The diets were formulated at IGV GmbH, Getreidemittelverarbeitung, Potsdam (Germany). During feed preparation, the new feed mixture was heated again above 90 ° C. For Experiment II each aquarium was stocked with 100 post larvae. The feeding was kept unchanged for the entire 44 days. Four different diets from the same raw material, with the clay mineral/ algae additive added, were formulated at Research Diet Services BV, WijkbijDuurstede (the Netherlands), each in 3 different pellet sizes (300-500 µm; 500-700 µm; 700-1400 µm). The reference diet contained fish meal (40.3%), shrimp meal (10%), squid meal (6%), wheat (13.53%), wheat flour (25%), fish oil (2%), vitamin premix (1.5%), Lecithin (1%), salt (0.6%) and Cholesterol (0.07%). The three test diets were produced including 1% clay additive, 2% clay additive and 2% clay powder combined with 2%microalgae meal (Chlorella vulgaris). The test diets were analyzed according to their feed composition and energy contents (Table 1). Experiment III repeated Experiment II by testing the control diet, 2% clay powder and 4% of the clay additive (2%)/microalgae (2%) mixture for 32 days, using 150 post larvae of a new shrimp batch and a constant feeding regime. As a major difference, the new control group was kept in the best performing recirculation system from Experiment II (50:50 clay powder and microalgae mixture) and vice versa, excluding an experimental set up bias. Clay powder and microalgae The clay mineral was mined in Friedland in MecklenburgWest Pomerania (Germany) by FIM Biotech GmbH. It is registered as "FIMIX (100% Friedland Clay Mineral) under EG (VO) 1060/2013",currently registered as a montmorillonite-illite mixed layer clay mineral, with a positive EFSA opinion concerning product security and running efficacy tests. The Friedland clay minerals were formed in shallow marine environments as a result of sedimentation of volcanic tephra and eroded detrital material, and underwent early diagenesis. Due to complex formation history, the Montmorillonite Friedland clay minerals contain, among others, mixed-layered illite and muscovite, as well as kaolin [47]. It has a less montmorillonite content (40- 60%) compared with the bentonites (60-80%). Furthermore, it differs in its composition from all other clay mineral sources [48] with a mineral fingerprint, according to the following formula $(Na_{0.58}Ca_{0.03}Mg_{0.10}Al_{2.22}Fe^{3+}{}_{1.09}Fe^{2+}{}_{0.18}Mg_{0.59}Si_{7.80}Al_{0.20}O_{20}(OH)_4)$ [49] It consists of SiO_2 (57.9-59.5 % total weight, Al_2O_3(17.0-18.5%),

Table 1: Nutrient and fatty acid content of diets from Exp II & III

	Control	1% MMT[1]	2% MMT[1]	Mixture[2]
Nutrients	**Diet 1**	**Diet 2**	**Diet 3**	**Diet 4**
Moisture (%)	5.4	5.8	6	8
Ash (%)	10.3	11.4	12.1	12.4
Protein (%)	40.8	40.6	40.4	39.2
Crudefiber (%)	2.3	2.3	2.5	2.3
Fat (%)	7.6	7.3	7.6	7.3
Fatty acid (%)				
C14:0	0.00	0.00	6.65	6.17
C16:0	23.84	23.94	22.11	22.65
C16:1	6.97	8.74	7.65	7.34
C18:0	4.83	4.61	5.08	4.25
C 18:1 n9c	19.25	18.79	18.01	17.67
C 18:2 n6c	17.85	17.05	16.14	16.45
C 20:1	2.84	2.7	2.58	2.53
C 18:3 n3	2.84	2.57	2.02	1.88
C 20:2	2.85	1.83	1.9	1.91
C 22:1 n9	0.00	1.92	0.00	0.00
C 23:0	0.00	0.00	0.81	0.79
C 22:2	0.00	0.00	0.00	0.91
C 20:5 n3	8.62	8.45	8.03	8.1
C 22:6 n3	10.11	9.4	9.02	9.35
Energy (kJ kg^{-1})	17,765.9	17,468.5	17,290.4	16,459.7
Energy (%)	100	98.3	97.3	92.7

Abbreviations: Values are based on the analysis at LUFA, Rostock (Germany)[1] , MMT = Montmorillonite[2] , Mixture = for Exp II & III the clay/ algae mixture contained 2% clay and 2% algae powder Fe_2O_3 (5.9-7.0%), K_2O (2.8-3.5%), MgO (1.5-2.6%), Na_2O (0.9-1.5 %), TiO_2 (0.6-1.5), CaO (0.25-0.35), P_2O_5 (0.09-0.15), and others (8.9-10.5%), or 23% Si, 9.6% Al, 3.8% Fe, 0.9% Mg, 0.2% Ca, 0.7% Na, 1.7% K, and 0.5% Ti. Other characteristics are the particle size of 90% ≤ 80 µm or 50% ≤ 25 µm, pH of 7.3, conductivity of 346 µS, cation exchange capacity of 38 meq/100 g, and exchangeable cations of Ca, K, Mg, Na.

The microalgae Chlorella vulgaris originated form a photobioreactor at a power station of the city Senftenberg (Vattenfall GmbH, Germany). FIM Biotech GmbH has developed a low-cost process, directly binding harvested Chlorella vulgaris dilution by using a defined Friedländer clay mineral, resulting in ready-toadd mineral/algae composite as a feed additive. The protein-rich biomass, in our case produced from CO_2 -containing flue gases from a conventional power plant, resulted from sedimentation under the presence of magnesium, followed by a drying and downsizing process. The clay/algae mixture was suitable as a shrimp feed additive, with the clay mineral already approved in Europe as a feed additive material [41].

Data collection and statistical analyses

At the beginning of each experiment, initial weights of the shrimps were taken. During the run of the experiments, random samples of 35 individuals from each aquarium were measured every two weeks. At the end of each experiment all shrimps were weighed and measured. Weight gain, specific growth rate (SGR, percentage weight gain per day), the feed conversion ratio (FCR) and survival were calculated for each group of shrimps. The data were analyzed by using One-Way ANOVA. Differences of means were evaluated for significance by the range test of TUKEY-HSD ($p<0.05$) for homogeneous variances and by the DUNNETT-T3 ($p<0.05$) test for inhomogeneous variances respectively. Calculations were performed with the SPSS 20.0 software package [42].

RESULTS

The optimal stocking density for the system used in this study was determined from the prestudy. The best survival rates were observed at stocking densities of 100-150 shrimps per tank or 200-300 post larvae per m^2

Experiment I

The results for this 60 days feeding trial are given in Table 2. During this experiment, the mean initial body weight of 0.047 g increased to 2.22-2.65 g. Best growth (final weight) was observed with the control (2.65 g), whereas treatment 1%, 2% and the mixture diet showed no significant final weights from 2.22-2.32 g. Total weight gain was best in the control group with 265.43 g (100%), and not significant with the 2% treatment with 261.69 g (98.6%, Figure 1). Furthermore, for all other growth parameters, the control group achieved the best results (the treatment feed had additional feed processing compared with the control). The final biomass and total weight gain were significant higher than in the 1%-clay mineral treatment and higher than in the clay/algae mixture treatment. The treatment with 2% montmorillonite showed no significant differences with the control group. The SGR in the 2%-treatment-group was lower than in the other 3 groups, with 6.43% d^{-1} growth per day compared to 6.73% d^{-1} in the control group, 6.50% d^{-1} in the 1% treatment group and 6.47% d^{-1} in the mixture treatment group. The lowest FCR was also observed in the control group (1.07), ranging between 1.2- 1.3 in the three tested groups.

In contrast to the respective growth parameters, highest survival was observed in the 2% clay mineral treatment-group (79.75%), although there were no significant differences to the control group (68.10%) and to the clay/algae-group (68.75%, Figure 2). The 1% treatment-group achieved the lowest survival (56.25%). The physical and chemical water parameters showed no significant differences among the feeding groups during this experiment except salinity with significant slightly differences between all groups (Table 3). The size distribution of the final shrimp weight differed between the different treatment groups in all three experiments (Figs. 3-5). In experiment I the median, minimum and maximum weight (g) in the control, 1% clay, 2% clay and 2% algae/2% clay treatment groups was 2.68 (0.13, 5.50), 2.36 (0.4, 4.2), 2.28 (0.24, 4.04), and 2.26 (0.75, 4.30), respectively. The most even size distribution was observed for the 2% algae/2% clay treatment group (Figure 3), resulting in a maximum weight difference of 5.37 g (±0.84) for the control, compared with 3.8 g (±0.7) in the

Table 2: Shrimp growth and survival parameters during 3 different feeding experiments, means (± SD), groups with not sharing a superscript in the same column are significantly different ($p<0.05$, $n=4$) for Experiment I, mixture-treatment, the clay/algae mixture contained 1.5% clay and 0.5% algae powder, for Experiment II and III the clay/algae mixture contained 2% clay and 2% algae powder.

EXP I						
group	**survival [%]**	**final weight [g]**	**final biomass [g]**	**total weight gain [g]**	**SGR[1] [% d^{-1}]**	**FCR[2]**
Control	68.10^{ab}±9.43	2.65^{a}±0.84	270.34^{a}±25.14	265.43^{a}±25.79(100%)	6.73^{a}±0.10	1.07^{a}±0.06
1%-Treatment	56.25^{a}±9.11	2.32^{b}±0.70	196.24^{b}±23.34	190.38^{b}±23.70(74.2%)	6.50^{ab}±0.06	1.34^{b}±0.12
2%-Treatment	79.75^{b}±5.91	2.22^{b}±0.61	265.64^{a}±6.80	261.69^{a}±9.40(98.6%)	6.43^{b}±0.10	1.20^{ab}±0.06
Mixture-Treatment	68.75^{ab}±11.00	2.29^{b}±0.60	234.83^{ab}±23.95	231.67^{ab}±25.54(87.3%)	6.47^{ab}±0.22	1.23^{ab}±0.10
EXP II						
group	**survival [%]**	**final weight [g]**	**final biomass [g]**	**total weight gain [g]**	**SGR[1] [% d^{-1}]**	**FCR[2]**
Control	33.00^{a}±10.61	2.85^{a}±1.66	67.60^{a}±19.10	62.50^{a}±19.10 (100%)	5.79^{ab}±0.76	1.15^{ab}±0.13
1%-Treatment	16.25^{a}±5.85	3.23^{a}±2.20	32.34^{a}±26.09	27.24^{a}±26.09(43.7%)	3.75^{a}±1.56	2.09^{b}±1.03
2%-Treatment	32.00^{a}±11.05	3.34^{a}±1.29	73.45^{a}±33.70	68.35^{a}±33.70(110%)	5.89^{b}±1.00	1.02^{ab}±0.23
Mixture-Treatment	54.75^{b}±7.80	3.18^{a}±1.15	151.25^{b}±4.67	146.15^{b}±4.67 (234%)	7.70^{b}±0.07	0.76^{a}±0.01
EXP III						
group	**survival [%]**	**final weight [g]**	**final biomass [g]**	**total weight gain [g]**	**SGR[1] [% d^{-1}]**	**FCR[2]**
Control	26.00^{a}±10.46	1.14^{a}±0.48	29.54^{a}±8.10	27.58^{a}±8.10(100%)	14.08^{a}±0.48	1.37^{a}±0.43
2%-Treatment	41.50^{a}±15.86	0.95^{b}±0.38	39.37^{a}±7.45	37.41^{a}±7.45 (135.5%)	13.55^{a}±0.65	0.93^{a}±0.16
Mixture-Treatment	45.00^{a}±9.90	0.89^{b}±0.36	40.15^{a}±5.81	38.19^{a}±5.81(138.4%)	13.25^{a}±0.38	0.94^{a}±0.17

Abbreviations: [1] SGR = Specific growth rate [% d^{-1}] = (ln final weight-ln initial weight) x t^{-1} x 100; t = 60 days for Exp I, 44 days for Exp II and 32 days for Exp III,[2] FCR = Feed conversion ratio = feed consumed [g] x weight gain $[g]^{-1}$.

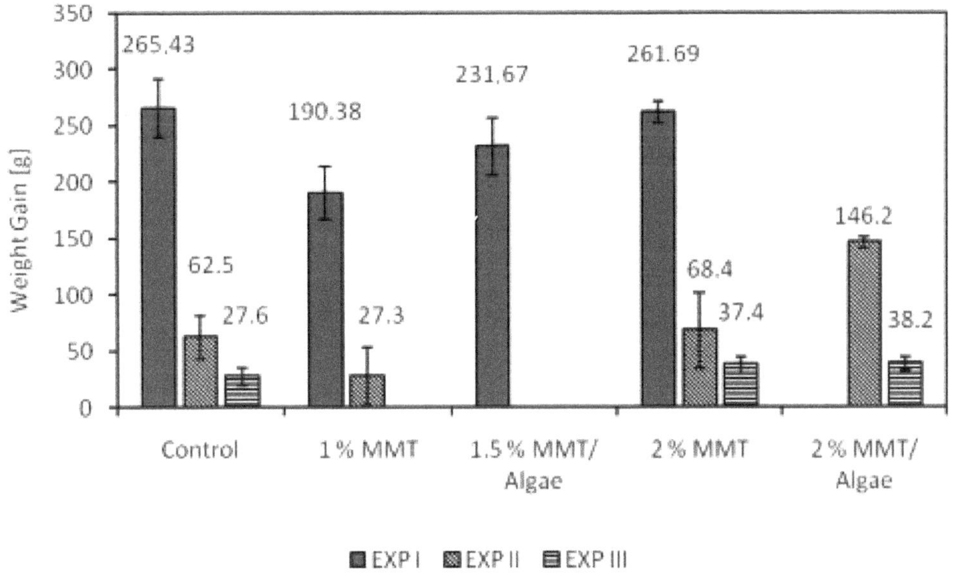

Figure 1 Weight gain [g] of L. vannamei during Experiments I-III

1% clay treatment group, 3.8 g (±0.61) in the 2% clay treatment group, and 3.55 g (±0.60) in the mixture treatment group.

Experiment II Experiment II was carried out for 44 days and results are given in Table 2. During the acclimatization phase in a separate holding tank, we recorded a problem with the biofilter resulting in ammonia (0.8 mg L^{-1}) and nitrite (0.135 mg $L^{-1)}$ peaks, with possible negative effects onto shrimp post larvae survival in the following cultivation in the recirculation units. During the run of the experiment, water parameters were stable, with only slightly differences between the four treatment groups (Table 3).

Compared with Experiment I, the shrimp larvae had a significantly lower survival rate (Table 2). Comparing the four treatment groups, the clay/algae-treatment-group had a significant higher survival rate (54.75%) than those shrimps fed with the control diet (33.00%). The 1%-clay mineral treatment (16.25%) was lower, and the 2% clay mineral treatment (32.00%) group showed no significant better survival to the control group. All treatment groups (3.18-3.34 g) achieved a higher mean final weight than the control group (2.85 g), even though there were no significant differences. The shrimps fed the clay/algae mixture treatment diet had a significantly higher final biomass (151.25 g) and weight gain (234%) compared with the control group

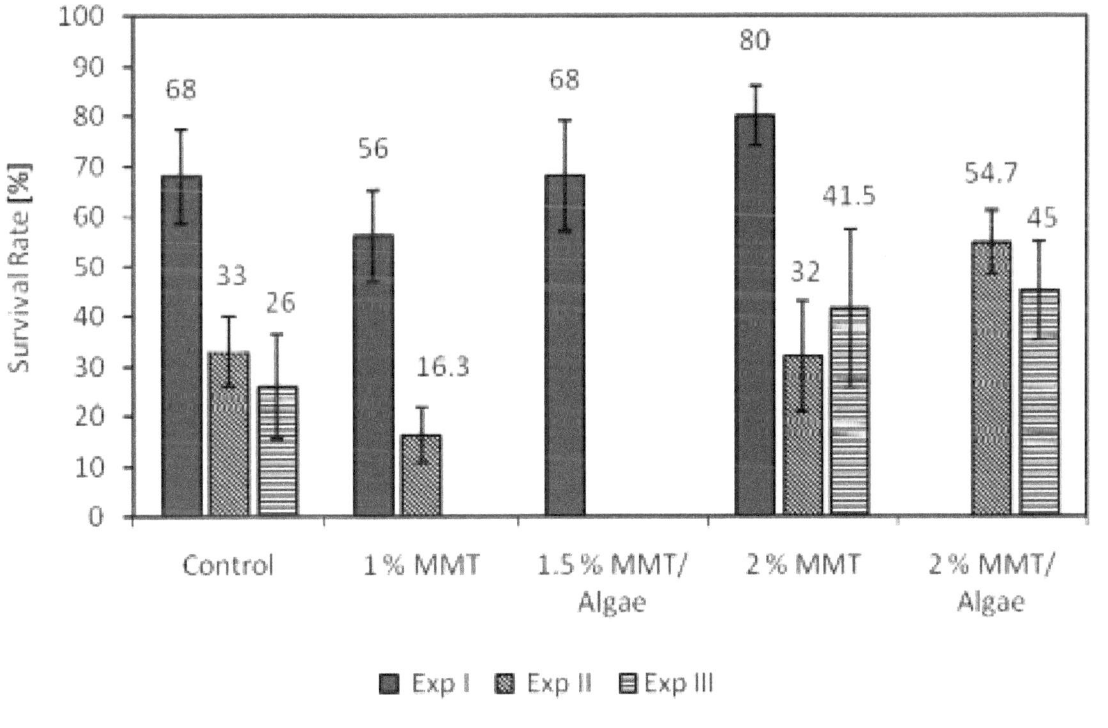

Figure 2 Survival [%] of L. vannameiduring Experiments I-III.

Table 3: Water parameters during Experiments I-III, means (± SD). Groups with not sharing a superscript in the same column are significantly different ($p<0.05$, $n=4$)

EXP I							
Parameter	Oxygen [mg L^{-1}]	pH	Temperature [°C]	Salinity [‰]	Ammonia [mg L^{-1}]	Nitrite [mg L^{-1}]	Nitrate [mg L^{-1}]
Control	$6.75^{a}\pm0.24$	$7.86^{a}\pm0.46$	$30.69^{a}\pm0.76$	$13.90^{a}\pm0.18$	$0.05^{a}\pm0.01$	$0.10^{a}\pm0.08$	$15.23^{a}\pm7.70$
1% Clay	$6.91^{a}\pm0.22$	$8.09^{a}\pm0.39$	$31.23^{a}\pm0.56$	$13.63^{b}\pm0.26$	$0.05^{a}\pm0.02$	$0.10^{a}\pm0.07$	$14.81^{a}\pm6.75$
2% Clay	$6.78^{a}\pm0.37$	$7.75^{a}\pm0.41$	$31.08^{a}\pm0.67$	$14.18^{c}\pm0.31$	$0.05^{a}\pm0.03$	$0.10^{a}\pm0.05$	$14.70^{a}\pm5.78$
Mixture	$6.95^{a}\pm1.07$	$7.84^{a}\pm0.41$	$31.12^{a}\pm0.57$	$14.83^{d}\pm0.15$	$0.06^{a}\pm0.03$	$0.11^{a}\pm0.08$	$16.65^{a}\pm8.63$
EXP II							
Parameter	Oxygen [mg L^{-1}]	pH	Temperature [°C]	Salinity [‰]	Ammonia [mg L^{-1}]	Nitrite [mg L^{-1}]	Nitrate [mg L^{-1}]
Control	$7.80^{a}\pm0.20$	$7.82^{a}\pm0.09$	$30.34^{a}\pm0.42$	$13.70^{a}\pm0.35$	$0.07^{a}\pm0.02$	$0.02^{a}\pm0.01$	$2.94^{a}\pm0.85$
1% Clay	$7.70^{ab}\pm0.19$	$7.95^{b}\pm0.05$	$30.77^{a}\pm0.61$	$13.78^{a}\pm0.19$	$0.05^{b}\pm0.02$	$0.02^{a}\pm0.01$	$2.53^{a}\pm0.44$
2% Clay	$7.61^{ab}\pm0.17$	$7.91^{ab}\pm0.07$	$30.86^{a}\pm0.50$	$14.08^{ab}\pm0.27$	$0.05^{b}\pm0.01$	$0.02^{a}\pm0.00$	$3.03^{a}\pm1.68$
Mixture	$7.52^{b}\pm0.26$	$7.84^{ab}\pm0.18$	$30.72^{a}\pm0.42$	$14.33^{b}\pm0.06$	$0.06^{ab}\pm0.02$	$0.02^{a}\pm0.01$	$3.20^{a}\pm1.21$
EXP III							
Parameter	Oxygen [mg L^{-1}]	pH	Temperature [°C]	Salinity [‰]	Ammonia [mg L^{-1}]	Nitrite [mg L^{-1}]	Nitrate [mg L^{-1}]
Control	$7.75^{a}\pm0.13$	$7.96^{a}\pm0.14$	$30.64^{a}\pm0.85$	$14.67^{a}\pm0.12$	$0.01^{a}\pm0.01$	$0.01^{a}\pm0.00$	$2.42^{a}\pm0.79$
2% Clay	$8.04^{b}\pm0.16$	$7.76^{b}\pm0.19$	$29.91^{a}\pm1.01$	$15.37^{b}\pm0.16$	$0.01^{a}\pm0.01$	$0.01^{a}\pm0.00$	$2.08^{a}\pm0.89$
Mixture	$7.74^{a}\pm0.15$	$7.92^{ab}\pm0.14$	$30.23^{a}\pm0.74$	$14.59^{a}\pm0.13$	$0.03^{a}\pm0.03$	$0.01^{a}\pm0.00$	$2.22^{a}\pm0.97$

(100%). Shrimp larvae fed with the 2% clay mineral treatment diet had a slightly higher final biomass (73.45 g) and weight gain (110%) than the control group and the 1% treatment group (32.34 g; 43.7%), but without any significant differences. This was in contrast to the feed energy content, that was highest in the control feed compared with the 1% clay (-1.7%), 2% clay (-2.7%) and the 2% clay/2% algae (-7.3%) treatments (Table 1). Taking into account the energy contents, the adjusted biomass gain based on energy input was 113% (2% clay mineral), 44.5% (1% clay mineral) and 251% (2% clay/2% algae) compared with the control. The size distribution of the final shrimp weight differed between the different treatment groups also at the end of Experiment II (Figure 4). The median, minimum and maximum weight in the control, 1% clay, 2% clay and 2% algae/2% clay treatment groups was 2.43 (0.14, 8.43), 2.96 (0.18, 8.43), 3.19 (0.67, 6.21), and 3.31 (0.14, 6.21), respectively. The most even

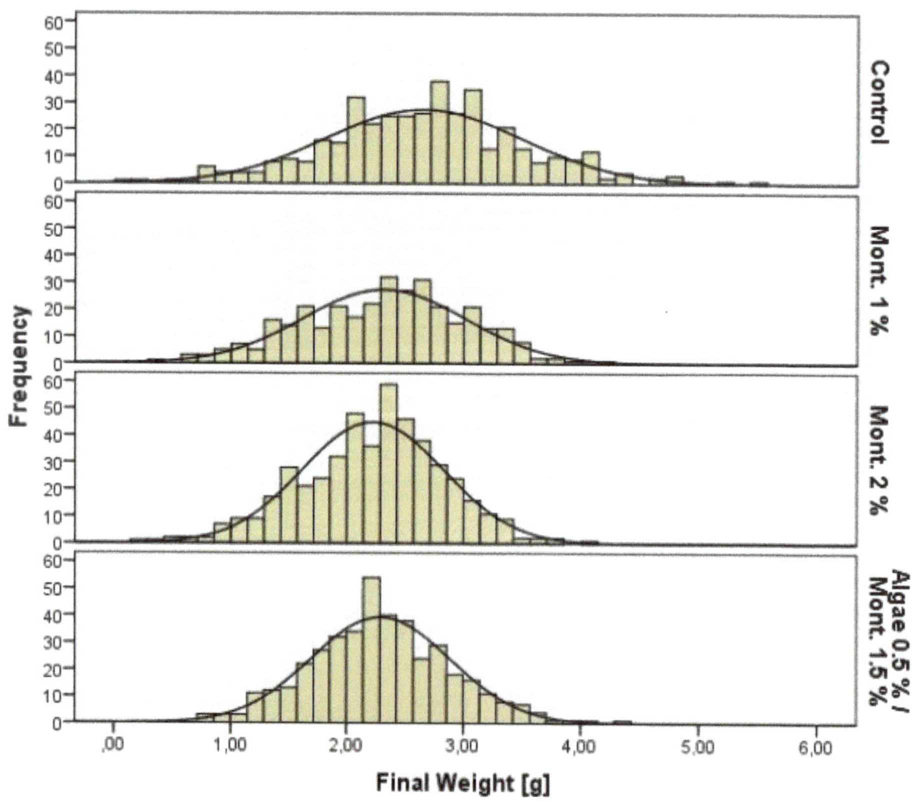

Figure 3 Size distribution histogram of shrimp final weight [g] and different treatment groups of clay in Experiment I

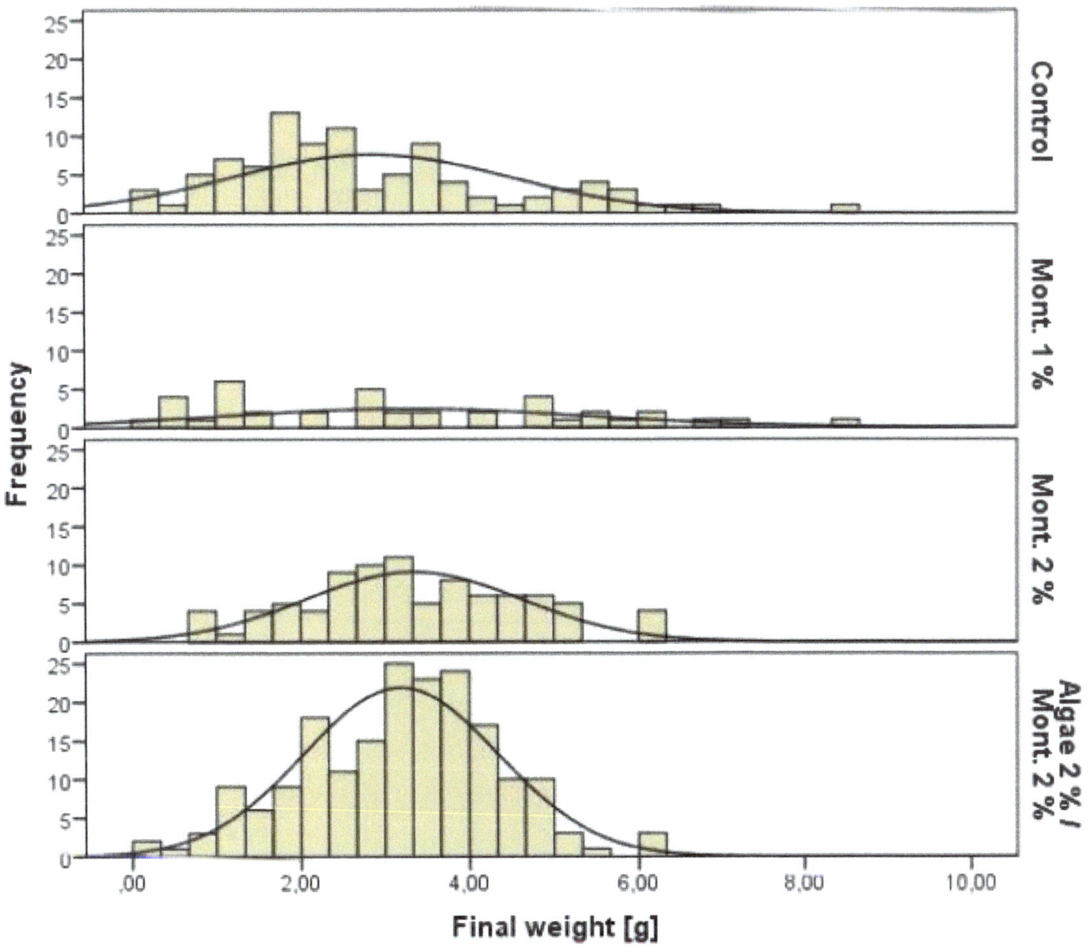

Figure 4 Size distribution histogram of shrimp final weight [g] and different treatment groups of clay in Experiment II.

size distribution was observed for the 2% clay treatment group, resulting in a maximum weight difference of 5.54 g (±1.29), followed by the 2% algae/2% clay (6.07, ±1.15), 1% clay (8.25, ±2.20) treatment group and the control (8.29, ±1.66, Figure 4).

Experiment III

The results of Experiment III showed significant difference only between the control treatment groups with a greater final weight of 1.14 g compared to the other diets (Table 2). However, the 2% clay mineral treatment (41.50%) and the clay/algae mixture treatment (45.00%) groups had a higher survival rate than the control group (26.00%). As a result of the higher survival rate and shrimp number in the aquaria, the same feed intake and slightly reduced energy content, the mean final weight of the shrimps was lower in both treatment groups than in the control. However, weight gain (surviving shrimps x individual final weight) was 30-35% higher in both treatment groups compared with the control. The SGR slightly better results in the control group compared to the 2% treatment and the clay/algae mixture (Table 2). The control had a higher FCR of 1.37 compared with a FCR of 0.93 for the 2% treatment group and 0.94 for mixture treatment group. The size distribution of the final shrimp weight differed between the three treatment groups in Experiment III (Figure 5). The median, minimum and maximum weight in the control, 2% clay and 2% algae/2% clay treatment groups was 1.01 (0.44, 3.20), 0.87 (0.19, 2.12), and 0.84 (0.22, 2.60), respectively. The most even size distribution was observed for the 2% clay treatment group (Figure 5), resulting in a maximum weight difference of 1.93 g (±0.38), compared with in the other treatment group (2.36 g, ±0.36) and the control (2.76 g, ±0.47).

DISCUSSION

Clay minerals are commonly used as a feed ingredient in land based animal husbandry, e.g. chicken-, pig- or cow- farming. As a feed additive, they have been demonstrated to have positive effects onto the treated animals, e.g. clay minerals can improve weight gain, feed efficiency and egg production in poultry farming. A significant increase in average weight gain was observed by adding 1-3% bentonite to the feed [30]. Similar results under 1-2% treatments with sodium bentonite were found by [31]. They monitored a higher feed intake and weight gain and lower FCR in the treated animals. Other studies noticed an increase in average daily weight gain and growth performance in pig farming. A slight increase in feed intake, weight gain and feed conversion for growing pigs by the use of 0.15% montmorillonite was reported by [34]. In a second study, they repeated these results for weanling pigs by using 0.2% montmorillonite [35]. This challenges the application of these feed additives also for fish and shrimp aquaculture. In the cultivation of aquatic animals clay was used to increase turbidity and to reduce the bacteria level in the production water of cod larvae, and observed indirect positive effects onto growth and survival [39]. Under the clay treatment the larvae had a slightly higher dry weight and a 20-25% better survival.

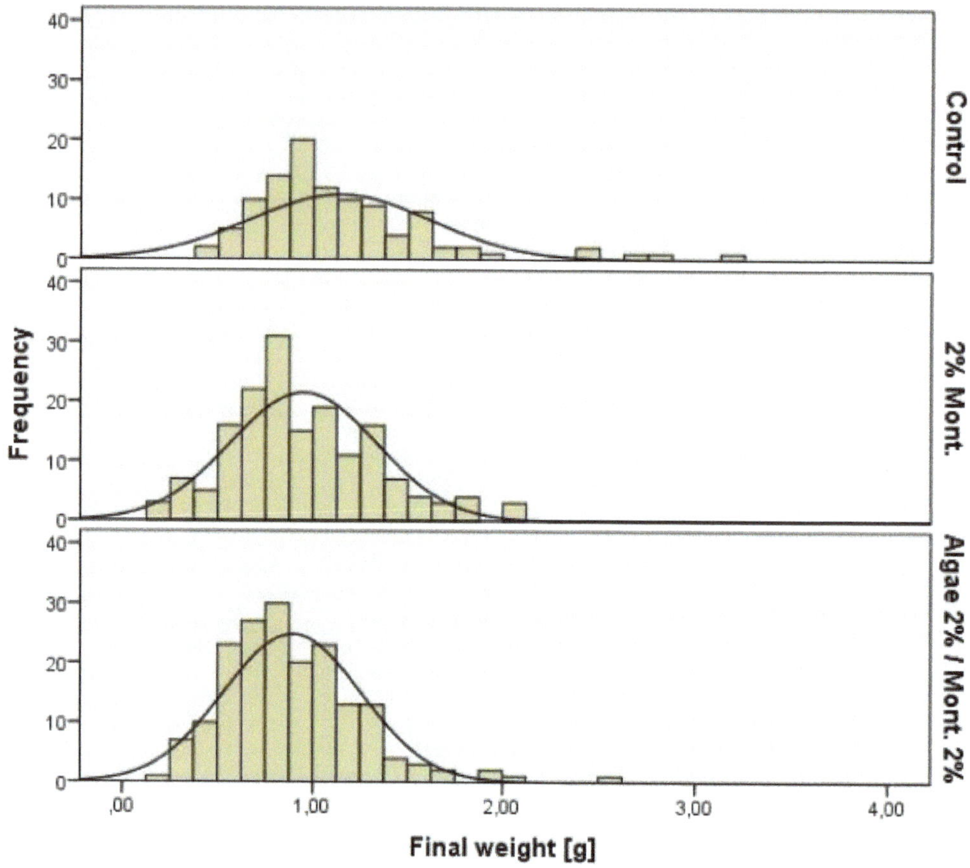

Figure 5 Size distribution histogram of shrimp final weight [g] and different treatment groups of clay in Experiment III.

The application of clay minerals in rainbow trout aquaculture was studied with direct positive effects onto growth and feed conversion [40]. The major outcome of his study was that the efficiency of the additive depended on the source of the clay mineral as well as the given amount, resulting in an increase in weight gain of 10% at a clay content in the trout feed of 2 and 5%. This is in the range of the results in the present study, where the Friedland clay mineral demonstrated a positive effect onto Litopenaeus vannamei survival, final total weight (110%) and FCR at 2% (with or without algae additive) in three independent trials (Table 2). On the other hand, the application of 1% clay mineral reduced the shrimp performance in the present trials compared with the control. Furthermore, the effects of montmorillonite clay minerals were analyzed onto Nile tilapia (Oreochromis niloticus) growth with the use of 0.15% montmorillonite in the fish diets [37]. A slight enhancement of growth performance and survival was observed. According to our results, this might be caused by an underestimate of the needed amount (0.15 vs. 2%) or the kind of clay mineral that was used. Clay minerals have many medical and therapeutically effects onto animal and human health [24,25,51], due to desirable physical and physico-chemical properties, such as high adsorption capacity, specific surface area, swelling capacity, and reactivity to acids [24,52]. They are used in the pharmaceutical industry in many different forms, and administered either orally or topically e.g. as antiseptics, disinfectants, and even antiinflammatories [25]. Montmorillonite has been demonstrated to have positive effects onto the intestinal microflora and antibacterial activities [53,54]. The influence of montmorillonite on the intestinal microflora, digestibility and digestive enzyme activities in the Nile Tilapia, Oreochromis niloticus was studied by [38]. They recorded an improvement of growth performance, reduction of intestinal aerobic bacterial counts and effects onto the composition of the intestinal microflora. Furthermore, they recorded an improvement of the digestibility of dry matter and crude protein, and improved enzyme activities of total protease, amylase, lipase and alkaline phosphatase. The antibacterial activity of the montmorillonite resulted in positive enzyme activities, promoting digestion and adsorption of dietary nutrition [38]. The specially characterized Friedland montmorillonite– illite/muscovite mixed layer clay mineral used in the present study has similar production and health promoting effects in animal husbandry. As a feed additive, this specific clay mineral, already known as a highly efficient mycotoxin binder, increased body weight and feed intake in pig farming, milk production and feed intake in cow farming, and induced a significant increase in body weight gain (approx. 10%) and feed intake in chicken farming [43-45]. A significant reduction of inflammatory bowel diseases was reported through in situ perfusion of an extract of this clay mineral with rats [26]. According to [27], Friedland clay minerals are a highly efficient phosphate binder, and assumed to have positive effects onto human kidney diseases. The use of clay minerals was reported to reduce Ammonium from wastewater [55]. The present study demonstrates, based on higher survival rates, such positive effects of this clay mineral also onto an invertebrate aquaculture species, and under recirculation aquaculture and experimental conditions. Possible negative effects such as nutrient binding capacities have not been tested so far.

We could observe significant different growth performance, survival rates, SGR, FCR and final size distribution of the tested shrimps under Friedland clay mineral application. The initiate stocking density with 200-300 post larvae m^{-2} was high, the glass aquaria without major hiding possibilities and a high ammonium load at the beginning of the second experiment represented non-favorable conditions for the shrimps during the run of the experiments. Experiment I showed better survival (79.75%) but less weight gain (262 g) in the 2% montmorillonite treatment group, compared with 68% and 265 g in the control. This is in contrast to the treatment of the test diets that were heated above 90 ° C compared with the control diet, possibly affecting its nutritional value. Consequently, the control diet without feed additive and higher energy content should have performed better than the test diets including the feed additive. This is a good explanation for the lower final weight and the higher FCR observed for the three test groups. The total final weight gain of the 2% treatment group is caused by the highest survival rate during this experiment. This results of the first experiment is supported by the other two experiments, where the water conditions where nonfavorable (Experiment II) for the post larvae survival. Shrimps with the feed additive treatment demonstrated significant better survival, though the energy content of the feed was reduced between 1.7 and 7.3%. Especially after the high ammonium values during the acclimatization phase, shrimps treated with the clay/algae feed reached 234% of the weight gain compared with the control group, and an obvious lower FCR. This was repeated in Experiment III, with the best growth performance under treatment with 2% montmorillonite with or without microalgae powder. Though the three independent experiments achieved different shrimp performances, based on a different feed treatment before the experiments and water quality during the weaning phase, the same trend was observed. We can conclude that under three different scenario using different imported charges of post larvae, survival rates and total weight gain increased under treatment of 2% Friedland montmorillonite. On the other hand, the 1% feed additive showed a lower performance compared with the control, underlining the statement by [40] that not only the origin but also the concentration effects the growth performance in rainbow trout. Also in aquatic invertebrates, digestibility of nutrients might be supported through the clay feed additive, as described by earlier authors [37,38,53] and also for the higher vertebrates [34,35,43-45]. An important observation is the reduced size range of the clay mineral treatment groups after the three independent experiments, where the 2% clay additive performed best, resulting in the most uniform size distribution of the shrimps (Figures. 3-5). During experiment II, the 2% clay/2 % algae feed additive performed best while the 2% clay had a more similar size range after Experiments I and III. Taking into account the higher survival rates and the observation that weak shrimps where directly consumed by the other animals in the aquaria, less cannibalism might be the reason for this result. High stocking densities of shrimps cause increasing interactions between the single shrimp individuals, resulting in cannibalism and increasing mortality rates [56-58]. In our experiments, the stocking densities of 200-300 Shrimps per m^2 are comparable to super intensive aquaculture conditions. In addition, the glass aquaria without substrate are an extreme environment for the shrimps, leaving them especially vulnerable to cannibalism, thus resulting in a high size distribution and a low survival. Under such super intensive conditions it is necessary to maintain a healthy stock with unsusceptible and resistant animals, possibly promoted by the Friedland clay minerals with and without Chlorella vulgaris microalgae. The applied Chlorella vulgaris microalgae originated from a bioreactor attached to a power station fed with CO_2 . Algae as a feed additive in aquaculture resulted in positive effects onto the cultivated organisms, such as promoting general health, antibacterial, antifungal and antiviral effects or the supplementation with nutrients, vitamins and other ingredients [39,59-61]. Shrimps are natural algae grazers in shrimp ponds and different algae have been reported to be of importance for the cultivation of larval shrimps [62,63]. The Friedland clay minerals can be used for an efficient and low cost harvesting of microalgae suspension. We could demonstrate that a combination of the C. vulgaris microalgae with the beneficial Friedland clay mineral is applicable for shrimp cultivation, additionally making a direct use of bioreactor produced microalgae for a healthy shrimp aquaculture possible. It is recommended that also other microalgae species and their combinations with the Friedland clay minerals should be tested in future studies.

CONCLUSION

White leg shrimp (Litopenaeus vannamei) post larvae, 12 days post hatch (PL 1), were fed with different diets including Friedland montmorillonite– illlite/muscovite mixed layer clay minerals as a feed additive with or without the addition of Chlorella vulgaris green algae. Positive effects on shrimp growing parameters were found in three independent experiments, especially with an application of 2% clay minerals. The addition of 1% clay minerals in the diets had no positive effects or reduced the performance compared with the control. Addition of microalgae in the diets (mixture with clay mineral) showed comparable effects onto shrimp growth to the 2% montmorillonite treatment. Friedland montmorillonite clay minerals as a feed ingredient resulted in a higher biomass gain after the run of the experiments. Though having a reduced individual weight gain compared with the control also caused by a reduced energy content of the feed, higher survival rates and a more even size distribution of L. vannamei resulted in the best performance of the shrimps with 2% montmorillonite. The more homogenous size heterogeneity of shrimps in combination with Chlorella vulgaris and montmorillonite (2%/2%) also indicated a positive effect of the feed onto shrimp cohorts under the experimental super intensive conditions. We affiliate this with a suppressed cannibalism in the aquaria. In conclusion, Friedland montmorillonite illite/ muscovite mixed layer clay minerals have beneficial effects onto White leg shrimp (Litopenaeus vannamei) survival and performance at a concentration of 2% with and without addition of Chlorella vulgaris green algae.

ACKNOWLEDGEMENTS

The present study was financially supported through FIM Biotech GmbH (Shrimp Health I and II) and the Chair of Aquaculture and Sea-ranching, University of Rostock.

REFERENCES

1. FAO. World Wide Web electronic publication. 2014.
2. Wyban JA, Swingle JS, Sweeney JN, Pruder GD. Development and commercial performance of high health shrimp using specific pathogen free (SPF) brood stock Penaeus vannamei. Orlando. Proceedings of the Special session on Shrimp Farming, The World Aquaculture Society. Aquaculture.1992; 92: 254-260.
3. Moriarty DJW. Diseases control in shrimp aquaculture with probiotic bacteria. In: Microbial Biosystems: New Frontiers. Proceedings of the 8th International Symposium on Microbial Ecology (ed. by Bell C.R., Brylinsky M.& Johnson-Green P). Atlantic Canada Society for Microbial Ecology, Halifax, Canada. 1999.
4. Nhan DT, Wille M, Schryver PD, Defoirdt T, Bossier P, Sorgeloos P. The effect of poly ß-hydroxybutyrate on larviculture of the giant freshwater prawn Macrobrachiumrosenbergii. Aquaculture. 2010; 302: 76-81.
5. Newaj-Fyzul A, Al-Harbi AH, Austin B. Review: Developments in the use of probiotics for disease control in aquaculture. Aquaculture. 2014; 431: 1-11.
6. Rairat T, Chuchird N, Limsuwan C. Effect sangrovit WS on growth, survival and prevention of Vibrio harveyi in rearing of Pacific white shrimp (Litopenaeus vannamei). Kasetsart University Fisheries Research Bulletin. 2013; 37: 19-29.
7. Holmström K, Gräslund S, Wahlström A, Poungshompoo S, Bengtsson BE, Kautsky N. Antibiotic use in shrimp farming and implications for environmental impacts and human health. International Journal of Food Science and Technology. 2003; 38: 255-266.
8. Cabello FC. Heavy use of prophylactic antibiotics in aquaculture: a growing problem for human and animal health and for the environment. Environ Microbiol. 2006; 8: 1137-1144.
9. Defoirdt T, Boon N, Sorgeloos P, Verstraete W, Bossier P. Alternatives to antibiotics to control bacterial infections: luminescent vibriosis in aquaculture as an example. Trends Biotechnol. 2007; 25: 472-479.
10. Defoirdt T, Sorgeloos P, Bossier P. Alternatives to antibiotics for the control of bacterial disease in aquaculture. Curr Opin Microbiol. 2011; 14: 251-258.
11. Burridge L, Weis JS, CabelloF, Pizarro J, Bostick K. Chemical use in salmon aquaculture: A review of current practices and possible environmental effects. Aquaculture. 2010; 306: 7-23.
12. Felix N, Sudharsan M. Effect of glycine betaine, a feed attractant affecting growth and feed conversion of juvenile freshwater prawn Macrobrachiumrosenbergii. Aquaculture Nutrition. 2004; 10: 193- 197.
13. Huang X, Zhou H, Zhang H. The effect of Sargassum fusiforme polysaccharide extracts on vibriosis resistance and immune activity of the shrimp, Fenneropenaeus chinensis. Fish Shellfish Immunol. 2006; 20: 750-757.
14. Nimrat S, Boonthai T, Vuthiphandchai V. Effects of probiotic forms, compositions of and mode of probioticadministration on rearing of Pacific white shrimp (Litopenaeus vannamei) larvae and post larvae. Animal Feed Science and Technology. 2011; 169: 244-258.
15. Rico A, Phu TM, Satapornvanit K, Min J, Shahabuddin AM, Henriksson PJG et al. Use of veterinary medicines, feed additives and probiotics in four major internationally traded aquaculture species farmed in Asia. Aquaculture. 2013; 412-413: 231-243.
16. New MB. A review of dietary studies with shrimp and prawns. Aquaculture. 1976; 9: 101-144.
17. Gräslund S, Holmström K, Wahlström A. A field survey of chemicals and biological products used in shrimp farming. Mar Pollut Bull. 2003; 46: 81-90.
18. Davis DA, Samocha TM, Bullis RA, Patnaik S, Browdy C, Stokes A et al. Practical diets for Litopenaeus vannamei, (Boone. 1931): working towards organic and/or all plant production diets. In: Avancesen Nutricion Acuicola VII. Memoriasdel VII SimposiumInternacional de NutricionAcuicola. 16-19 Noviembre, 2004. Hermosillo, Sonora, Mexico.
19. Jaime-Ceballos B, Villarreal H, Garcia T, Pérez-Jar L, Alfonso E. Effect of Spirulina platensis meal as feed additive on growth, survival and development in Litopenaeus schmittishimp larvae. Revista de Investigaciones Marinas. 2005; 26: 235-241.
20. 20.Patnaik S, Samocha TM, Davis DA, Bullis RA, Browdy CL. The use of HUFA-rich algal meals in diets for Litopenaeus vannamei. Aquaculture Nutrition. 2006; 12: 395-401.
21. Hanel R, Broekman D, GraafSde, Schnack D. Partial Replacement of Fishmeal by Lyophylized Powder of the Microalgae Spirulinaplatensis in Pacific White Shrimp Diets. The Open Marine Biology Journal. 2007; 1: 1-5.
22. Ju ZY, Forster IP, Dominy WG. Effects of supplementing two species of marine algae or their fractions to a formulated diet on growth, survival and composition of shrimp (Litopenaeus vannamei). Aquaculture. 2009; 292: 237-243.
23. Ju ZY, Deng DF, Dominy W. A defatted microalgae (Haematococcuspluvialis) meal as a protein ingredient to partially replace fishmeal in diets of Pacific white shrimp (Litopenaeus vannamei, Boone, 1931). Aquaculture. 2012; 354: 50-55.
24. Carretero MI, Pozo M. Clay and non-clay minerals in the pharmaceutical industry Part I. Excipients and medical applications. Applied Clay Science. 2009; 46: 73-80.
25. Carretero MI, Pozo M. Clay and non-clay minerals in the pharmaceutical and cosmetic industries Part II. Activeingredients. Applied Clay Science. 2010; 47: 171-181.
26. Breitrück A, Sparmann G, Mitzner S, Kerkhoff C. Establishment of a novel extracorporeal bowel model to study luminal approaches to treat inflammatory bowel disease. Disease Models & Mechanisms. 2013; 6: 1487-1493.
27. Fraunhofer. Treating chronic kidney disease using clay minerals. Research News of the Fraunhofer-Gesellschaft. 2014; 01.

28. 28.Huwig A, Freimund S, Käppeli O, Dutler H. Mycotoxin detoxication of animal feed by different adsorbents. Toxicol Lett. 2001; 122: 179-188.

29. Quisenberry JH. The Use of Clay in Poultry Feed. Clays and Clay Minerals. 1968; 16: 267-270.

30. Tauqir NA, Nawaz H. Performance and Economics of Broiler Chicks Fed on Rations Supplemented with Different Levels of Sodium Bentonite. International Journal of Agriculture & Biology. 2001; 3: 149-150.

31. Salari H, Kermanshahi H, Moghaddam HN. Effect of Sodium Bentonite and Comparison of Pellet vs Mash on Performance of Broiler Chickens. International Journal of Poultry Science. 2006; 5: 31-34.

32. Colling GF, Thomasson SA, Ku PK, Miller ER. Sodium Bentonite in Swine Diets. Journal of animal science. 1980; 50: 272-277.

33. Lindemann MD, Blodgett DJ, Kornegay ET, Schurig GG. Potential ameliorators of aflatoxicosis in weanling/ growing swine. J Anim Sci. 1993; 71: 171-178.

34. Hu CH, Xia MS, XuZR, Xiong L. Effects of Copper-bearing Montmorillonite on Growth Performance and Digestive Function of Growing Pigs. Asian-Australas. J. Anim. Sci. 2004; 17: 1714-1720.

35. Hu CH, Gu LY, Luan ZS, Song J, Zhu K. Effects of montmorillonite-zinc oxide hybrid on performance, diarrhea, intestinal permeability and morphology of weanling pigs. Animal Feed Science and Technology. 2012; 177: 108-115.

36. Walz LS, White TW, Fernandez JM, Gentry LR, Blouin DC, Froetschel MA. Effects of fish meal and sodium bentonite on daily gain, wool growth, carcass characteristics, and ruminal and blood characteristics of lambs fed concentrate diets. Journal of Animal Science. 1998; 76: 2025-2031.

37. Hu CH, Xu Y, Xia MS, Xiong L, Xu ZR. Effects of Cu2+-exchanged montmorillonite on growth performance, microbial ecology and intestinal morphology of Nile tilapia (Oreochromis niloticus). Aquaculture. 2007; 270: 200-206.

38. Hu CH, Xu Y, Xia MS, Xiong L, Xu ZR. Effects of Cu2+-exchanged montmorillonite on intestinal microflora, digestibility and digestive enzyme activities of Nile tilapia. Aquaculture Nutrition. 2008;14: 281- 288.

39. Attramadal KJK, Tøndel B, Salvesen I, Øie G, Vadstein O, Olsen Y. Ceramic clay reduces the load of organic matter and bacteria in marine fish larval culture tanks. Aquacultural Engineering. 2012; 49: 23-34.

40. Smith RR. Recent advances in nutrition: clay in trout diets. Salmonid. 1980; 4 : 16-18.

41. FIM Biotech (pers. comm.). FIM Biotech GmbH, Am Kupfergraben 6a, D-10117 Berlin. Germany.

42. SPSS Inc. Statistic Software Package, Chicago, IL, USA. 2013.

43. Männer K. Efficacy and tolerance of supplemented FIMIX E 558 (Bentonit-Montmorillionit E 558) in broiler chickens from 1 to 35 d of age. Tolerance study of FIMIX E 558 (code: GM 4/13), Report to FIM Biotech GmbH, Institute of Animal Nutrition, University of Berlin. 2013a.

44. Männer K. Tolerance of supplemented FIMIX E 558 (BentonitMontmorillionit E 558) in dairy cattle during a 56-day-feeding period. Tolerance study of FIMIX E 558 (code: C 1/13), Report to FIM Biotech GmbH, Institute of Animal Nutrition, University of Berlin. 2013b.

45. Männer K. Tolerance of supplemented FIMIX E 558 (BentonitMontmorillionit E 558) in multiparous sows during one complete reproductive cycle. Tolerance study of FIMIX E 558 (code: SL 4/13), Report to FIM Biotech GmbH, Institute of Animal Nutrition, University of Berlin. 2013c.

46. Bautista MN, Lavilla-Pitogo CR, Subosa PF. Aflatoxin B, Contamination of Shrimp Feeds and its Effect on Growth and Hepatopancreas of Pre-adult Penaeusmonodon. Journal of the Science of Food and Agriculture. 1994; 65: 5-11.

47. Ostrowski-Meissner HT, LeaMaster BR, Duerraq EO, Walsh WA. Sensitivity of the Pacific white shrimp, Penaeusvannamei, to aflatoxin B1. Aquaculture. 1995; 131: 155-164.

48. Henning KH, Kasbohm J. Mineral composition and genesis of finegrained sediments of Quaternary and Praequartenary in East Germany with special emphasis on "Friedlander Clay Minerals". Institute of Geological Sciences, Ernst-Moritz-Arndt-University of Greifswald, Annual Meeting of the DTTG. 1998; Report 6: 147-167 [in German].

49. FIM Biotech. FIM Biotech GmbH, Am Kupfergraben 6a, D-10117 Berlin, Germany. 2012.

50. Kumpulainen S, Kiviranta L. Mineralogical and Chemical Characterization of Various Bentonite and Smectite-Rich Clay Materials, Finland, Clays in Natural & Engineered Barriers for Radioactive Waste Confinement, 4TH International Meeting-March 2010 NANTES, FRANCE, P/GEO/PC/06

51. Slamova R, Trckova M, Vondruskova H, Zraly Z, Pavlik I. Clay Minerals in animal nutrition. Applied Clay Science. 2011; 51: 395-398.

52. Ayari F, Srasra E, Trabelsi-Ayadi M. Characterization of bentonitic clays and their use as adsorbent. Desalination. 2005; 185: 391-397.

53. Droy-Lefain MT, Drouet Y, Schatz B. Sodium glycodeoxycholate and spinability of gastrointestinal mucus: protective effect of smectite. Gastroenterology. 1985; 88: 1369.

54. Holešová S, Valášková M, Hlavác D, Madejová J, Samlíková M, Tokarský J, et al. Antibacterial kaolinite/ urea/chlorhexidine nano composites: Experiment and molecular modelling. Applied Surface Science. 2014; 305: 783-791.

55. Eturki S, Ayari F, Jedidi N, Ben Dhia H. Use of Clay Mineral to Reduce Ammonium from Wastewater. Effect of Various Parameters. Surface Engineering and Applied Electrochemistry. 2012; 48 (3): 276-283.

56. Reid B, Arnold CR. The Intensive Culture of the Penaeid Shrimp Penaeusvannamei Boone in a Recirculating

Raceway System. Journal of the World Aquaculture Society. 1992; 23: 146-153.

57. Williams AS, Davis DA, Arnold CR. Density-Dependent Growth and Survival of Penaeussetiferus and Penaeusvannamei in a Semi-Closed Recirculating System. Journal of the World Aquaculture Society. 1996; 27: 107-112.

58. Main KL, Van Wyk P. Farming Marine Shrimp in Freshwater Systems: An Economic Development Strategy for Florida - Final Report. Harbor Branch Oceanographic Institution, Florida Department of Agriculture and Consumer Services, USA. 1999.

59. Mustafa MG, Nakagawa YH. A review: dietary benefits of algae as an additive in fish feed. The Israeli Journal of Aquaculture-Bamidgeh. 1995; 47: 155-162.

60. Guedes AC, Malcata FX. Nutritional value and uses of microalgae in aquaculture, In: Aquaculture, Dr. Zainal Muchlisin (Ed.). 2012; 59-78.

61. Priyadarshani I, Rath B. Commercial and industrial applications of micro algae-a review. J. Algal Biomass Utln. 2013; 3: 89-100.

62. Otoshi CA, Montgomery AD, Look AM, Moss SM. Effects of Diet and Water Source on the Nursery Production of Pacific White Shrimp Litopenaeus vannamei. Journal of the World Aquaculture Society. 2001; 32: 243-249.

63. Nunez M, Lodeiros C, De Donato M, Graziani C. Evaluation of microalgae diets for Litopenaeus vannamei larvae using a simple protocol. Aquaculture International. 2002; 10: 177-187.

CHAPTER 10

DIETARY SUPPLEMENTATION OF DUNALIELLA SALINA ON GROWTH PERFORMANCE AND BODY COMPOSITION OF INDIAN WHITE SHRIMP FENNEROPENAEUS INDICUS (H MILNE EDWARDS)

Osamah A Ahmad, Adnan J Salama and Sambhu Chithambaran

Keywords: Dunaliella Salina;Fenneropenaeus Indicus; White Shrimp;Dietary

INTRODUCTION

The production of natural feed additives for the aquaculture industry is a thriving sector. These substances are being used for several purposes including the enhancement of the immune systems of farmed shrimp, promoting growth, attaining the desired flesh and skin pigmentation, as well as improving the organoleptic properties of the farmed product[1,2]. Astaxanthin is the major carotenoid responsible for the pink-red pigmentation of many fish and shrimp species[3]. Carotenoids are a source of pro-vitamin A which increase survival rate, weight gain and disease resistance against white spot syndrome virus in shrimp[4,5]. Carotenoids not only control ammonia level and resistance to stress but also improve hepato-pancreatic function, cholesterol and polyunsaturated fatty acids from oxidation and provide antioxidant protection[6-8]. Among the carotenoids, astaxanthin was found to be more effective than β-carotene as an antioxidant and a better agent to destroy free radicals than other carotenoids[9,10].

Dunaliella salina is a microalgae occurring naturally in a number of locations worldwide. In the marine environment, *D. salina*appears green, however, in conditions of high salinity and light intensity, the microalgae turns red due to the production of protective carotenoids in the cells[11]. The majority of harvested micro-algae are currently being sourced from one of the most pristine environments in the world – a remote coastal salt lagoon in Western Australia[12]. *D. salina* is harvested without any harmful solvents or chemicals and the carotenoids (highly prized anti-oxidant pigments responsible for the red color) are then extracted for use in pharmaceuticals, cosmetics, nutritional supplements, aquaculture feeds and food coloring[13]. *D.salina* paste has a wide range of applications as feed stock. As a direct feed for filter feeders such as corals and sponges, as an enrichment for live feed organisms such as *Artemia* and *Rotifers*, especially for marine ornamental fish where the coloration is an important consideration, as an enrichment in fresh food (Mussels, Pipis, etc.) for crustacean larvae and brood stock, as a natural source of carotenoids in diets for shrimps and fish[14,15]. Considering the importance of *D. salina* as a feed additive in aquaculture, the present

Citation: Chithambaran, S., et al. Dietary supplementation of Dunaliella salina on growth performance and body composition of Indian white shrimp, Fenneropenaeus indicus (H. Milne Edwards). (2015) J Marine Biol Aquacult 1(1): 1- 5.

study was conducted to evaluate the effect of *D. salina* on growth, pigmentation and body composition in Indian white shrimp, *Fenneropenaeus indicus* an ideal candidate species for coastal aquaculture practice in the Kingdom.

RESULTS

Growth Performance: Data on growth performance of shrimp are presented in Table 1 and Figure 1. A diet containing D. salina at 2.0% (T3) showed superior growth when compared to control and other treatments. ANOVA shows that there is significant difference ($p < 0.05$) on weight between control and treatments. High SGR (%) was observed in T3 followed by T4, T2 and T1 and it was low in T5. Biomass was high in T3 compared to control and other treatments. Highest survival was recorded in Control followed by T1, T3, and T4 and it was lowest in T2.

Table 1: Growth, survival and biomass of shrimp fed *D. salina* diets

Parameters	Control M ± SD	T1 M ± SD	**T2 M ± SD**	**T3 M ± SD**	**T4 M ± SD**	**T5 M ± SD**
Initial length (cm)	9.2 ± 0.1	9.2 ± 0.1	9.2 ± 0.1	9.2 ± 0.1	9.2 ± 0.1	9.2 ± 0.1
Initial weight (g)	4.4 ± 0.1	4.4 ± 0.1	4.4 ± 0.1	4.4 ± 0.1	4.4 ± 0.1	4.4 ± 0.1
Final length (cm)	13.6 ± 0.9	13.6 ± 0.7	13.5 ± 0.8	14.3 ± 0.7	14.2 ± 0.6	13.3 ± 0.9
Final weight (g)*	15.5 ± 3.6^{a}	16.1 ± 2.8^{a}	16.7 ± 2.8^{ab}	17.6 ± 2.9^{b}	17.2 ± 3.4^{b}	15.1 ± 3.0^{a}
Net length gain (cm)	4.4 ± 0.6	4.4 ± 0.9	4.3 ± 0.8	5.1 ± 0.5	5.0 ± 0.6	4.1 ± 0.9
Net weight gain (g)	11.1 ± 1.6	11.7 ± 1.2	12.3 ± 0.9	13.2 ± 1.3	12.8 ± 31.7	10.7 ± 1.0
SGR (%)	0.83 ± 0.3	0.85 ± 0.3	0.85 ± 0.4	0.89 ± 0.3	0.88 ± 0.4	0.82 ± 0.3
Survival (%)*	78.3 ± 2.6^{d}	63.3 ± 3.1^{c}	48.3 ± 2.9^{a}	60 ± 2.0^{c}	58.5 ± 2.8^{b}	58.3 ± 4.0^{b}
Biomass (g)	730.3 ± 12.1	610.6 ± 9.8	471.4 ± 10.3	632.0 ± 13.2	600.4 ± 11.1	530.1 ± 9.4

M-Mean; SD- Standard Deviation*$p < 0.05$; $^{a, b, c, d}$. Means with the same superscript do not differ from each other (Duncan's test)

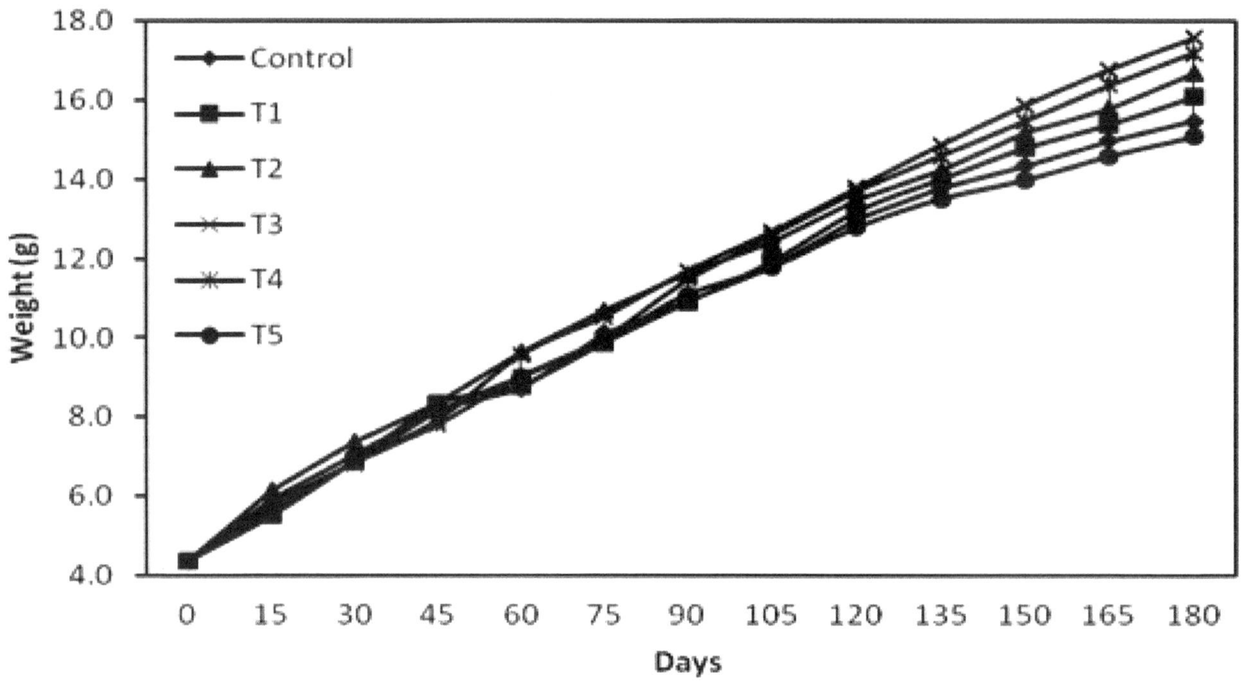

Figure 1: Growth performance of *F. indicus* fed *D. salina* diets

Water Quality Parameters: Data on physical and chemical water quality parameters observed during the culture period is depicted in Figure 2, 3. These parameters were found to be within the range suitable for shrimp growth and there was no significant difference ($p > 0.05$) on parameters between control and treatment tanks.

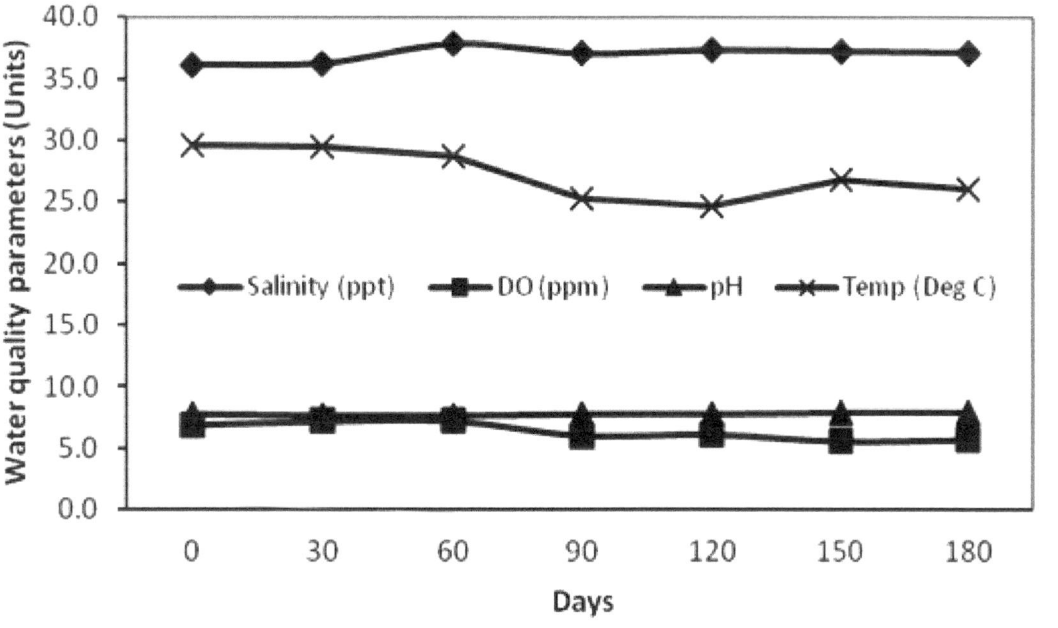

Figure 2: Physical water quality parameters during the culture period

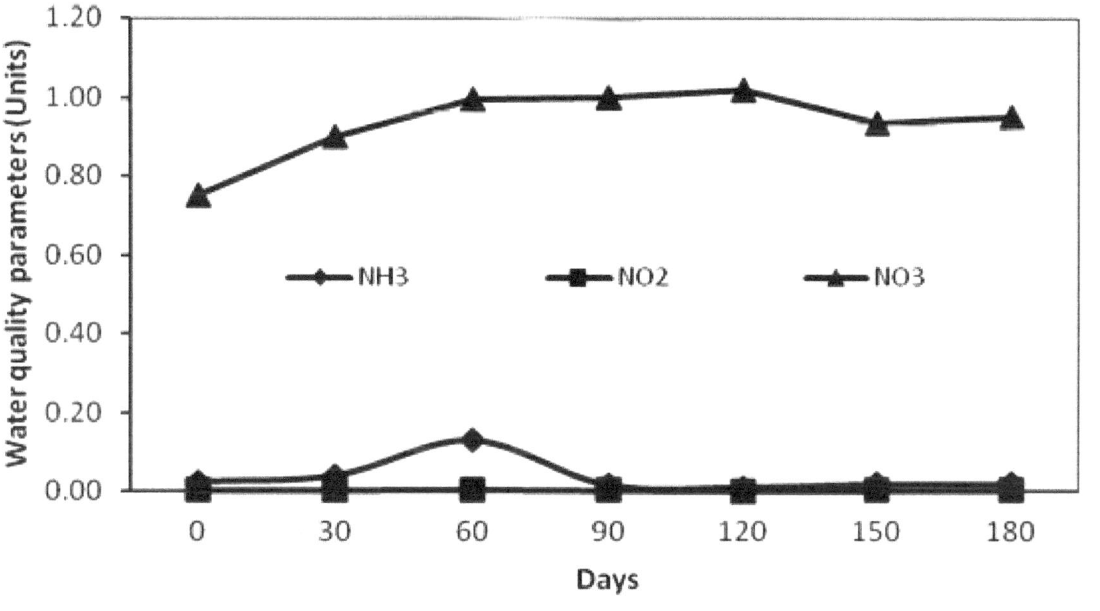

Figure 3: Chemical water quality parameters during the culture period

Feed Utilization: Details of feed utilization study are presented in Table 2. Protein Efficiency ratio and apparent protein digestibility were found to be high in T3 when compared to control and other treatments. Feed conversion ratio was low in T3. Significant difference ($p < 0.05$) in feed conversion efficiency was observed between control and T3 but there was no significant difference in feed consumption, excretion, assimilation and protein digestibility between control and treatments.

Table 2: Feed utilization of shrimp fed *D. salina* diets

Parameters	Control M ±SD	T1 M ± SD	T3 M ± SD	T3 M ± SD	T4 M ± SD	T5 M ± SD
Feed consumption (g)^{NS}	1 ± 0.0	1 ± 0.0	1 ± 0.0	1 ± 0.0	1 ± 0.0	1 ± 0.0
Excretion (g)^{NS}	0.1 ± 0.1	0.1 ± 0.0	0.1 ± 0.1	0.1 ± 0.0	0.1 ± 0.0	0.1 ± 0.0

Assimilation (g)^{NS}	0.7 ± 0.1	0.7 ± 0.1	0.7 ± 0.1	0.7 ± 0.1	0.7 ± 0.1	0.7 ± 0.1
FCE (%)*	2.90 ± 0.3^{c}	2.40 ± 0.4^{b}	2.20 ± 0.3^{b}	1.80 ± 0.5^{a}	2.20 ± 0.3^{b}	2.10 ± 0.4^{a}
Protein efficiency ratio^{NS}	2.57 ± 0.7	2.28 ± 0.6	2.28 ± 0.6	3.14 ± 0.7	2.85 ± 0.6	2.00 ± 0.4
Protein digestibility (%)^{NS}	78.9 ± 1.8	79.4 ± 1.5	80.0 ± 2.2	82.0 ± 2.6	81.1 ± 1.8	78.9 ± 1.7

*$p < 0.05$; NS- $p > 0.05$; M- Mean; DS- Standard Deviation; FCE-Feed conversion efficiency$^{a, b, c}$. Means with the same superscript do not differ from each other (Duncan's test)

Physical Quality Of Shrimp: Details on physical quality of shrimp after culture are presented in Table 3. Soft shell and loose shell percentage was low in *D. salina* fed shrimps compared to control and the observed difference between control and treatments was significant ($p < 0.01$). Shrimp colour and taste were found to be increased in higher dose of *D. salina* fed shrimps.

Table 3: Physical quality of shrimp fed *D. salina* diets

Parameters	Control M ± SD	T1 M ± SD	T2 M ± SD	T3 M ± SD	T4 M ± SD	T5 M ± SD
Soft Shell (%)*	4.2 ± 0.6	3.0 ± 0.3	6.9 ± 2.1	5.6 ± 1.1	2.9 ± 0.9	5.0 ± 2.1
Loose Shell (%)*	27.7 ± 2.3^{c}	20.7 ± 3.1^{a}	24.1 ± 1.9^{b}	22.2 ± 2.5^{b}	20.0 ± 2.8^{a}	20.0 ± 1.8^{a}
Hard Shell (%)*	68.1 ± 2.5^{a}	76.3 ± 3.4^{b}	69.0 ± 3.6^{a}	72.2 ± 2.6^{b}	77.1 ± 1.6^{b}	75.0 ± 2.9^{b}
Shrimp colour (cooked)	light	light	light	Dark	Dark	Dark
Taste	Bland	Bland	Light sweet	Light sweet	Light sweet	Light sweet

*$p < 0.05$; M- Mean; DS- Standard Deviation$^{a, b, c}$. Means with the same superscript do not differ from each other (Duncan's test)

B -Carotene Analysis Data on β-carotene analysis is shown in Table 4. The average β-carotene level in *D. salina* is found to be 0.5%. The residual β-carotene content in shrimp carcass after experiment is found to be negligible.

Table 4: β-carotene content in *D. salina*, feeds and carcass

Parameters	Con-trol	T1	T2	T3	T4	T5
β-carotene in D.salina (%)	0.5	0.5	0.5	0.5	0.5	0.5
β-carotene in Feed (%)	0	0.25	0.5	1.0	2.0	7.0
β-carotene in carcass (%)	0	0.000057	0.000063	0.00003	0.000044	0.000046

Proximate Composition Analysis: Results on proximate composition analysis of shrimp carcass are shown in Table 5. Dry matter, moisture, protein, lipid, ash and nitrogen free extract (NFE) contents did not vary significantly ($P > 0.01$) between control and treatments.

Table 5: Proximate composition of *F. indicus* fed *D. salina* diets

Parameters	Control M ± SD	T1 M ± SD	T2 M ± SD	T3 M ± SD	T4 M ± SD	T5 M ± SD
Dry matter (%)^{NS}	90.1 ± 00	89.9 ± 0.1	89.7 ± 0.3	90.5 ± 0.2	89.9 ± 0.1	89.9 ± 0.4

Moisture (%)NS	9.9 ± 00	10.2 ± 0.1	10.3 ± 0.3	9.6 ± 0.2	10.2 ± 0.1	10.1 ± 0.4
Protein (%)NS	71.2 ± 0.4	69.2 ± 0.5	69.4 ± 0.6	68.7 ± 1.0	69.6 ± 0.3	69.8 ± 1.3
Lipid (%)NS	4.9 ± 0.6	5.9 ± 00	4.6 ± 00	5.1 ± 0.3	4.8 ± 0.1	4.9 ± 0.1
Ash (%)NS	11.5 ± 00	11.7 ± 0.1	12.1 ± 00	11.6 ± 0.1	11.8 ± 0.4	12 ± 0.4
NFE (%)	2.6 ± 0.9	3.1 ± 0.5	3.6 ± 0.3	4.1 ± 1.6	3.7 ± 0.4	3.1 ± 0.3

NS $p > 0.01$; M-Mean: SD- Standard Deviation

DISCUSSION

Results of the present study show that dietary incorporation of *D. salina* brought significant effect on growth performance in white shrimp, *F. indicus* and a dose at 2% inclusion in diet exhibited high growth rate compared to control. Similar growth enhancement was also reported in tiger shrimp, Penaeus monodon when *D. salina* extract was added into the supplementary feed[3,4]. Supplementation of *D. salina* was not only improved growth, but also enhanced health, immunity and disease resistant in tiger shrimp. Growth promoting effect of *D. salina* was also reported in pacific white shrimp, Litopenaeus vannamei when carotenoid compounds were supplemented[2,5]. Therefore, it is suggested that the enhanced growth observed in *F. indicus* is due to incorporation of *D. salina* in the feed. Studies show that the dose of *D. salina* required for growth enhancement in tiger prawn and pacific white shrimp are varied[3,4]. These finding clearly reveal that the growth promoting effect of *D. salina* is dose dependent and species specific.

Several reports suggest that crustaceans have the metabolic activity to introduce structural modifications into carotenoids that they obtain from their diet, in particular, by introducing hydroxy groups at C(3) and C(3') and keto groups at C(4) and C(4'). Thus, canthaxanthin (β, β-carotene- 4,4'-dione), zeaxanthin (β, β-carotene- 3,3'-diol) and even β-carotene can undergo metabolic conversion into astaxanthin[16,17]. Therefore, it can be suggested that growth promoting effect of *D. salina* found on *F. indicus* may be due to the presence of β-carotene content in it.

Water quality parameters recorded during the study was found to be within the suitable range required for the growth of white shrimp[18]. The declining trend of water temperature during culture period was due to the start of winter season. Low concentrations of nitrate observed during culture period suggest the oxidation of ammonia[19]. According to Avnimelech, et al[20], only about 25% of the feed nutrients are converted into harvestable products hence contributing to high nitrogen residues in tank water, especially total ammonia nitrogen (TAN), which is the sum of both ammonia and ammonium which will adversely affect shrimp growth. This may be a reason for the elevated ammonia level recorded during later phase of culture in present study.

Result of the feed utilization study shows that incorporation of *D. salina* in diets did not affect feed consumption and assimilation. However, protein digestibility was found to be increased slightly in fast growing shrimp and this may be due to the better digestion of food. Proximate composition analysis of meat shows that there is no significant difference in shrimp meat quality which indicates that incorporation of *D. salina* did not influence meat quality of white prawn. Similar studies have also been reported in shrimp when feed additives are supplemented through diets[21,22].

In shrimp culture industries, color is one of the major factors, which determines the market price of black tiger shrimp (*Penaeus monodon*) in the international market[23]. Supplementation of carotenoid pigments into diets has been demonstrated to yield shrimp colour and higher pigments in shrimp[15-17]. Moreover, carotenoid pigments have positive effect on the immunological and stress response[24]. Several studies have been made to find alternative sources of astaxanthin and other carotenoids such as yeast, *Phaffia sp.* and many species of algae[25,26]. In the present study also *D. salina* diet fed shrimp showed better color than that of the control. Therefore, it is strongly opined that >*D. salina* has the ability to improve shrimp colour in white shrimp, *F. indicus*.

It is considered that penaeid shrimp cannot biosynthesize carotenoids from mevalonic acid, but can alter dietary carotenoids by oxidation and deposit them in their tissues Tanaka, et al[27] reported on the metabolism of carotenoids in kuruma shrimp and suggested that some of the dietary carotenoid pigments such as astaxanthin, β-carotene, isocryptoxanthin, echinenone, canthaxanthin, phoenicoxanthin, zeaxanthin and 4-ketozeaxanthin were converted into astaxanthin in shrimp body. Especially, astaxanthin was the most effective substance for pigmentation in shrimp when compared with β-carotene and canthaxanthin[28]. However, Chien et al[17] studied the pigmentation of the black tiger shrimp by feeding the shrimp with diets containing different carotenoid sources, e.g., β-carotene, *Spirulina*, *Phaffia* yeast and krill oil. A marked increase of carotenoid content in the carapace was observed in the group fed with *Spirulina* supplemented diets and suggested that zeaxanthin, which is the major carotenoid in*Spirulina*, was rapidly converted to astaxanthin. Moreover, 125 mg/kg of synthetic β-carotene or 125–175 mg/kg β carotene extracted from *D. salina* have been reported as a pigment sources in black tiger shrimp and demonstrated that black tiger shrimp has the metabolic ability to converted β-carotene into astaxanthin[3]. Based on above observations, it is suggested that *F. indicus*has the ability to convert the carotenoid content in *D. salina* and this may be a reason for the better colour development in *F. indicus*.

From previous reports, dietary carotenoids were converted into astaxanthin and deposited in the shrimp body in free form by the association with protein and exists as carotenoprotein and esterifed forms which are predominantly a mono ester and di ester of long-chain fatty acids[29]. However, the incorporation of astaxanthin esters in test diets seemed less effective than the free form as reported by[17]. The results of the present study shows that *D. salina* incorporated diet fed shrimp had better colour than that of the control shrimp. Hence it is suggested that the carotene content in *D. salina* plays an important role on pigmentation in white shrimp as in the case of other shrimps.

Result of the present study show that shrimp body composition was not influenced by *D. salina* incorporation. The light sweet taste of treatment shrimp can be correlated to the high saline conditions of the Red Sea water used for culture. HPLC analysis on residual β-Carotene content in meat of experimental shrimp shows that presence of β-carotene content at very low level and it did not show significant difference from that of the control. The enhanced color noticed in the shrimp fed *D. salina* did not change during cooking and it is suggested that the developed color will not change while cooking. In conclusion, it is suggested that carotene content in *D. salina* play a significant role in enhancing growth and coloration in white shrimp, *F. indicus*.

METHODS

The study was conducted for a period of 180 days in fiber glass tanks (500L) at the University Fish Farm at Obhur, Jeddah. There were control and treatment for the study and both were laid out in a completely randomized design. Healthy and uniform size juvenile (4.4 ± 0.1; 9.2 ± 0.1 cm) produced at the Farm Hatchery were stocked at the rate of 20 pieces/tank. Dry powder of *D. salina* was procured from National Aquaculture Group (Naqua), Jeddah, Saudi Arabia and incorporated at 0.5, 1.0, 2.0, 4.0 and 8.0% in the diet and designated as treatments T1, T2, T3, T4 and T5 respectively. *D. salina* free diet was considered as control. A standard fishmeal based pellet feed (Naqua, Jeddah) having 35% protein in diet was selected for *D. salina* incorporation and the details of feed ingredients are shown in Table 6. Required quantity of *D. salina* (powder) was thoroughly mixed with soaked feed and extruded out through a pelletizer having 2m in the die. The pellets were then oven dried at 60°C and kept it in airtight containers for storage. Feeding was done at 5% of the biomass daily at 6:00 AM, 1 PM and 6 PM. Water exchange was done at the rate of 20% every day. Shrimp were sampled fortnightly to record their total length and weight and the feed quantity was re adjusted after every sampling. Specific growth rate was calculated as Loge W2-Log_eW1/T2-T1 (where W2 is the weight of shrimp at time T2 and W1 is the weight of shrimp at time T1).Water quality parameters such as temperature, dissolved oxygen, pH and salinity (HAAC, HQ40d multiply portable device meter) were recorded daily. Ammonia (unionized), nitrates (NO_3) and nitrites (NO_2) were recorded (HAAC, device colorophotometer – DR 3900) weekly. On 180th day, the experiment was terminated and total biomass and survival was calculated.

Table 6: Feed ingredients in the supplementary feed

Ingredients	Incorporation (%)
Fish meal	18.00
Soya meal	40.00
Wheat	31.78
Fish oil	3.50
Sea weed powder	0.50
Mono sodium phosphate	0.38
Mono calcium phosphate	0.50
Yeast autolyzate	0.50
Soya lecithin oil	2.53
Anti-oxidant	0.10
Anti-mould	0.15
Binder	1.00
Vitamin premix	0.35
Mineral premix	0.25
DL Methionine	0.27
Vitamin C	0.14
Lysine (Amino acid)	0.05
Total (g)	100

A short term laboratory experiment was conducted in fiber glass tanks (50 liter) using 10 shrimps (10.0 ± 1.2 g) for 30 days to assess the feed intake, conversion efficiency, and apparent protein digestibility of the control and treatment shrimp. For this, shrimp were collected from the experimental tanks and provided with weighed amount of feed for 2 hours daily between 7.00 am and 7:00 pm. The unconsumed feed and fecal matter were siphoned out separately after 1hour of feeding. The pooled, dried and weighed faecal matter was used for nutrient analysis. Feed conversion efficiency and nutrient digestibility were calculated as: Feed conversion efficiency (FCE %) = Wet weight gain (g) / Feed consumed (g) x 100; Apparent protein digestibility (%) = Protein in feed - Protein in excreta / Protein in feed x100.

A panel of experts evaluated the physical quality of shrimp such as colour, loose shell, soft shell, hard shell and taste after the experiment. Proximate analysis of shrimp carcass was analyzed for the estimation of protein, lipid, glycogen, fiber and ash[30]. Experimental feed, shrimp samples and dry powder of *D. salina* were subjected for the estimation of β-carotene content by High Pressure Liquid Chromatography (HPLC) [31] and the samples were analyzed at IDAC lab, Riyadh, Saudi Arabia. One way analysis of variance (ANOVA) was employed to find out the statistical difference in growth, water quality parameters and physical quality of shrimp between control and treatment[32].

ACKNOWLEDGMENT

The author, Osamah A Ahmad is gratefully thankful for the research grant awarded by KACST (AT-34-232) for post graduate studies. Authors are thankful to the Dean Faculty of Marine Sciences and the Managing Director, Naqua for the facilities provided.

REFERENCES

1. Wang, X., Wille, N.R., Wadstrom, T. Astaxanthin-rich algal meal and vitamin C inhibit *Helicobacter pylori* infection in Balb/cA mice. (2000) Antimicrob Agents Chemother 44(9): 2452- 2457.
2. Arredondo-Figueroa, J.L., Pedroza-Islas, R., Ponce-Palafox, J.T., et al. Pigmentation of Pacific white shrimp (*Litopenaeus vannamei*, Boone, 1931) with esterified and saponified carotenoids from red chili (*Capsicum annuum*) in comparison to astaxanthin. (2003) Rev Mex Ing Quim vol 2(2): 101- 108.
3. Boonyaratpalin, M., Thongrod, S., Supamattaya, K., et al. Effect of β-carotene source, *Dunaliella salina*, and astaxanthin on pigmentation, growth, survival and health of *Penaeus monodon*. (2001) Aquacult Res 31(s1): 182- 190.
4. Supamattaya, K., Kiriratnikoma, S., Boonyaratpalin, M., et al. Effect of a *Dunaliella* extract on growth performance, health condition, immune response and disease resistance in black tiger shrimp (*Penaeus monodon*). (2005) Aquaculture 248(1-4): 207- 216.
5. Linan-Cabello, M.A., Paniagua-Michel, J., Zenteno-Savin, T. Carotenoids and retinal levels in captive and wild shrimp,*Litopenaeus vannamei*. (2003) Aquacult Nutr 9(6): 383- 389.
6. Pan, C.H., Chien, Y.H., Hunter, B. The resistance to ammonia stress of*Penaeus monodon* Fabricius juvenile fed diets supplemented with astaxanthin. (2003) J Exp Mar Biol Ecol 297(1): 107- 118.
7. Mcnulty, H., Jacob, R.F., Mason, R.P. Biologic activity of carotenoids related to distinct membrane physicochemical interactions. (2008) Am J Cardiol 101(10A): 20D- 29D.
8. Britton, G. Structure and properties of carotenoids in relation to function. (1995) FASEB J 9(15): 1551- 1558.
9. Nielsen, B.R., Mortensen, A., Jorgensen, K., et al. Singlet versus triplet reactivity in photo degradation of c40 carotenoids. (1996) J Agric Food Chem 44(8): 2106- 2113.
10. Palozza, P., Barone, E., Mancuso, C., et al. The protective role of carotenoids against 7-keto-cholesterol formation in solution. (2008) Mol Cell Biochem 309(1-2): 61- 68.
11. Oren, A. A hundred years of *Dunaliella* research, 1905-2005. (2005) Saline Syst 1: 2- 15.
12. Tafreshi, A.H., Shariati, M. *Dunaliella* biotechnology: methods and applications. (2009) J Appl Microbiol 107(1): 14- 35.
13. Ben-Amotz, A., Polle, J.E.W., Subba Rao, D.V. The algae *Dunaliella*: Biodiversity, Physiology, Genomics and Biotechnology. (2009) New Hampshire: Science Publishers.
14. Stottrup, J.G., McEvoy, L.A. Live Feeds in Marine Aquaculture. (2003) Oxford: Blackwell Publishing Ltd.
15. Dufosse, L., Galaup, P., Yaron, A., et al. Microorganisms and microalgae as sources of pigments for food use: a scientific oddity or an industrial reality. (2005) Trends Food Sci Technol 16(9): 389- 406.
16. Liao, W.L., Nur-E-Borhan, S.A., Okada, S., et al. Pigmentation of cultured black tiger prawn by feeding with a *Spirulina*supplemented diet. (1993) Nippon Suisan Gakk 59(1): 165- 169.
17. Chien, Y.H., Jeng, S.C. Pigmentation of kuruma prawn, *Penaeus japonicus* Bate, by various pigment sources and levels and feeding regimes. (1992) Aquaculture 102(4): 333- 346.
18. Wickins, J.F. The tolerance of warm water prawn to recirculated water. (1976) Aquaculture 9(1): 19- 37.
19. Cohen, J.M., Samocha, T.M., Fox, J.M., et al. Characterization of water quality factors during intensive raceway production of juvenile *Litopenaeus vannamei* using limited discharge and biosecure management tools. (2005) Aquacult Eng 32(3-4): 425-442.
20. Avnimelech, Y., Ritvo, G. Shrimp and fish pond soils, processes and management. (2003) Aquaculture 220(1-4): 549- 567.
21. Nakagawa, H., Gomez-Diaz, G. Usefulness of *Spirulina sp.* meal as feed additive for giant freshwater prawn, *Macrobrachium rosenbergii*. (1995) Aquacult Sci 43(4): 521- 526.
22. Jaime-Ceballos, B., Villareal, H., Garcia, T., et al. Effect of *Spirulina platensis* meal as feed additive on growth,

survival and development in *Litopenaeus schmitti* shrimp larvae. (2005) Rev Investig Mar 26(3): 235- 241.

23. Latscha, T. Carotenoids in aquatic animal nutrition. (1991) Proceeding of the Aquaculture feed processing and Nutrition Workshop, Bankok, Thailand 19-25: 68-78
24. Chien, Y., Pan, C., Hunter, B. The resistance to physical stresses by *Penaeus monodon* juveniles fed diets supplemented with astaxanthin. (2003) Aquaculture 216(1-4): 177- 191.
25. Sanderson, G.W., Jolly, S.O. The value of *Phaffia* yeast as a feed ingredient for salmonid fish. (1994) Aquaculture 124(1-4): 193-200.
26. Boonyaratpalin, M., Supamattaya, K., Borisuth, C. The immune system in black tiger shrimp, *Penaeus monodon* Fabricius: VIII. Effect of astaxanthin on blood parameters, immune system and disease resistance in black tiger shrimp (*Penaeus monodon Fabricius*). (2000) Songklanakarin J Sci Technol 22: 633- 639.
27. Tanaka, Y., Matsuguchi, H., Katayama, T., et al. The biosynthesis of astaxanthin-XVIII. The metabolism of carotenoids in the prawn, *Penaeus japonicus* Bate. (1976) B Jpn Soc Sci Fish 42(2): 197-202.
28. Estermann, R. Biological functions of carotenoids. (1994) Aquaculture 124(1-4): 219-222.
29. Yamada, S., Tanaka, Y., Sameshima, M., et al. Pigmentation of prawn (Penaeus japonicus) with carotenoids: I. Effect of dietary astaxanthin, β-carotene and canthaxanthin on pigmentation. (1990) Aquaculture 87(3-4): 323-330.
30. AOAC. Official method of analysis of the Association of official analytical chemists. Horwitz, W., Ed. 17th edition. (2000) Washington, DC.
31. Van Vliet, T., Van Schaik, F., Van Schoonhoven, J., et al. Determination of several retinoids, carotenoids and E-vitamins by high-performance liquid chromatography. Application to plasma and tissues of rats fed a diet rich in either beta-carotene or canthaxanthin. (1991) J Chromatogr 553(1-2): 179-186.
32. Snedecor, G.W., Cochran, W.G. Statistical methods. (1967) Oxford and IBH Publishing Company, Calcutta, 593.

CHAPTER 11 PREDICTING FUTURE THERMAL HABITAT SUITABILITY OF COMPETING NATIVE AND INVASIVE FISH SPECIES: FROM METABOLIC SCOPE TO OCEANOGRAPHIC MODELLING

Stefano Marras, Andrea Cucco, Fabio Antognarelli, Ernesto Azzurro, Marco Milazzo, Michel Bariche, Momme Butenschön, Susan Kay, Massimiliano Di Bitetto, Giovanni Quattrocchi, Matteo Sinerchia and Paolo Domenici

Keywords: Conservation physiology distribution modelling fish physiology global warming invasive species thermal habitat suitability

INTRODUCTION

Sea surface temperatures are predicted to continue rising globally throughout the 21st century, with dramatic consequences for marine ecosystems in the near future (IPCC, 2013). Rapid poleward shifts of isotherms were recorded in recent decades, specifically in the northern hemisphere (Burrows et al., 2011), ultimately contributing to changes in the composition of marine communities (Harley et al., 2006). Direct effects of temperature changes include a shift in species distribution (i.e. range contractions or extensions) towards higher latitudes and depths (Perry et al., 2005) and an increase in extinction rates of many marine organisms (Ben Rais Lasram et al., 2010). It was argued that one of the first processes to cause extinction or relocation to cooler waters is the reduction in aerobic performance in warming conditions (Pörtner and Knust, 2007). However, species do not exist in isolation but as part of large communities of interacting species; recent studies revealed the importance of taking into account species interactions as biotic multipliers of climate change effects (Zarnetske et al., 2012; Blois et al., 2013; Milazzo et al., 2013). Therefore, the relative changes in the performance of interacting species are expected to provide useful insights to explain invasion success (Azzurro et al., 2014) and to predict future changes in species distribution (Pörtner and Knust, 2007; Pörtner and Farrell, 2008).

In the Mediterranean Sea, warming conditions are believed to be facilitating the arrival and spread of tropical invaders at an unprecedented rate (Hiddink et al., 2012). This basin can be considered as a model area for investigation of the different thermal windows of interacting species and the potential ecological consequences of climate change. At present, almost 90 fish species (Belmaker et al., 2013; Golani et al., 2002) coming from the Red Sea (hereafter referred as Lessepsian migrants) have entered the Mediterranean Sea through the Suez

Citation: Marras S, Cucco A, Antognarelli F, Azzurro E, Milazzo M, Bariche M, Butenschön M, Kay S, Di Bitetto M, Quattrocchi G, Sinerchia M, Domenici P (2015) Predicting future thermal habitat suitability of competing native and invasive fish species: from metabolic scope to oceanographic modelling. Conserv Physiol 3(1): cou059; doi:10.1093/conphys/cou059

Canal. This represents the main invasion route for alien species (Galil, 2009; Zenetos et al., 2010, 2012). This phenomenon, once confined to Levantine waters (Por, 1990), is proceeding with an apparent accelerating rate of introductions (Galil, 2009). Moreover, it has recently started to extend westwards through the Sicilian Channel (Quignard and Tomasini, 2000), which separates the western and the eastern sectors of the basin. This channel has long been considered the ultimate limit to the spread of Lessepsian fish species to the Western Mediterranean (Quignard and Tomasini, 2000). However, in the last two decades, the number of tropical and subtropical fishes recorded in this area has dramatically increased (Azzurro et al., 2013a), and a few of them recently reached the western sectors of the basin (e.g.Fistularia commersonii, Azzurro et al., 2012). An increasing number of studies have used ecological modelling for predicting the future distribution of native and invasive fishes in the Mediterranean Sea (Ben Rais Lasram et al., 2010; Albouy et al., 2012, 2013) and have suggested that the projected climate change might cause a significant loss of climatically suitable habitat for endemic fish species (Albouy et al., 2013).

The increasing success of the tropical invasive species may be related in part to the fact that their thermal window of aerobic performance is likely to be shifted towards warmer temperatures than that of the temperate indigenous species. Here, we investigated the relationship between temperature and aerobic performance in a pair of herbivorous fish species that occupy a similar ecological niche: the invasive marbled spinefoot (Siganus rivulatus) and its native counterpart, the salema (Sarpa salpa). These two species are among the few herbivorous fishes occurring in the Mediterranean (Azzurro et al., 2007) and they may compete with each other (Golani, 1993), especially in conditions of limiting trophic resources, such as reduced algal biomass (Sala et al., 2011). The marbled spinefoot entered the Mediterranean through the Suez Canal in the 1920s and established a large population along the African coast around the 1940s (Golani et al., 2002). It has also spread through the Sicilian Channel, via the African coasts (Golani et al., 2002), but has not yet entered the Western Mediterranean (Zenetos et al., 2010; Golani, et al., 2002). The native salema is very common and widespread along the Mediterranean coasts (Verlaque, 1990), but in the easternmost sectors of the Mediterranean, this species has probably been outcompeted by the marbled spinefoot (Bariche et al., 2004; Galil, 2009), and a competitive superiority of marbled spinefoot vs. salema has been hypothesized (Bariche et al., 2004). Furthermore, other Siganid species (e.g. Siganus luridus) have recently extended their distribution, colonizing the islands of the Sicilian Channel (Schembri et al., 2012; Azzurro and Andaloro, 2014). Expected changes in the relative dominance of these two herbivorous species might have important ecological consequences in the near future, because Siganids have the potential to graze algal resources intensively and cause large-scale impact on natural habitats and local food chains (Sala et al., 2011; Vergés et al., 2014a).

Here, we aim to estimate how changes in water temperature influence the physiological performances (i.e. metabolic scope, MS) of both marbled spinefoot and salema. Metabolic scope represents the scope within which aerobic activities, such as swimming, feeding or growth, must be accommodated (Fry, 1971). As such, MS is a suitable metric for analysing the energetics of habitat selection (Fry, 1971) and is related to the ecological success of a given species (Pörtner and Knust, 2007). Thus, MS can provide a bioenergetic basis for studying how environmental constraints govern the spatial and temporal distribution pattern of fishes (Cucco et al., 2012a). We used this information to model the thermal habitat suitability (THS, a dimensionless index based on the relationship between MS and temperature) in relationship to current and future environmental conditions, taking the whole Mediterranean Sea into account, and focusing on the Sicilian Channel (Fig. 1) because of its biogeographical relevance for the westward progress of this invasion. A multidisciplinary approach combining a physiological empirical model and oceanographic numerical models was employed to evaluate THS of marbled spinefoot vs. salema, and to allow predictions of future fish distributions.

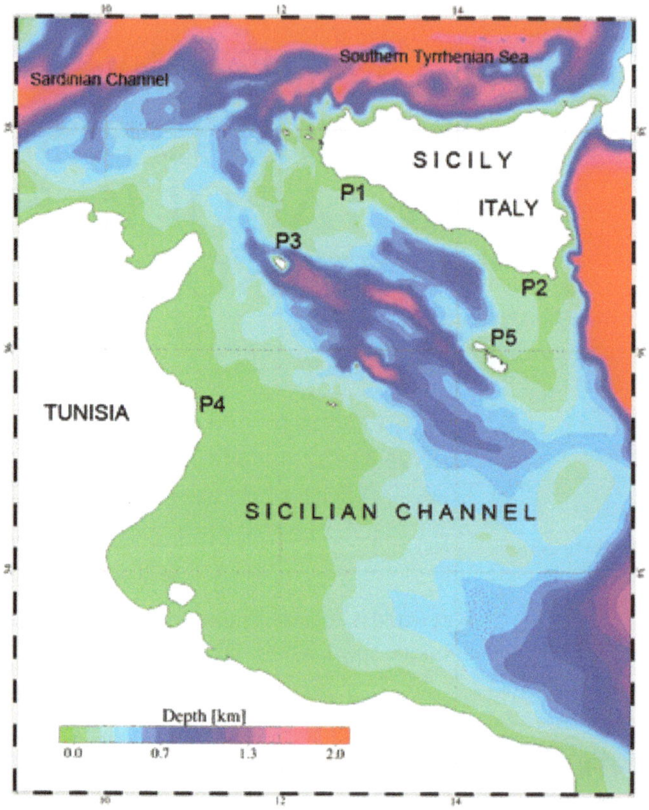

Figure 1: Sicilian Channel, study area. P1 to P5 show the locations of the stations adopted for a time series analysis of the NMS in the regional-scale model.

MATERIALS AND METHODS

Area of investigation

The Mediterranean Sea is divided in two main sub-sectors, the western and the eastern basins, which are connected through the shallower Sicilian Channel (Manzella et al., 1988; Astraldi et al., 1999). The exchanges of water with the Atlantic Ocean, the Black Sea and the Red Sea are controlled through the shallow straits of Gibraltar, the Bosphorus and the Suez Channel, respectively. The average surface temperature ranges roughly between 16 and 20°C in the winter and between 24 and 28°C in the summer, with generally higher temperatures (by ~2°C) in the eastern basin. Mediterranean isotherms are rapidly shifting west- and northwards, and average sea surface temperature is predicted to increase by a further 2.3–2.6°C or even 3.1°C by the end of this century (Somot et al., 2008).

The water circulation of the Sicilian Channel is characterized by the presence of two main current patterns: a south-eastern surface flow, which transfers the Atlantic water (AW) from the western to the eastern sub-basin; and a north-westward sub-surface current in the opposite direction, which exchanges the intermediate Levantine water (LIW) between the eastern and western sub-basins (Malanotte-Rizzoli, 2001). Ten years (1983–1992) of monthly averaged sea surface temperature maps reveal that during the climatological winter, the isotherms run zonally from the Sicilian Channel to Cyprus. This generates a bimodal (north–south oriented) regimen of the surface temperature within the Sicilian Channel (Marullo et al., 2007). The climatological summer yields southern thermal fronts, especially within the Sicilian Channel, where a strong coastal upwelling appears along the Sicilian coastlines. Two regions with different thermal regimens dominate the Sicilian Channel throughout the year: a warmer region along the African coasts and a colder region along the southern coasts of Sicily.

Overview of methods

Physiological empirical models and oceanographic numerical modelling were used to make predictions of how changes in environmental conditions may affect the thermal habitat suitability of the native salema and the invasive marbled spinefoot in the Mediterranean Sea. Two different climate scenarios were considered: a present-day projection (RA, hereafter 'present day') corresponding to the past decade from 2001 to 2009; and a near future scenario, 2040–2049, based on a national enterprise scenario driven by atmospheric conditions of an A2 AR4 SRES projection (A2, hereafter 'future scenario'). The overall approach consists of the following steps: (i) measurement of metabolic scope in the two fish species and defining the equation relating MS and temperature; (ii) modelling the oceanographic conditions to predict the changes in temperature and other hydrodynamic variables in the selected coastal waters in relationship to the present-day projection and the future scenarios; and (iii) coupling the physiological relationships with the oceanographic model to run the simulations of the present-day projection and the future scenarios.

Measuring the metabolic scope of salema and marbled spinefoot

A total of 72 wild juvenile marbled spinefoot (total length 11.8±1.3 cm; mass 19.1±6.4 g at the time of testing) were captured along the Lebanese coast, offshore of Beirut. They were transported via air cargo to the IAMC-CNR laboratory in Oristano (Sardinia), where they were transferred to a circular tank (1.5 m diameter) supplied with recirculated natural sea water maintained at constant temperature (20±0.3°C) and salinity (35.1±0.2‰), under a natural photoperiod. A total of 72 wild juvenile salema (total length 11.1±1.1 cm; mass 19.4±5.8 g at the time of testing) were captured in the Oristano Gulf, Sardinia, Italy and transported to the IAMC-CNR laboratory, where they were maintained in the same way as the marbled spinefoot. One month later, groups of 12 individuals of each species were transferred to rectangular 100 l aquaria (a total of 12 aquaria, six per species, each one containing 12 individuals). The six different aquaria for each species were used to acclimate each group of 12 fish to six different water temperatures (i.e. 17, 20, 23, 26, 29 and 32°C). Final acclimation temperature was reached by increasing (or decreasing, depending of the final acclimation temperature) the water temperature by 1°C per day. Fish were left for 1 month at the final acclimation temperature before a single experiment. Note that none of S. salpa individuals survived at acclimation temperatures higher than 26°C. Fish were fed daily with commercial pellets, and feeding was suspended for at least 24 h before testing.

Metabolic rate was measured in individual animals. Fish were removed from the holding tank and introduced into a round tank (50 l), where they were chased to exhaustion. Fish were considered exhausted when they did not respond to stimulation (Roche et al., 2013). Immediately after exhaustion, fish were placed in one of four static respirometers (0.5 l; Loligo Systems, Tjele, Denmark) immersed in an outer tank. Instantaneous oxygen uptake (MO_2, in milligrams of O_2 per kilogram per hour) was measured by intermittent flow respirometry (Steffensen, 1989) once every 30 min. Temperature was kept constant for the duration of the experiment. Metabolic scope was assessed in non-limiting oxygen conditions (i.e. in normoxia corresponding to an oxygen saturation >80%). Water flow from the external bath through the respirometers was driven by an external pump that was set to turn on and off for alternating 15 min periods. This allowed decreases in water oxygen content to be measured every 15 s for 15 min while the respirometer was in the closed state. The respirometer was then flushed with aerated water for 15 min. The oxygen consumption during each closed phase was calculated using linear least-squares regression (excluding the first and last 2 min of each closed phase). Fish could not see each other nor interact while in the separate respirometers during the measurement of oxygen consumption. Water oxygen levels were measured with optodes (Oxy-4 mini; PreSens Precision Sensing GmbH, Regensburg, Germany) and associated software (Pre-Sens Oxy 4v2), and MO_2 was then calculated (Marras et al., 2010, 2013; Killenet al., 2011, 2012). Fish were left in the tank for 24 h, then removed, measured for length, width and weight and placed in a common holding tank different from the one used for acclimation. Measurements of background microbial respiration in the system were taken before and after the oxygen measurement in the respirometers.

The first 15 min (excluding the first and last 2 min as described above) of the oxygen consumption measurements obtained immediately after the introduction of individual fish into the respirometer were used to determine the maximal metabolic rate (MMR), which is the maximal aerobic metabolic rate measured in non-limiting conditions. Standard metabolic rate (SMR), which supports maintenance activities, such as ventilation or osmoregulation, and corresponds to the oxygen consumption of a resting, fasted and non-maturing fish, was measured using the 15th percentile method (Chabot and Claireaux, 2008). Metabolic scope (MS, i.e. the energy available to an individual to accomplish all of its tasks, including swimming, digestion etc.) was obtained as the difference between MMR and SMR as follows:

$$\mathrm{MS} = \mathrm{MMR} - \mathrm{SMR},$$

(1)

(1)where MMR (in milligrams of O_2 per kilogram per hour) and SMR (in milligrams of O_2 per kilogram per hour) are based on the following equations:

$$\mathrm{MMR} = d\left(\frac{T_m - T}{T_m - T_{act}}\right)^{\omega} \times \exp\left[-\omega\left(\frac{T_m - T}{T_m - T_{act}}\right)\right]$$

(2)

$$\mathrm{SMR} = a \times \exp(b \times T)$$

(3)

where T_{act} is the temperature at which MMR (MO_{2max} in the study byLefrançois and Claireaux, 2003) is maximized and T_m is the temperature at which MMR is equal to zero. The values for T_{act}, T_m and the remaining empirical parameters (a, b, ω and δ) were obtained as the result of the fitting procedure between the MMR, SMR experimental data and the parametric functions described by equations (2) and (3).

The numerical models and the simulation set-up for present-day and future scenarios

Two different numerical approaches were followed to simulate the changes of the thermal habitat between the present-day projection and the future scenario at Mediterranean basin scale and at sub-regional scale for the Sicilian Channel area. At the basin scale, the Proudman Oceanographic Laboratory Coastal Ocean Modelling System [POLCOMS;Holt and James, 2001) was implemented. This model is a three-dimensional, baroclinic, primitive equations model using terrain following σ-coordinates in the vertical direction. The implementation of the Mediterranean Sea model extends over the whole basin, with an open boundary at the Gibraltar Strait and a closed boundary at the Bosphorus Strait. The model domain was discretized using a finite difference mesh with a spatial resolution of 0.1° for the horizontal directions and 40 levels with a variable intralevel distance for the vertical direction. At the regional scale, a high-resolution three-dimensional hydrodynamic numerical model (SHYFEM3D, Umgiesser and Bergamasco, 1995; Umgiesser et al., 2004;Cucco et al., 2009, 2012b) based on the finite element method was applied. SHYFEM3D solves the shallow water equations integrated over each layer in their formulations with water levels and transports. SHYFEM3D uses finite elements for horizontal spatial discretization, withz-layers for vertical discretization. The SHYFEM3D implementation covers the area comprised between 8.5° and 16.5° longitude and 31.2° and 39.5° latitude and includes the whole Sicilian Channel, part of the Sardinian Channel and Southern Tyrrhenian Sea. The model domain was discretized by an unstructured finite element mesh with a spatial resolution varying between 300 m along the coastlines and 10 km for the inner part of the channel and for the eastern and western domain boundaries. Forty levels were used for the vertical direction.

The data providing atmospheric and ocean boundary condition forcing of the POLCOMS model simulations were based on present-day reanalysis data, modified by reference to data from the ECHAM5 GCM for the future projections (Jungclaus et al., 2006; Roeckner et al., 2006). For the present-day model runs (2000–2009), atmospheric reanalysis data were taken from ERA-Interim data (http://www.ecmwf.int/en/research/climate-reanalysis/era-interim) for the meteorological forcing and from GLORYS2 reanalysis (www.mercator-ocean.fr) for the ocean boundary conditions. The spatial resolution is about 0.7° for ERA-Interim and 0.25° for GLORYS2, with a temporal resolution of at least 24 h for all variables. River flow data were taken from the Global NEWS database (Seitzinger et al., 2005). A delta method was used to create forcing data for the future scenario, in order to retain the spatial and temporal resolution of the reanalysis data. The future scenario data set was generated from the present-day values by applying a modification of the same size as the difference between the 2000s and 2040s given by the GCM. Changes in river forcings were derived from the results of the ELME project (Langmeadet al., 2007).

A downscaling technique was adopted to simulate the hydrodynamics of the Sicilian Channel for the two decades of the present-day projection and future scenario. The approach consisted of using the three-dimensional oceanographic data [water temperature (T), water salinity (S), water currents (Vel) and water levels (η)] produced by the POLCOMS model basin-scale application, as well as the meteorological data sets previously described, as open boundaries and forcing conditions for the SHYFEM3D model runs. A relaxation procedure was adopted to constrain the SHYFEM3D model solution to the imposed T, S and Vel values along the whole lateral boundary and for an inner buffer zone extending 25 km inside the domain (Cucco et al., 2012b). No river loads were included as lateral boundary conditions. This is because the investigated area is characterized by an arid and semi-arid climate, with no major river inflows.

Coupling physiological relationships with modelling to assess thermal habitat suitability

The results of the present-day and future model runs were processed to obtain the temporal and spatial variability of the daily water temperature obtained by averaging the values computed between the sea surface and a depth of 30 m for the sub-regional scale analysis, and only at the surface for the basin-scale analysis. The data sets obtained consisted of horizontal distributions of the daily temperature values during the two climatological conditions computed both from the basin-scale model results and from the sub-regional model results. The temperature values were used as input data for computing the spatial and temporal variability of the MS of the two fish species for the present-day projection and the future scenario at both basin and regional scales. The mean monthly temperatures for the present-day projection and the future scenario in the Sicilian Channel ranged between 13.1 and 31.2°C.

The differences in the aerobic performance for a defined thermal environment between the two fish species were computed by defining the thermal habitat suitability (THS) and the coexistence factor (CF). The THS is a dimensionless index that was computed, for each element of the model domain, as the time average of the MS values normalized with respect to the maximal computed value MS_{max} (NMS):

$$\mathrm{THS}_i = \frac{\sum_{t=0}^{N}(\mathrm{MS}_{i,t})}{N \times \mathrm{MS}_{\max}}. \tag{4}$$

The THS computed for the whole simulated period (N) and for both fish species ranged between zero and one, providing a degree of suitability within each element of the model domain (i) in relationship to the two fish species.

Considering NMS:

$$\mathrm{NMS}_i = \frac{\mathrm{MS}_i}{\mathrm{MS}_{\max}}, \tag{5}$$

computed for each element of the domain and for both fish species, a theoretical threshold (T_{MS}) of NMS for high aerobic performance was identified based on the current distribution of the two fish species and the corresponding thermal habitat suitability values. The coexistence factor was calculated from the QMS, which is a binary quantity calculated by associating to the elements i a value of one when THS > T_{MS} and a value of zero when NMS < T_{MS}. The QMS provided a semi-quantitative evaluation of the environmental fitness for both fishes. The coexistence factor was defined as the difference between the QMS computed for the marbled spinefoot and those computed for the salema. Coexistence factor ranges between minus one and one and reads as:

$$\mathrm{CF}_i = \mathrm{QMS}_i^S - \mathrm{QMS}_i^R, \tag{6}$$

$$\begin{aligned} \mathrm{QMS}_i^S &= 0 \quad \mathrm{THS}_i \le T_{\mathrm{MS}}, \\ \mathrm{QMS}_i^S &= 1 \quad \mathrm{THS}_i > T_{\mathrm{MS}}, \\ \mathrm{QMS}_i^R &= 0 \quad \mathrm{THS}_i \le T_{\mathrm{MS}}, \\ \mathrm{QMS}_i^R &= 1 \quad \mathrm{THS}_i > T_{\mathrm{MS}}, \end{aligned}$$

where the superscripts R and S stand for marbled spinefoot and salema, respectively. A negative and a positive value of CF indicate environmental conditions favouring marbled spinefoot and salema, respectively, whereas CF = 0 indicates areas where fish species are favoured equally by the thermal conditions.

RESULTS

Metabolic response of salema and marbled spinefoot

The curves shown in Fig. 2 describe the relationship between MS and water temperature for both native and invasive species. The values of the empirical parameters obtained by the fitting procedure are shown in Table1, for each species. Metabolic scope increased as a function of temperature until it reached a maximal value. The optimal temperature for aerobic scope (T_{opt}) was defined as the temperature at which the MS was maximum. Beyond T_{opt}, MS decreased at a certain rate until it reached zero. Critical temperature (T_{crit}) was defined as the temperature at which MS was equal to zero. Metabolic scope curves in the salema and marbled spinefoot were described by different values of T_{opt} and T_{crit}. Metabolic scope increased in the salema up to a maximal value of 709 mg O_2 kg^{-1} h^{-1} at 21.8°C (i.e. T_{opt}), whereas at higher temperatures MS decreased to zero at T = 28.7°C (i.e. T_{crit}). In line with their T_{crit}, none of the salema individuals survived at the acclimation temperatures of 29 and 32°C. In the marbled spinefoot, MS increased up to a maximal value of 850 mgO_2 kg^{-1} h^{-1} at 29.1°C (i.e. T_{opt}), whereas at higher temperatures it decreased to zero at T = 37.5°C (i.e. T_{crit}). The normalized MS as a function of water temperature is shown in Fig. 2b for both species.

Table 1. List of the parameters and corresponding values used to estimate Eq. (2) and (3).

Parameters	Units	Values for *S. salpa*	Values for *S. rivulatus*
a	mg O_2 $kg^{-1}h^{-1}$	190.2	176.7
b	$°C^{-1}$	0.067	0.050
ω		1.4	1.3
δ	mg O_2 $kg^{-1}h^{-1}$	6931.1	6762.2
T_m	°C	32.8	42.7
T_{act}	°C	24.3	32.2

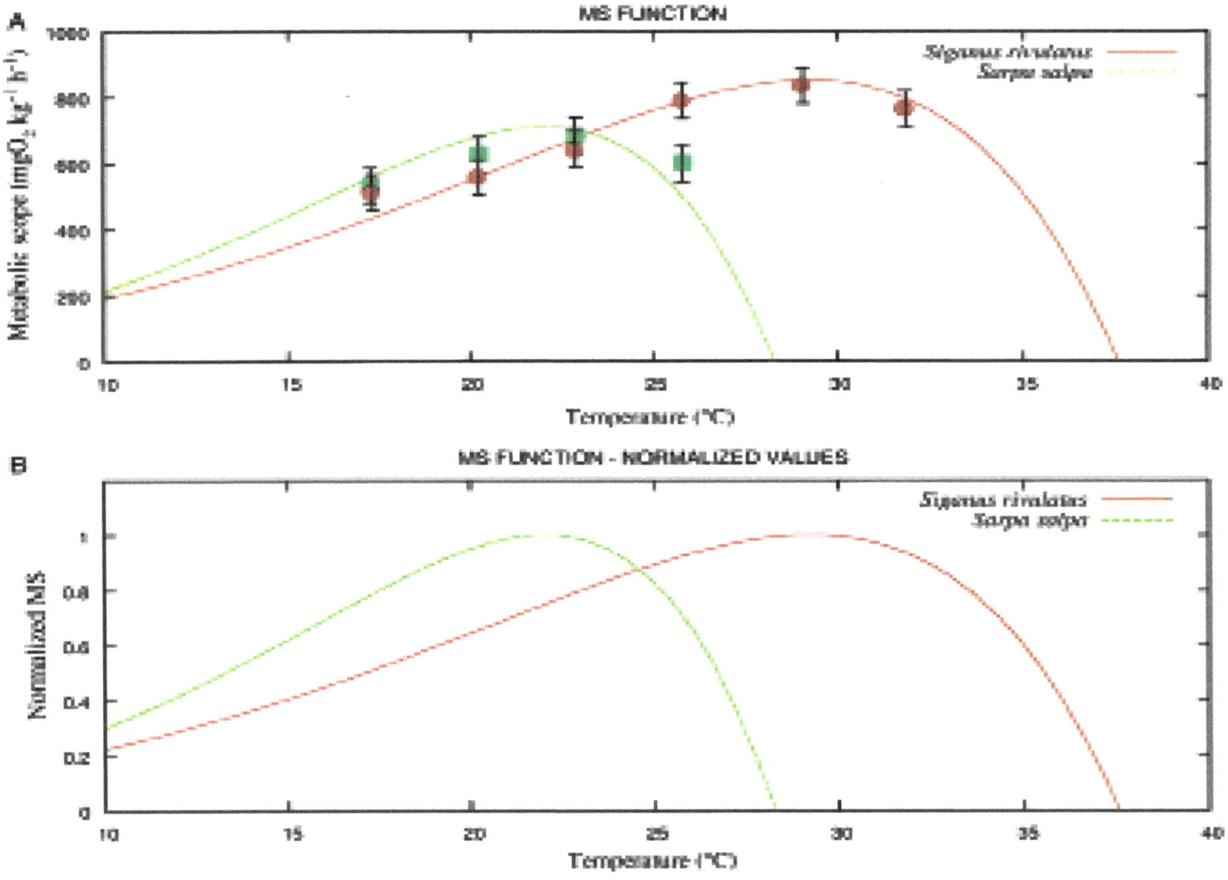

Figure 2: (A) Metabolic scope (MS) curve as a function of temperature in marbled spinefoot (Siganus rivulatus) and salema (Sarpa salpa). Green squares represent the MS values measured in salema and red circles represent the MS values measured in marbled spinefoot (means± SD). Note that none of the S. salpaindividuals survived at acclimation temperatures higher than 26°C. (B) Normalized MS as a function of temperature in marbled spinefoot and salema.

The accuracy of the computed empirical function was verified by comparing the MS obtained from the empirical measurements of MMR and SMR, carried out during the whole set of laboratory experiments in both fish species, and the predicted ones (MS_{pre}). A linear regression analysis based on equation (1) was performed (forcing the regression line through the origin), and the following significant relationship was found:

$$MS_{obs} = (1.01 \pm 0.026) \times MS_{pre}, \quad (7)$$

with $P<0.0001$, $N=106$ and $r^2=0.80$. The linear relationship obtained was characterized by a slope not significantly different from one ($P=0.50$), indicating an acceptable level of accuracy in the empirical model when predicting MS in relationship to T.

Thermal habitat predictions

The model results were analysed to obtain the variability of the daily average temperature at the basin and regional levels for two climatological windows, representative of the decade 2000–2009 (present-day projection) and the decade 2040–2049 (future scenario), respectively. At the basin scale, a general rise in average temperature up to 1°C with an average increase of around 0.6°C was found between the present-day projection and the future scenario. The highest increase in temperature, corresponding to 1.8°C, was found in the Northern Adriatic Sea during the winter period. Lower increments, between 0.3 and 0.7°C, were computed for the central and deeper areas of the eastern and western sub-basins, throughout the whole year. At the regional scale, in the Sicilian Channel, the surface temperature varied considerably from the southern, warmer part of the channel to the northern, colder one. According to the present-day projection results, the southern area was characterized by high variability in surface temperature [with the lowest values during the winter (between 14.5 and 20.7°C, monthly averaged temperatures) and the highest values in summer (between 21.5 and 30.1°C, monthly averaged temperatures)]. This was especially true in the shallow water banks in front of Tunisia. For this area, the increase in surface temperature between present-day projection and future scenario ranged between 0.3 and 1.3°C, with the maximal increment found during the autumn and the minimal increment during the spring. The northern area was characterized by low variability of the surface temperature. This was due both to the presence of Atlantic Water, which restrains the seasonal changes of temperature along the Sicilian coastal areas, and to the higher latitudes. The lowest temperatures were computed during the winter (ranging between 13.1 and 18.5°C, monthly averaged temperatures) with maximal and minimal values found in

December and January, respectively. During the summer, the computed monthly averaged temperature varied between 17.5 and ~26.5°C, with the maximum found in August and the minimum in June. Along with the lower variability, the results obtained by the future scenario revealed a marked increase in the surface temperature in the southern part of the basin. The increase in temperature varied between 0.2 and 1.5°C, with the maximum value during the autumn and the minimum in the spring.

Physiological–oceanographic coupling at basin scale

At basin scale, the MS and NMS of the two fish species were computed using surface water temperature data obtained by a simulation run (MOD1) for the present-day projection and the future scenario. The distributions of the THS obtained from the present-day projection and the future simulated scenario are shown in Fig. 3. The present-day projection results show a clear distinction between the values obtained for the eastern and the western parts of the basin, with high THS values found in the former. This is in line with the existing data on distribution, which suggest that the current thermal habitat of the eastern basin is suitable for the marbled spinefoot, while in the western basin the THS values are as low as 50–60% of the maximum. In contrast, most of the Mediterranean Sea shows relatively high THS for the salema, in line with the current widespread distribution of this species (data for salema, www.fishbase.org; data for the marbled spinefoot, Azzurro et al., 2013b). Based on the THS isoline including 95% of the current distribution of marbled spinefoot and salema, a threshold below which the thermal habitat conditions become unsuitable in each species (T_{MS}) was identified as 0.62 of the maximal THS. The future scenario results show an increase in the extension of areas thermally suitable for the marbled spinefoot, such that virtually all of the Mediterranean Sea shows THS values >62% of the maximum. Conversely, for the salema, the thermally suitable area decreases in size, with large areas showing values around 50–60% of the maximal THS along the Lebanese and Tunisian–Libyan coasts.

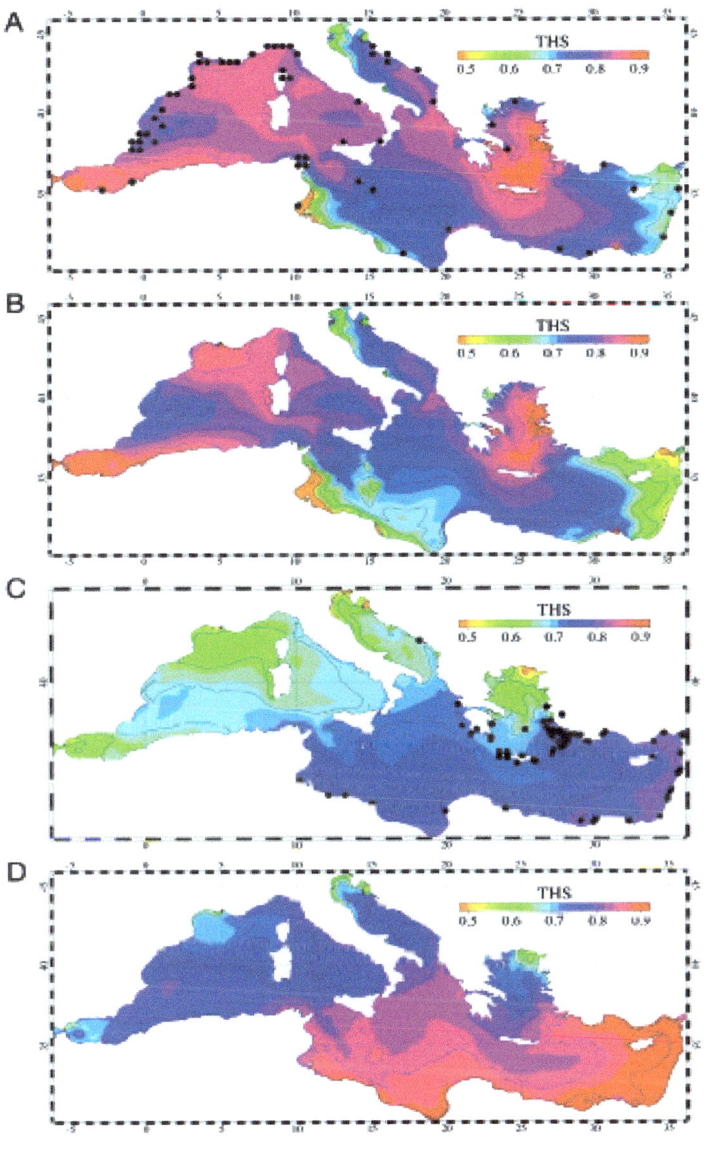

Figure 3: Thermal habitat suitability (THS) computed for the whole Mediterranean Sea from the basin-scale model results. (A) Thermal habitat suitability of the salema computed from the present-day simulation results. (B) Thermal habitat suitability of the salema computed from simulation results of the future scenario. (C) Thermal habitat suitability of the marbled spinefoot computed from present-day projection results. (D) Thermal habitat suitability of the marbled spinefoot computed from the simulation results of the future scenario. Black dots represent the sites where the fish species were observed.

Physiological–oceanographic coupling at regional scale

The numerical approach used at the regional scale allowed the reproduction of the hydrodynamic conditions in more detail. The computation of the NMS for the two fish species was carried out at different depths (i.e. between 5 and 30 m), within the depth range of distribution of the two species [i.e. salema 0–70 m; marbled spinefoot 0–60 m (Louisy, 2005)]. The time series of NMS of the two species computed at five locations selected in the Sicilian Channel are shown in Fig. 4 for the present-day projection and Fig. 5 for the future scenario. For the present-day projection, the NMS of the salema showed a strong decrease in the summer. Conversely, the pattern of NMS for the marbled spinefoot showed an increase in NMS during the summer months, together with a higher overall NMS in southern than in northern sites. The future scenario showed a general increase of NMS in the marbled spinefoot in all stations and at all depths. In contrast, the NMS of the salema showed a general reduction, especially during the summer. The spatial distributions of THS for the present-day projection and the future scenario are shown in Fig. 6. For the salema, the areas of high THS value tended to decrease in size in the future scenario compared with the present-day projection, especially in the southern sites, while the opposite is true for the marbled spinefoot. This trend is confirmed by comparing the coexistence factor between the present-day projection and the future scenario (Fig. 7). The future scenario showed an increase of the areas where the CF<0 (i.e. areas where THS of the marbled spinefoot is higher than that of the salema) and where CF=0 (i.e. areas where THS is favourable for both fish) and a decrease of areas with CF>0 (i.e. areas where THS of the salema is higher than that of the marbled spinefoot). Note that where CF=0, the QMS of both fish species were always equal to one so that no cases of THS unfavourable for both fish species were found. Specifically, when comparing the present-day projection with the future scenario, the area of suitability increased from 4 to 7% for CF<0 and from 66 to 82% for CF=0, while it decreased from 30 to 11% for CF>0.

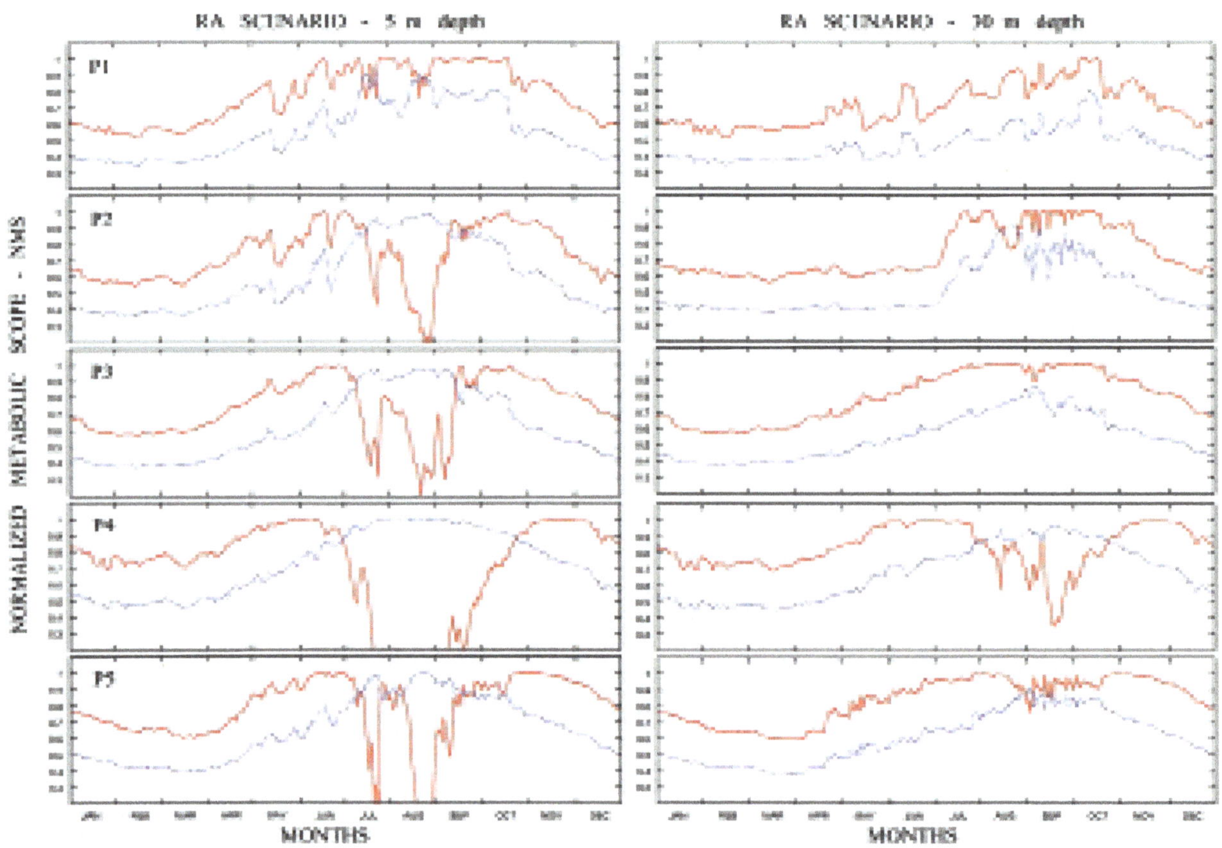

Figure 4: Present-day projection. Temporal variability of the daily normalized MS computed for salema (red line) and for marbled spinefoot (blue line) at different depths for the five selected sites in the Sicilian Channel, throughout the year (averaged over 10 years). Left panel refers to results at 5 m depth and right panel at 30 m depth.

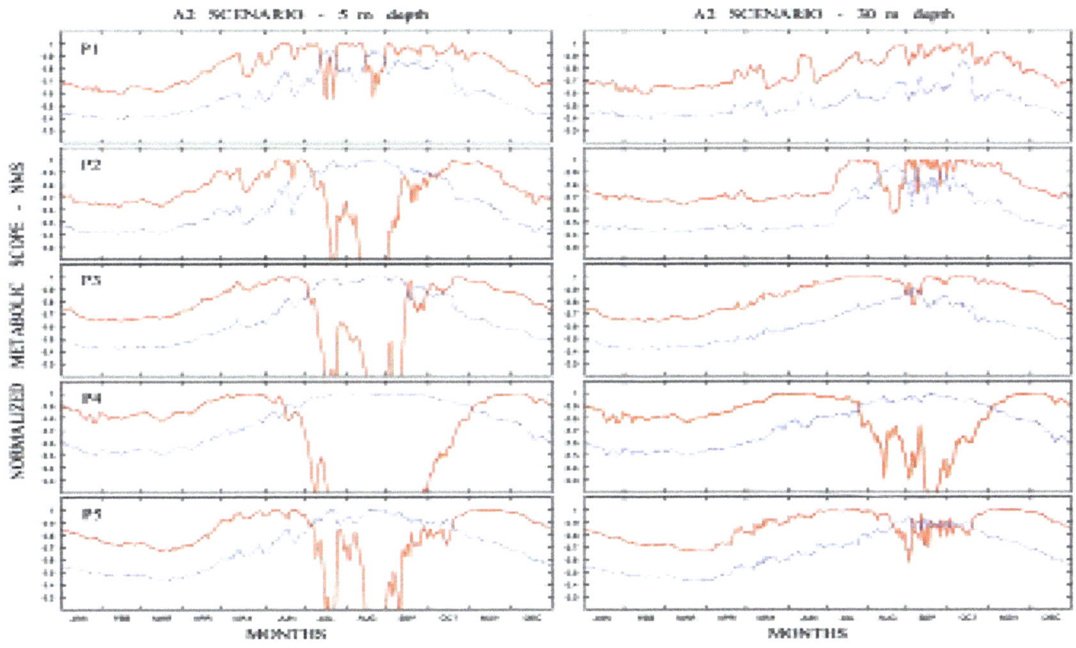

Figure 5: Future scenario. Temporal variability of the daily normalized MS computed for the salema (red line) and for the marbled spinefoot (blue line) at different depths for the five selected sites in the Sicilian Channel, throughout the year (averaged over 10 years). Left panel refers to the results at 5 m depth and right panel at 30 m depth.

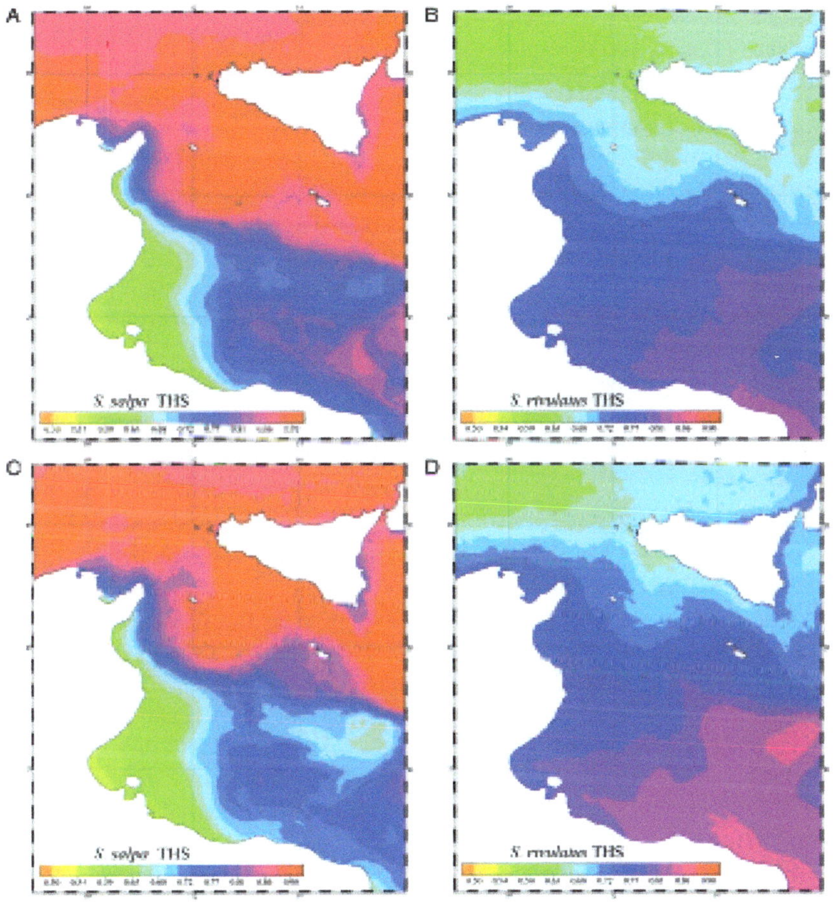

Figure 6: Thermal habitat suitability (THS) of salema and marbled spinefoot computed for the present-day projection and the future scenarios. (A) and (B) Thermal habitat suitability of salema and marbled spinefoot, respectively, computed from present-day projection simulation results. (C) and (D) Thermal habitat suitability of salema and marbled spinefoot, respectively, computed from future scenario simulation results.

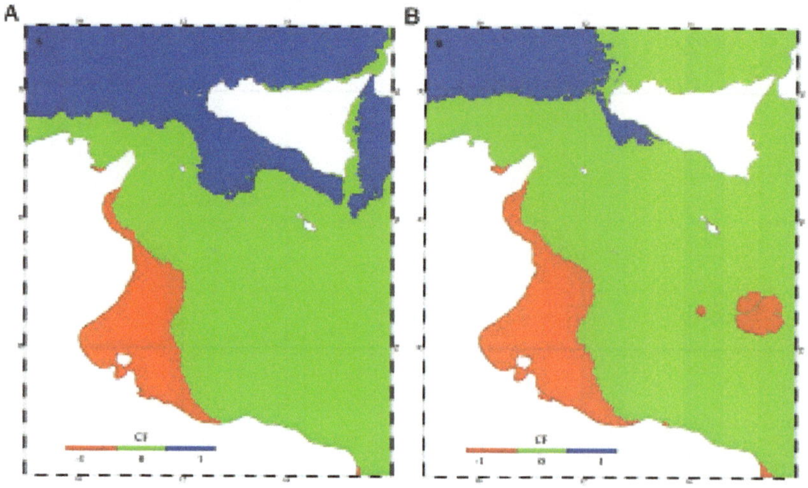

Figure 7: (A) Coexistence factor (CF) computed for the present-day projection. (B) Coexistence factor computed for the future scenario.

DISCUSSION

Our work represents the first attempt at comparing the physiological performance of a native fish species with that of its invasive counterpart integrated with an oceanographic modelling approach. For the present-day scenario, our basin-scale model showed higher THS for the invasive marbled spinefoot in the Eastern compared with the Western Mediterranean Sea. According to our basin-scale model, over the next 40 years the THS of the marbled spinefoot is expected to increase in the Western Mediterranean Sea, reaching values currently found in the Eastern Mediterranean, where the species is already well established (Golani, 1993).

The relationship between aerobic scope and temperature is in agreement with previous work that explored the effects of temperature on the physiological performance of siganids (Saoud et al., 2008). Saoud and colleagues found that growth rate [a variable known to be affected by aerobic scope (Pörtner and Knust, 2007; Khan et al., 2014)] was maximized in marbled spinefoot kept at 27°C (Saoud, et al., 2008). However, no data are available on the effect of temperature on the physiological performance of the salema. Interestingly, the metabolic scope of Mugil cephalus (a coastal fish species commonly found near the surface within a shallow depth range; Louisy, 2005) show similar, though slightly higher, values T_{opt} to those found here for the salema, i.e. ~25°C (Cucco et al., 2012a).

The thermal habitat suitability of both marbled spinefoot and salema largely matched the current distribution patterns of the two species. Currently, the marbled spinefoot occupies most of the Eastern Mediterranean but has not yet entered the western basin. Interestingly, the only part of the eastern basin where the marbled spinefoot has not been reported is the northern Aegean Sea (Zenetos et al., 2010; Golani et al., 2002; Vergés et al., 2014b). Our model showed relatively low THS values for this area, similar to those found in the Western Mediterranean Sea. The competitive superiority of the marbled spinefoot in the Eastern Mediterranean Sea when compared with salema may also be due to its greater adaptability to fluctuating environmental conditions (Bariche et al., 2004). Juvenile marbled spinefoot were found to be highly tolerant to variations in temperature and salinity, oxygen deficiency and overcrowding (Ben Tuvia et al., 1973; Popper and Gundermann, 1975).

According with the predictions based on the IPCC scenarios (IPCC, 2013), over the next 40 years the thermal properties of the Western Mediterranean could become suitable for the marbled spinefoot, if it reaches similar temperature values currently occurring in the Eastern Mediterranean. At the same time, large sectors of the eastern basin (e.g. Lebanese and Libyan coast) could become thermally unsuitable for the salema, which could progressively restrict its distribution to the western basin. Hence, the marbled spinefoot is expected to move into the western basin and compete with the salema. In the eastern basin, we predict that it will slowly replace the salema in several areas, as has already happened in the coastal waters of Lebanon (Bariche et al., 2004). The salema was still an abundant species in the 1930s, when a modest population of marbled spinefoot occurred along the coast of Lebanon and Syria. The marbled spinefoot population has increased significantly since then, and competition between the two species was noted in the 1960s in the same area (reviewed by Kalogirou et al., 2012). Considering the absence of salema along the coast of Lebanon in recent observations (Bariche et al., 2004), the displacement process of the native fish can be estimated to have occurred within the past five to seven decades.

The documented spread of the marbled spinefoot in the Eastern Mediterranean has led to important consequences for the composition and structure of subtidal benthic communities. This has caused algal forests to turn into 'barrens', where canopy algae abundance, benthic biomass and species richness are dramatically reduced. Recent work has found functional differences in the feeding behaviour of the salema and the marbled spinefoot (Vergés et al., 2014b). Although the salema showed a higher overall feeding rate in terms of biomass consumption, this temperate fish species feeds exclusively on adult microalgae, while the marbled spinefoot feeds on both adult microalgae and epilithic algal matrix, which, in most cases, contains microalgal recruits. Therefore, the reduction in the epilithic algal matrix biomass, caused by the feeding habits of the tropical marbled spinefoot, can severely affect the biomass and biodiversity of the temperate reefs (Vergés et al., 2014b). This does not only cause a direct reduction in the algal abundance and diversity, but also implies a loss of critical habitat for understory algae and many invertebrates that live in the algal canopies (Vergés et al., 2014b), which are the prey for the majority of carnivorous fish species.

The Sicilian Channel area may be considered the 'door' to the western basin; therefore, we have focused our attention on this area, for which a high-resolution model was run. Our results suggest that the thermal habitat of the southern part of the Sicilian Channel will be increasingly more suitable for the marbled spinefoot than for the salema (Figs 6 and 7), and a population increase of the invasive species in these sectors (which include a large part of the Tunisian coasts) could represent a significant step for the westward progress of the invasion. In areas such as Malta and south-eastern Sicily, the thermal suitability for marbled spinefoot and salema is predicted to reach similar values by 2050 (Fig. 7). Moreover, the northern part of the Sicilian Channel was predicted to become thermally suitable for both species (Fig. 6). Therefore, the invasion of the marbled spinefoot could potentially expand northward as a result of warming seawater conditions. Conversely, in south-western Sicily, THS could remain relatively unsuitable for the marbled spinefoot.

Our findings could have implications for a variety of questions in ecology, invasive biology and climate change (Pörtner and Peck, 2010; Jørgensen et al., 2012). To predict the invasive potential of Lessepsian fish species, biotic processes such as interspecific competition (Azzurro et al., 2014) need to be taken into account, in addition to the direct response to abiotic variables (i.e. temperature increase, induced by global warming; Killen et al., 2013). The present model is, admittedly, simplistic. In a warming scenario, species distribution shifts are likely to depend on a number of factors in addition to metabolic scope; these include food availability, predation and competition, as well as other abiotic factors, such as oxygen levels and salinity. However, temperature changes in the Mediterranean Sea could act as a critical factor influencing invasion success by affecting the physiological mechanisms responsible for invader superiority in a global warming scenario (Ben Rais Lasram et al., 2010; Albouy et al., 2012,2013). Further work on our model could be directed at including additional variables, such as food availability, thereby increasing the predictive power of the model. Furthermore, a useful addition to the present study would be the individual-based analyses of species' realized functional niches, aimed at quantifying the possible functional overlap between the marbled spinefoot and the salema in various thermal conditions. An example is found in species-specific microhabitat utilization patterns of herbivores (Brandl and Bellwood, 2014). In addition, behavioural work could be performed to test the possibility that competitive displacement may exacerbate the effect of temperature on fish performance, leading to habitat relocation of the 'loser' species, as found in other competing fish pairs (Milazzo et al., 2013).

At present, different models have been used to assess fish distribution and abundance. One of the most common is the bio-envelope model, which examines the relationship between key climatic variables and species distributions, mostly based on historical distributional data (Ben Rais Lasram et al., 2010; Albouy et al., 2012, 2013). Here, we propose the use of a physiology-based model to predict fish distribution. This type of model relies on solid, reproducible laboratory experiments, and uses metabolic scope as a proxy for the ecosystem thermal suitability in different climate scenarios. In this respect, it provides an advance in the use of mechanistic, cause–effect approaches for understanding the effect of environmental changes on habitat suitability. This approach might provide a physiologically driven tool for understanding and predicting the distribution and relative abundance of marine fish.

Funding

This work was supported by the European Community's Seventh Framework Programme (FP7/2007–2013) under Grant Agreement No. 266445 for the project Vectors of Change in Oceans and Seas Marine Life, Impact on Economic Sectors (VECTORS) and by the RITMARE Project, funded by the Italian Ministry of Education, Universities and

REFERENCES

1. Albouy C, Guilhaumon F, Araújo MB, Mouillot D, Leprieur F (2012) Combining projected changes in species richness and composition reveals climate change impacts on coastal Mediterranean fish assemblages. Glob Change Biol 18: 2995–3003.
2. Albouy C, Guilhaumon F, Leprieur F, Ben Rais Lasram F, Somot S, Aznar R, Velez L, Le Loc'h F, Mouillot D (2013) Projected climate change and the changing biogeography of coastal Mediterranean fishes. J Biogeogr 40: 534–547.
3. Astraldi M, Balopoulos S, Candela J, Font J, Gacic M, Gasparini GP, Manca B, Theocharis A, Tintoré J (1999) The role of straits and channels in understanding the characteristics of Mediterranean circulation. Progr Oceanogr 44: 65–108.
4. Azzurro E, Andaloro F (2014) A new settled population of the lessepsian migrant Siganus luridus (Pisces: Siganidae) in Linosa Island—Sicily Strait. J Mar Biol Assoc UK 84: 819–821.
5. Azzurro E, Fanelli E, Mostarda E, Catra M, Andaloro F (2007) Resource partitioning among early colonizing Siganus luridus and native herbivorous fish in the Mediterranean: an integrated study based on gut-content analysis and stable isotope signatures. J Mar Biol Assoc UK 87: 991–998.
6. Azzurro E, Soto S, Garofalo G, Maynou F (2012) Fistularia commersonii in the Mediterranean sea: invasion history and distribution modeling based on presence-only records. Biol Invas 5: 977–990.
7. Azzurro E, La Mesa G, Fanelli E (2013a) The rocky-reef fish assemblages of Malta and Lampedusa islands (Strait of Sicily, Mediterranean Sea): a visual census study in a changing biogeographical sector. J Mar Biol Assoc UK 93: 2015–2026.
8. Azzurro E, Soto S, Bariche M, Fanelli W, Maynou F (2013b) Exotic fish species in the Mediterranean Sea: analysis of occurrence records. Rapp Comm Int Mer Médit 40: 508.
9. Azzurro E, Tuset V, Lombarte A, Maynou F, Simberloff D, Pérez A, Sole R (2014) External morphology explains the success of biological invasions. Ecol Lett 17: 1455–1463.

10. Bariche M, Letourneur Y, Harmelin-Vivien M (2004) Temporal fluctuations and settlement patterns of native and Lessepsian herbivorous fishes on the Lebanese coast (eastern Mediterranean). Environ Biol Fishes 70: 81–90.
11. Belmaker J, Parravicini V, Kulbicki M (2013) Ecological traits and environmental affinity explain Red Sea fish introduction into the Mediterranean. Glob Change Biol 19: 1373–1382.
12. Ben Rais Lasram F, Guilhaumon F, Albouy C, Somot S, Thuiller W, Mouillot D (2010) The Mediterranean Sea as a 'cul-de-sac' for endemic fishes facing climate change. Glob Change Biol 16: 3233–3245.
13. Ben Tuvia A, Kissil GW, Popper D (1973) Experiments in rearing rabbitfish Siganus rivulatus in sea water. Aquaculture 1: 359–364.
14. Blois JL, Zarnetske PL, Fitzpatrick MC, Finnegan S (2013) Climate change and the past, present, and future of biotic interactions. Science 341: 499–504.
15. Brandl SJ, Bellwood DR (2014) Individual-based analyses reveal limited functional overlap in a coral reef fish community. J Anim Ecol 33: 421–430.
16. Burrows MT, Schoeman DS, Buckley LB, Moore P, Poloczanska ES, Brander KM, Brown C, Bruno JF, Duarte CM, Halpern BS, et al. (2011) The pace of shifting climate in marine and terrestrial ecosystems. Science 334: 652–655.
17. Chabot D, Claireaux G (2008) Environmental hypoxia as a metabolic constraint on fish: the case of Atlantic cod, Gadus morhua. Mar Pollut Bull 57: 287–294.
18. Cucco A, Umgiesser G, Ferrarin C, Perilli A, Canu DM, Solidoro C (2009) Eulerian and lagrangian transport time scales of a tidal active coastal basin. Ecol Model 220: 913–922.
19. Cucco A, Sinerchia M, Lefrancois C, Magni P, Ghezzo M, Umgiesser G, Perilli A, Domenici P (2012a) A metabolic scope based model of fish response to environmental changes. Ecol Model 237–238: 132–141.
20. Cucco A, Sinerchia M, Ribotti A, Olita A, Fazioli L, Sorgente B, Perilli A, Borghini M, Schroeder K, Sorgente R (2012b) A high-resolution real-time forecasting system for predicting the fate of oil spills in the Strait of Bonifacio (western Mediterranean Sea). Mar Pollut Bull 64: 1186–1200.
21. Fry FE (1971) The effect of environmental factors on the physiology of fish. In Hoar WS, Randall DJ, eds, Fish Physiology, Vol. VI. Academic Press, New York, NY, USA, pp 1–98.
22. Galil BS (2009) Taking stock: inventory of alien species in the Mediterranean Sea. Biol Invas 11: 359–372.
23. Golani D (1993) Trophic adaptations of Red Sea fishes to the eastern Mediterranean environment – review and new data. Israel J Zool 39: 391–402.
24. Golani D, Orsi-Relini L, Massuti E, Quignard JP (2002) CIESM Atlas of Exotic Species in the Mediterranean. Vol 1. Fishes, CIESM, Monaco, 256 p.
25. Harley CDG, Hughes AR, Hultgren KM, Miner BG, Sorte CJB, Thornber CS, Rodriguez LF, Tomanek L, Williams SL (2006) The impacts of climate change in coastal marine systems. Ecol Lett 9: 228–241.
26. Hiddink JG, Ben Rais Lasram F, Cantrill J, Davies AJ (2012) Keeping pace with climate change: what can we learn from the spread of Lessepsian migrants? Glob Change Biol 18: 2161–2171.
27. Holt JT, James ID (2001) An s coordinate density evolving model of the northwest European continental shelf: 1. Model description and density structure. J Geophys Res 106: 14015–14034.
28. IPCC (2013) Climate Change 2013: the Physical Science Basis. Working group 1 contribution to the fifth assessment report of the International Panel on Climate Change, International Panel on Climate Change, Cambridge, UK and New York, NY, USA.
29. Jørgensen C, Peck MA, Antognarelli F, Azzurro E, Burrows T, Cheung WWL, Cucco A, Holt RE, Huebert KB, Marras S, et al. (2012) Conservation physiology of marine fishes: advancing the predictive capacity of models. Biol Lett 8: 900–903.
30. Jungclaus JH, Keenlyside N, Botzet M, Haak H, Luo J-J, Latif M, Marotzke J, Mikolajewicz U, Roeckner E (2006) Ocean circulation and tropical variability in the coupled model echam5/mpi-om. J Climate 19: 3952–3972.
31. Kalogirou S, Azzurro E, Bariche M (2012) The ongoing shift of Mediterranean coastal fish assemblages and the spread of non-indigenous species. In Lameed AG ed, Biodiversity Enrichment in a Diverse World. InTech, Rijeka, Croatia, pp 263–280.
32. Khan JR, Pether S, Bruce M, Walker SP, Herbert NA (2014) Optimum temperatures for growth and feed conversion in cultured hapuku (Polyprion oxygeneios)—is there a link to aerobic metabolic scope and final temperature preference? Aquaculture 430: 107–113.
33. Killen SS, Marras S, McKenzie DJ (2011) Fuel, fasting, fear: routine metabolic rate and food deprivation exert synergistic effects on risk taking in individual juvenile European seabass. J Anim Ecol 80: 1024–1033.
34. Killen SS, Marras S, Ryan MR, Domenici P, McKenzie DJ (2012) A relationship between metabolic rate and risk-taking behaviour is revealed during hypoxia in juvenile European sea bass. Funct Ecol 26: 134–143.
35. Killen SS, Marras S, Metcalfe NB, McKenzie DJ, Domenici P (2013) Environmental stressors alter relationships between physiology and behaviour. Trends Ecol Evol 28: 651–658.
36. Langmead O, McQuatters-Gollop A, Mee LD (Eds) (2007) European Lifestyles and Marine Ecosystems: Exploring Challenges for Managing Europe's Seas. University of Plymouth Marine Institute, Plymouth, UK.
37. Lefrançois C, Claireaux G (2003) Influence of ambient oxygenation and temperature on metabolic scope and scope for heart rate of the sole (solea solea). Mar Ecol Prog Ser 259: 273–284.
38. Louisy P (2005) Guide d'Identification des Poisson Marins Europe et Mediterranee. Les Editions Eugen Ulmer, Paris, France.

39. Malanotte-Rizzoli P (2001) Currents systems in the Mediterranean Sea. Steele J, Thorpe S, Turekian K, Encyclopedia of Ocean Sciences. Elsevier, Oxford, UK, pp 605–612.

40. Manzella GMR, Gasparini GP, Astraldi M (1988) Water exchange between the eastern and western Mediterranean through the Strait of Sicily. Deep-Sea Res Part A Oceanogr Res Papers 35: 1021–1035.

41. Marras S, Claireaux G, McKenzie DJ, Nelson JA (2010) Individual variation and repeatability in aerobic and anaerobic swimming performance of European sea bass, Dicentrarchus labrax. J Exp Biol 213: 26–32.

42. Marras S, Killen SS, Domenici P, Claireaux G, McKenzie DJ (2013) Relationships among traits of aerobic and anaerobic swimming performance in individual European sea bass, Dicentrarchus labrax. PLoS ONE 8: e72815.

43. Marullo S, Buongiorno Nardelli B, Guarracino M, Santoleri R (2007) Observing the Mediterranean Sea from space: 21 years of Pathfinder-AVHRR sea surface temperatures (1985 to 2005). Re-analysis and validation. Ocean Sci 3: 299–310.

44. Milazzo M, Mirto S, Domenici P, Gristina M (2013) Climate change exacerbates interspecific interactions in sympatric coastal fishes. J Anim Ecol 82: 468–477.

45. Perry AL, Low PJ, Ellis JR, Reynolds JD (2005) Climate change and distribution shifts in marine fishes. Science 308: 1912–1915.

46. Popper D, Gundermann N (1975) Some ecological and behavioral aspects of siganids population in the Red Sea and the Mediterranean coast of Israel in relation to their suitability for aquaculture. Aquaculture 6: 127–141.

47. Por FD (1990) Lessepsian migration. An appraisal and new data. Bulletin de l'Institut Oceanographique (Monaco) 7: 1–10.

48. Pörtner HO, Farrell AP (2008) Physiology and climate change. Science 322: 690–692.

49. Pörtner HO, Knust R (2007) Climate change affects marine fishes through the oxygen limitation of thermal tolerance. Science 315: 95–97.

50. Pörtner HO, Peck MA (2010) Climate change effects on fishes and fisheries: towards a cause-and-effect understanding. J Fish Biol 77: 1745–1779.

51. Quignard JP, Tomasini JA (2000) Mediterranean fish biodiversity. Biologia Marina Mediterranea 7: 1–66.

52. Roche DG, Binning SA, Bosiger Y, Johansen JL, Rummer JL (2013) Finding the best estimates of metabolic rates in a coral reef fish. J Exp Biol 216: 2103–2110.

53. Roeckner E, Brokopf R, Esch M, Giorgetta M, Hagemann S, Kornblueh L, Manzini E, Schlese U, Schulzweida U (2006) Sensitivity of simulated climate to horizontal and vertical resolution in the ECHAM5 atmosphere model. J Climate 19: 3771–3791.

54. Sala E, Kizilkaya Z, Yildirim D, Ballesteros E (2011) Alien marine fishes deplete algal biomass in the eastern Mediterranean. PLoS ONE 6: e17356.

55. Saoud IM, Mohanna C, Ghanawi J (2008) Effects of temperature on survival and growth of juvenile spinefoot rabbitfish (Siganus rivulatus). Aquac Res 39: 491–497.

56. Schembri PJ, Deidun A, Falzon MA (2012) One Siganus or two? On the occurrence of Siganus luridus and Siganus rivulatus in the Maltese Islands. Mar Biodivers Rec 5: e71.

57. Seitzinger SP, Harrison JA, Dumont E, Beusen AHW, Bouwman AF (2005) Sources and delivery of carbon, nitrogen, and phosphorus to the coastal zone: an overview of global Nutrient Export from Watersheds (NEWS) models and their application. Glob Biochem Cycles 19: GB4S01.

58. Somot S, Sevault F, Déqué M, Crépon M (2008) 21st century climate change scenario for the Mediterranean using a coupled atmosphere–ocean regional climate model. Glob Planet Change 63: 112–126.

59. Steffensen JF (1989) Some errors in respirometry of aquatic breathers: how to avoid and correct for them. Fish Physiol Biochem 6: 49–59.

60. Umgiesser G, Bergamasco A (1995). Outline of a primitive equation finite element model. In Rapporti e Studi, Vol. XII. Istituto Veneto di Scienze, Lettere ed Arti, Venice, Italy, pp 291–320.

61. Umgiesser G, Canu D, Cucco A, Solidoro C (2004) A finite element model for the Venice lagoon. Development, set up, calibration and validation. J Marine Syst 51: 123–145.

62. Vergés A, Steinberg PD, Hay ME, Poore AG, Campbell AH, Ballesteros E, Heck KLJ, Booth DJ, Coleman MA, Feary DA, et al. (2014a) The tropicalization of temperate marine ecosystems: climate-mediated changes in herbivory and community phase shifts. Proc Biol Sci 281: 20140846.

63. Vergés A, Tomas F, Cebrian E, Ballesteros E, Kizilkaya Z, Dendrinos P, Karamanlidis AA, Spiegel D, Sala E (2014b) Tropical rabbitfish and the deforestation of a warming temperate sea. J Ecol 102: 1518–1527.

64. Verlaque M (1990) Relations entre Sarpa salpa (Linnaeus, 1758) (Teleostean, Sparidae) et les autres poissons brouteurs et le phytobentos algal Mediterraneen. Oceanologica Acta 13: 373–388.

65. Web of ScienceGoogle Scholar

66. Zarnetske PL, Skelly DK, Urban MC (2012) Biotic multipliers of climate change. Science 336: 1516–1518.

67. Zenetos A, Gofas S, Verlaque M, Çinar ME, García Raso E, Bianchi CN, Morri C, Azzurro E, Bilecenoğlu M, Froglia C, et al. (2010) Alien species in the Mediterranean Sea by 2010. A contribution to the application of European Union›s Marine Strategy Framework Directive (MSFD). Part I. Spatial distribution. Mediterranean Mar Sci 11: 381–493.

68. Zenetos A, Gofas S, Morri C, Rosso A, Violanti D, Garcia JE, Cinar ME, Almogi-Labin A, Ates AS, Azzurro E, et al. (2012) Alien species in the Mediterranean Sea by 2012. A contribution to the application of European Union›s Marine Strategy Framework Directive (MSFD). Part 2. Patterns in introduction trends and pathways. Mediterranean Mar Sci 13: 312–336.

CHAPTER 12 POTENTIAL USE OF INLAND SALINE GROUNDWATER FOR THE PRODUCTION OF LIVE FEEDS FOR COMMERCIAL AQUACULTURE PURPOSES

McDowall S, Awal S and Christie A

Keywords: Live feed; Rotifer; Copepod; Artemia; Brachionus plicatilis; Cyclops spp; Nannochloropsis oculata; SGW; ASW; Neubauer haemocytometer; Dryland salinity

INTRODUCTION

The production of live feeds is a crucial segment of the global aquaculture industry. In effect, aquaculturists seek to develop a simplified food web, which utilises algae and zooplankton to feed larval fish. Algae, which form the base of the food web, are primary producers, and are used as an enrichment diet for zooplankton such as rotifers, Artemia and copepods. The importance of zooplankton in aquaculture is well established and has been mentioned in considerable detail elsewhere [1-4]. Australia is blessed with vast marine water resources, though lacks freshwater sources that are suitable for aquaculture purposes. Unfortunately, much of Australia's arable lands are subject to secondary salinisation from the agricultural practices of land clearing, irrigation, and the replacement of deep-rooted native perennial vegetation with shallow-rooted annual crops [5]. Inefficient irrigation distribution systems, and the reduction in evapotranspiration that have resulted from land clearing have produced rising water-tables and salt mobilization, decreasing the fertility and economic return of the affected lands [6]. Historical evidence shows an upward trend in the financial costs and reduction in rural production from salinity. Prime Minister's Science, Engineering and Innovation Council of Australia (PMSEIC) [7] identified that the area damaged by salinity was estimated to be 4.5% of currently cultivated land, at an annual cost of $130 million in lost agricultural production, consisting of around $90 million in damage to valuable land resources, and at least $40 million in loss of environmental assets. PMSEIC [7] acknowledges that it is inevitable that the consequences caused by salinity will get worse because land use decisions have already been made and are counterproductive in trying to deal with this issue. Later, National Land and Water Resources Audit (NLWRA) [8] reported that Australia had approximately 5.7 million hectares of agricultural and pastoral land at high risk of being affected by secondary salinisation, with an estimated 17 million hectares at high risk by 2050. Further to these problems are ecological indications that Australia is experiencing increases in the salinity of underground bore water. Examples are the Deutgam (which serves the Werribee irrigation District) and the Koo Wee Rup (which serves the Westernport region) aquifers found by Boland & Toulmin [9] to be at risk of saline intrusion. This will limit the potential level of extraction for freshwater crop

Citation: McDowall S, Awal S, Christie A (2016) Potential use of Inland Saline Groundwater for the Production of Live Feeds for Commercial Aquaculture Purposes. J Aquac Mar Biol 4(1): 00071. DOI: 10.15406/jamb.2016.04.00071

use, but at the same time provides considerable opportunity for utilisation by aquaculture enterprises.

In response to this issue, research has been undertaken into the use of inland saline groundwater for the aquacultural production of finfish, crustaceans, molluscs and seaweed [10-17] but, significantly, little information has been produced on the production of commercially important live feed species for aquaculture in inland saline groundwater. With this in mind, the aim of this study is to test the hypothesis that commercially important live feed species can be successfully grown in inland saline groundwater by quantifying the growth patterns that are observed. This investigation also seeks to offer recommendations for diversification with regards to the use of saline water for rural farmers that are affected by irrigation or dryland salinity, as well as adding to information concerning the salt water preferences of aquaculture is for the culture of various live feed organisms, as listed in Allan [18], Pillay & Kutty [19] and Smith & Barlow [20]. This study takes these examples further and looks at the capacity of inland saline groundwater to be a viable and alternative medium in the production of live aquaculture food items.

MATERIALS AND METHODS

Four experiments, coded T1 (Artemia), T2 (Cyclopoid copepod, Cyclops spp), T3 (rotifer Brachionus plicatilis) and T4 (microalgae Nannochloropsis oculata) were conducted as part of this research. The algae used in this experiment (both to act as a test organism and as an enrichment diet for the zooplankton populations) were obtained pre-experiment from the CSIRO (Australian National Algae Collection Centre, Hobart, Tasmania). These were strains of the algae Dunaliella tertiolecta, Nannochloropsis oculata, Pavlova salina, and Isochrysis spp.

Research Design and Experimental Set Up

The experimental design in each Trial (1, 2, 3 & 4) featured trials run intriplicate in each of the three saline waters. Aquaria units were assembled in a pattern that was completely random. During the experiment, abiotic variables such as pH (Hanna Instruments, USA), ammonia-nitrogen(Aquarium Pharmaceuticals, USA), salinity (SR6-Salinity meter, Vitalsine, China), temperature, (Hanna Instruments, USA), dissolved oxygen (OxyGuard Handy Polaris Dissolved Oxygen Meter) levels and lighting (Electronics Light Meter No. 05132201) were measured and kept in a balanced condition so that the study highlighted the merits of the different waters. The starting point for salinity was 25ppt with a pH of 8.5. The individuals in the experiment were the live food species used for aquaculture. The treatment waters consisted of seawater, artificial sea water and saline groundwater. The response variable was the size and density of algal colonies and zooplankton at the end of the study.

Description of Saline Water and Chemical Properties

The saline groundwater with a salinity level of 25ppt was sourced from Donald (a small town in central Victoria approximately 283 km north west of Melbourne, Australia; location 36°22′13.07″S, 142°58′56.68″E). Sea water of 33ppt salinity was obtained from Seaholme (a coastal town approximately 12 km west southwest of Melbourne on the north western shore of Port Phillip Bay; location 37°52′11.94″S, 144°50′33.32″E). Artificial seawater was prepared using a commercial artificial seawater mix (Prodac Ocean Fish, Prodac International, Cittadella, Italy), with tap water being added to arrive at a salinity of 25ppt. This product was used because of the quoted 99% grade of purity, and because it is presumably representative of a typical commercial artificial seawater mix. Chemical composition of groundwater and seawater, namely Sodium (Na^+), Potassium (K^+), Calcium (Ca_2^+), Magnesium (Mg_2^+), Sulphate ($SO_4^{\ 2-}$) and Chloride (Cl^-) content, were determined with an Atomic Absorption Spectrometer (AAS) (Varian AA20, Melbourne), and quantitative analysis was further conducted by undertaking titrations where applicable and determination of total dissolved solids (TDS) was undertaken according to the protocols described by APHA-AWWA-WEF [21].

Counting method

Before counting, all samples were homogenised to ensure that random sampling error would not be minimised. This was done by giving the container a swirl and applying aeration to make sure that the biomass of algae or zooplankton were evenly distributed through the mix. Cellular density of algae was determined by counting three aliquots of cultures using a Neubauer Haemocytometer, while zooplankton samples were counted by collecting a 1mL sample at random, which was then distributed in droplet form onto a petri dish. Samples were then viewed through a dissection microscope (Leica, China) with counting assisted with a hand counter (Compass, Taiwan). Three samples consisting of 1mL each for a grand total of 3mL were then counted, and an average of the three readings weretaken and percentage error was determined.

Experimental procedure for Artemia salina (T1)

The method described in APHA- AWWA-WEF [21] was used to measure the moisture content of the Artemia eggs (Salt Creek Inc., Salt Lake City, Utah, USA). Hatching the Artemia eggs followed the techniques described in Daintith [22]. Nauplii were separated from the eggs using a homemade apparatus made from a 1.5lt plastic drink bottle and rubber tubing. With the bottle held in an inverted position and tubing pushed through the lid by 25mm, a concentrated batch of pure nauplii was collected. Algae feed was administered equally to all units over the experiment. The estimation of the Artemia biomass was established at 270,000 and distributed at 30,000 per aquaria. (Standard deviation = 1031.98, % of error 3.55).

Experimental procedure for cyclopoid copepod (Cyclops spp) (T2)

Copepod collection method

The population of copepods used in this experiment was obtained with the use of a zooplankton net with an 80μm mesh size (Australian Filter Specialists, NSW). Several sweeps across and through the surface layers at Seaholme in Port Phillip Bay was able to provide suitable biomass for the purposes of the experiment. The catch of copepods were then allowed to sit in a lightly aerated, dark coloured tub for one week prior to use in the experiment.

Two feeding diets were used for the copepods. Diet one was formulated from a feed used by Lee et al. [23]. Diet two consisted of Isochrysis spp algae. The combined use of two food types for copepods was adopted, as suggested by Rhodes & Boyd [24]. Estimation for the copepod biomass was established at approximately 81,000 individuals distributed at 9000 individuals per aquaria. (Standard deviation = 409.26, % of error 4.68).

Experimental procedure for rotifers Brachionus plicatilis (T3)

Rotifer cultures for this study were obtained through Fuji Fuels (a biofuels company based in Thomas town in the northern suburbs of Melbourne), and were maintained using miscellaneous techniques as described in other studies [22,25-26]. Rotifer biomass was estimated to be approximately 72,000 individuals, which were distributed at around 8,000 per aquaria (standard deviation=217.94, % of error 3.02). Algae Nannochloropsis oculata spp was used as the basic enrichment diet and inoculated into aquaria prior to the introduction of rotifers to ensure the presence of a readily available food source.

Experimental procedure for Nannochloropsis oculata (T4)

Nannochloropsis oculata was inoculated into all nine carboys at an initial count of 2.89 × 10-4ml-1 from pre- stock cell counts made using a compound light microscope (Prism/ Optical, Australia) with a Neubauer haemocytometer slide (Marienfeld, Germany).

Statistical analysis

All data were recorded for Hypothesis testing on Microsoft Excel (Seattle, Washington) spreadsheets (2007) for subsequent data analysis of by using analysis of variance (ANOVA) and F-test statistic at $\alpha = 0.05$.

RESULTS

The results of preliminary analysis of all waters utilized in these experiments are shown in Table 1.

Growth in Artemia salina (T1)

Figure 1 shows the growth characteristics of Artemia salina in the three saline waters and indicates that there are no significant differences between each of the waters being tested (P-value of 0.828, > α=0.05). Estimation of moisture content in Artemia cysts was found to have a mean value of 8.55%, with a mean dry weight of 0.4185g. Three samples were taken at random from the three waters and viewed through a dissection microscope at the same magnification. Figure 2a demonstrates the obvious visual differences in the growth rates of Artemia from each of the different water types that were tested as part of this investigation. Uniformly smaller bodies were observed from the artificial sea water aquaria. They evidently fed, as indicated by the food matter that was readily visible in the gut. The implication here is that the artificial sea water has inhibited the growth rate of individual Artemia but not fecundity and consequent rate of population increase.

The body sizes in the seawater tended to be more uniform and larger than the bodies depicted in the photo from the artificial sea water (Figure 2b). Being larger, the body colour tended to be darker, possibly as a consequence of having more food in the gut. There tended to be considerably more larger-sized bodies in the groundwater aquaria than the other two experimental water types that were tested, though the body sizes were not uniform and variation was considerable (Figure 2c). With such variety in body sizes being observed, it is possible that more population activity was occurring within these aquaria.

Growth in cyclopoid copepods (T2)

Initially all animals in each water type tested showed a sharp decrease in numbers. Figure 3 shows how the copepod biomass returned to measurable numbers after a very poor start, during which numbers were observed to decline significantly. The copepod densities that were observed between each of the waters being tested varied at a statistically highly significant level (P-value 0.0001, >α = 0.05). A series of photographs taken in the laboratory of Cyclopoid copepods revealed an interesting situation with regards to the coloration being adopted by the exoskeleton of these animals. When viewed under a microscope, (a Motic BA300 microscope with Modicam 2000 digital camera), a typically orange-brown colour (Figure 4a) that is characteristic of the body of this species was evident on the hollow exoskeleton, which represented the animal's first moult post-harvest from the sea. This can be compared with the colour of the post moult copepod (Figure 4b), where the pigment of its former colour has significantly changed to this glass-like appearance. In another situation (Figure 4c), a female copepod demonstrated that the animals had adapted to a changed environment as much as necessary to commence breeding. This situation may indicate that the Cyclopoid copepod may become more fecund with each generation, and that a certain level of domestication could be expected to develop that could lead to improved survival rates, but this is subject to confirmation by further experimentation.

Growth in rotifer Brachionus plicatilis (T3)

Figure 5 outlines the rate of rotifer growth in the three saline waters (P-value 0.00 < α=0.05). The statistical analysis for T3therefore demonstrated that the rotifer Brachionus plicatilis differed in terms of density depending on the water that it was being grown in. The low P-value with an F-statistic of 22.72 provides strong evidence against the H0 hypothesis.

Growth in Nannochloropsis oculata (T4)

The performance outline of the three saline waters is shown in Figure 6. Statistical analysis for T4 reveals a P-value of 0.067 >α = 0.05. This P-value for N.oculata means that the data are not statistically significant at the α = 0.05 level. However the 2.99 analysis of variance F statistic shows that there was a statistically significant difference in the mean values for saline groundwater over artificial sea water.

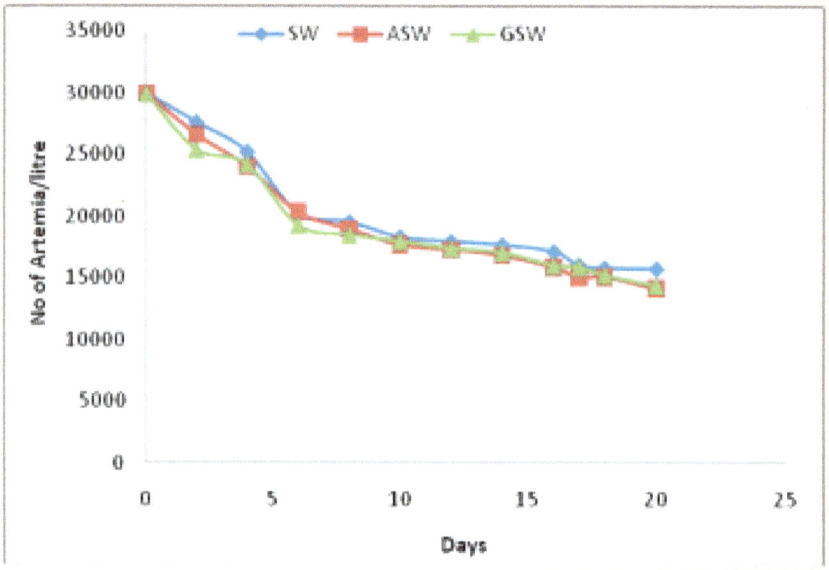

Figure 1: Growth of Artemia in Sea Water (SW), Artificial Sea Water (ASW) and Saline Ground Water (SGW).

Figure 2: (a) Smaller and lighter coloured Artemia grown in artificial sea water, (b) uniform body size of Artemia grown in sea water; (c) differing body sizes of Artemia grown in SGW.

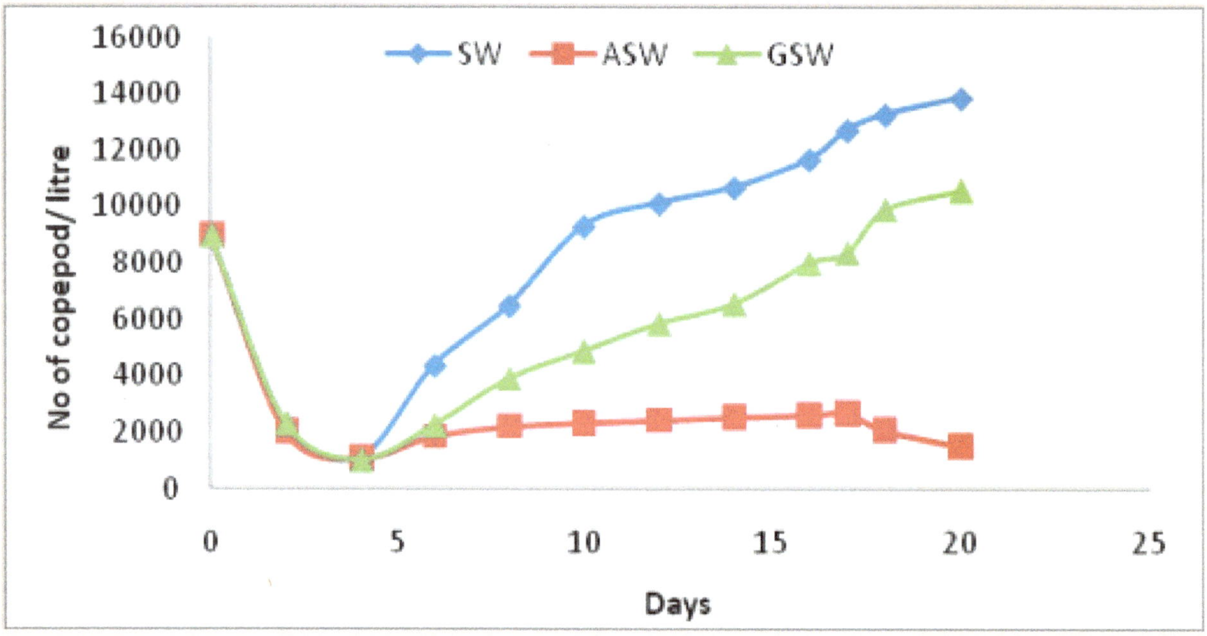

Figure 3: Growth of copepods in Sea Water (SW), Artificial Sea Water (ASW) and Saline Ground Water (SGW).

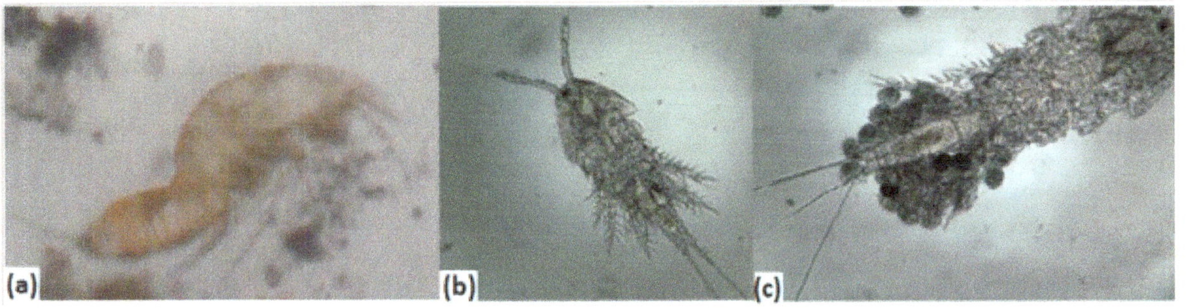

Figure 4: (a) Cyclopoid exoskeleton showing orange brown colour of wild harvested copepod. (b) Post moult Cyclopoid copepod from T2, saline groundwater; (c) Female copepod (from saline groundwater) with newly hatched naupli.

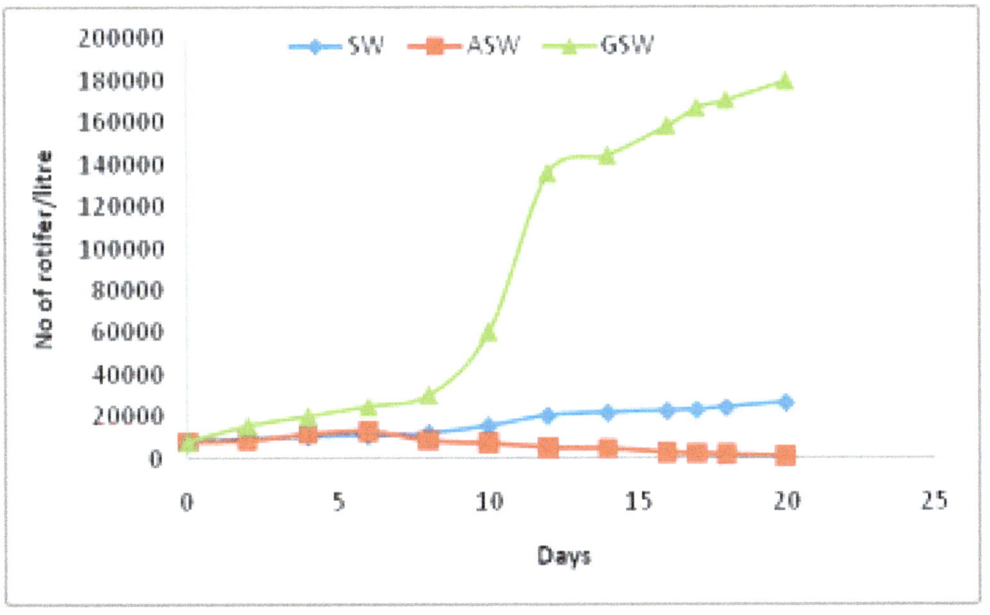

Figure 5: Growth of rotifer in Sea Water (SW), Artificial Sea Water (ASW) and Ground Saline Water (GSW).

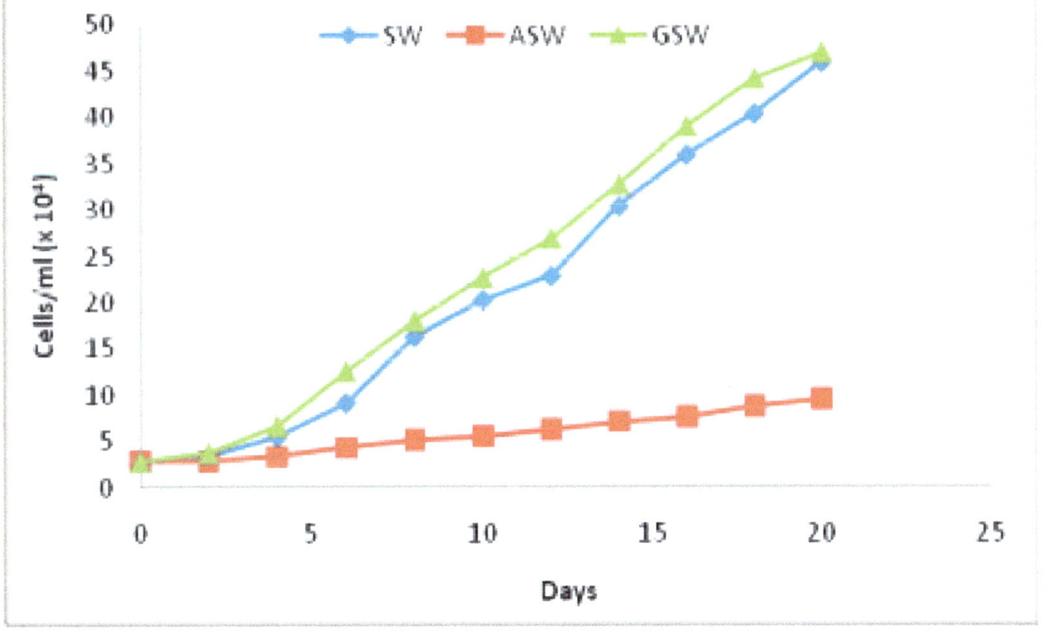

Figure 6: Growth of Nannochloropsis oculata spp in Sea Water (SW), Artificial Sea Water (ASW) and Ground Saline Water (GSW).

Table 1: Chemical analysis of Prodac Ocean Fish artificial sea water and inland saline groundwater, using Varian AA20 Atomic Absorption Spectrometer

Parameters	Sea	Artificial	SG/W
	Concentration (mg/L^{-1})	Concentration (mg/L^{-1})	Concentration (mg/L^{-1})
Na+	10930	10752	7580
Mg2+	1684	1295	1330
K+	410	390	90
Ca2+	370	416	293
SO_4^{2-}	3500	2701	2700
Cl^-	19640	19345	13600
% TDS	3.69	n/a	2.58

DISCUSSION

The first feed items taken by larval fish is often restricted by the gape of their mouths [27]; since food particles cannot exceed this size, even micro-pelleted diets may be unsuitable for fish and cannot be expected to provide them with the sustenance required to get them through the crucial early life stages [28-30]. The size of rotifers and early-instar Artemia and copepod species are generally small enough to be detected and devoured by larval fish species, explaining their attractiveness as a live feed diet for a multitude of fish species [31-33]. As this study has demonstrated, the breeding of live feed species can be achieved using saline water that just happens to be quite well suited to their needs. In the experiment T1 (featuring Artemia salina), darker orange/ brown coloured bodies were first noticed in the groundwater aquaria. As this colour difference between the waters was shortlived, the cause was thought to be more from a shorter lag period in the reproductive rate of the Artemia when placed in saline groundwater, as opposed to being from any specific ionic water parameters per se. Random photos, taken towards the end of the T1 Artemia trial, appeared to show the animal with a full gut, indicating that the animals had fed normally. These photos (Figure 2) also represent a point of difference in the average growth sizes seen from the different water types tested. Smaller and lighter coloured Artemia developed within the artificial seawater. Further study could be conducted on this point to see if this characteristic had any influence on a longer life span, as well as other important aspects, such as fecundity and overall productivity levels. Relatively uniform body sizes of Artemia were observed in the seawater aquaria. As the size of a larval fish's mouth is a major limiting factor in feeding, this uniform consistency in size becomes a considerable and desirable advantage in a live feed situation. Extra quantities of larger bodied Artemia were observed in the GW, which could potentially present issues with situations in which various instars are being offered to very small larvae, such as those produced by barramundi, Lates calcarifer, which are only 1.5mm long, and therefore require necessarily small (and preferably uniform) Artemia larvae upon which to feed [34-36]. This factor relates to the early observation of darker bodies and indicates that the Artemia have adapted somewhat sooner to the groundwater. The growth performance of copepods (Figure 2) and their initially poor performance, followed by an explosive population rebound, are in agreement with research carried out by Conseic et al. [1], who described various difficulties in copepod culture and survival, including the finding that greater success was initially found for second and third generation cultures. In this trial, a mistaken observation was made that the cultures had crashed because only dead orange/brown carcasses were found. One possible reason why the culture could have been thinned out so dramatically could have been cannibalistic tendencies, as indicated by Rhodes & Boyd [24] and Lee et al. [23]. At a later date, closer inspection revealed that while where there was much mortality; there were a reasonable number of hollow moulted exoskeletons. The authors' previous experiences with yabbies (the Australian native freshwater crayfish Cherax destructor) have shown how the fresh moult can be mistaken for the whole animal. Identification became obscure, as post-moult, the colour of the copepod body was near transparent or light grey, which displayed the internal organs in black. The experimental units were never shut down and as the copepods reappeared, population counts resumed. Even though the statistics say that there was no significant difference in using either of the salt water types, the preference of this study would be to use seawater. The reason for this is due to the delicate nature displayed by the animal in managing translocation from the catch-point and introduction into the culture water. The opinion of the authors from this study is that the population would be lost using either the groundwater or artificial seawater at these early stages. Comparable findings were made by Conseic et al. [1], which identify long generation times with some copepod species, as well as sensitivity to alterations in parameters such as hardness, salinity and temperature. Rotifer growth in seawater and artificial seawater were slow by comparison to the immediate acclimatization displayed in the GW aquaria. Growth in the groundwater developed at such a rate that all food was ingested overnight, leaving the water almost clear. Rising NH3 levels were a further indication of the increasing growth activity occurring in the groundwater aquaria, presumably due to the metabolic breakdown of protein and other substances that led to the formation and liberation of NH3 . While there was a point of difference in population numbers, no visible differences were observed in body size or colour of the animals from any of the units, after what would have been the second or third generations. In order to test and determine the suitability for supporting Nannochloropsis oculata growth for each water type, inoculations were staggered and time was given for the cultures to develop. After twelve days, sufficient green water had developed to determine biomass accurately. Growth in the artificial sea water units achieved half the productivity of the seawater units. Seawater performed consistently, whereas the growth produced from the GW was considerably better (the extent to which was somewhat surprising). Two main reasons for explaining the groundwater results would be that N. oculata is a suitable algae species in groundwater, while other candidate species, such as Dunaliella tertiolecta, Pavlova salina, and Isochrysis spp. may have to be tested first to ascertain suitability with this water type. Also, the combination of macro and micro nutrients of the groundwater were better assimilated by N. oculata. With the sea ater being used in its raw state, pollution within Port Phillip Bay could have also been a contributing factor, thought this would have to be tested and confirmed to be able to pass judgment on this issue. Aquaculture is earmarked as a growth industry for future sustainable food production and this project concentrates on an important element of that endeavor. Not only is this research relevant to regular practices within the aquaculture industry, but it aims to provide solutions for the use of inland areas that have been compromised by high salinity levels. Alternative farming methods are now a fact of life for many farming communities affected by dry land salinity. As the improvement of the land and water quality is often a slow and gradual process, some local farming communities have dwindled in numbers and health status due to economic hardships [37]. Aquaculture is heralded as a sustainable solution to providing income and industry growth. It is hoped that the results of this experiment and those that are similarly aligned could influence future planning with regards to aquaculture development. In Australia, as well as other parts of the world, land that is no longer suitable or viable for agricultural farming may be very suitable for aquaculture. The analysis of groundwater could yield vital information as to specific suitability towards a number of potential aquaculture harvests, whether they consist of algae, live feeds or finfish. The current study seeks to embrace the basic principles of ecological sustainability and biodiversity, as well as those of economic management, by demonstrating the ability to adapt to a changing world. It is hoped that the positive outcomes that result will influence various levels of aquaculture practice in future years. As a global example, Israel now has to deal with both dry land salinity and government-imposed water restrictions due to compromised land and water scarcity, situations that are exceedingly familiar to Australian farmers due to a fundamentally El-Nino/La Nina-based climatic system [38]. Many authors [8,10,38] indicate that this is becoming common in many parts of the world. Crop farmers invariably experience difficulties in sowing a crop, or produce crops of poor yield because of dry land salinity or increasing soil salinity that has stemmed from irrigation practices. One method of dealing with this problem would be the pumping and disposal of saline groundwater but this cost would be prohibitive to many small rural towns [39]. The topography of northern and southern Israel is somewhat similar to that of Mildura and Shepparton respectively, so there should be some scope for utilizing similar principles to develop aquaculture in this area [40]. Integration of aquaculture with agriculture has also become increasingly attractive [41], as there is a greater emphasis being placed on re-using resources such as land and water. This in turn leads to the development of guidelines for managing and controlling saline groundwater [20], which must be a major initiative well into the future. This study has demonstrated that there are potentially some very good opportunities for the utilisation of inland saline groundwater for live feed production and associated aquaculture activities. Clear advantages in the use of inland saline ground water for aquaculture have been proven in this study, with rotifers of the species Brachionus plicatilis and the algae Nannochloropsis oculata being particularly suited to the chemical composition of the water being used in the study. It could easily be argued that inland saline aquaculture clearly provides an option for diversification for agricultural holdings, but a sobering element that must be remembered is that the inland saline industry has been examined for suitability for the past two decades with only moderate success that has yet to be commercially realised to any great extent. Kolkovski [42] indicates that a good number of inland aquaculture farmers have already given up on inland saline aquaculture to look elsewhere, but it is hoped that by approaching aquaculture from the base of the food pyramid (i.e. With algae and zooplankton culture), more farmers can be encouraged to stay the course and develop hatcheries or grow out systems that are capable of supporting thriving enterprises. This way, the skills learnt from zooplankton production can be developed, and eventually enable a fish hatchery to be developed. This foundation then paves the way for the business to move in an upward direction. This can then develop into the growing out of fish stocks for markets. The overall majority of reports pertaining to inland saline aquaculture are about the production of fish or trialling suitable euryhaline fish species for this form of aquaculture. In other words, the building blocks of the industry were never established, as the prime focus appears to have only been on the end result. The fundamental reality of this top-down approach to building an industry is that no long term stability can be established, and that it has little or no diversity to fall back on in times of adversity. Few inland saline aquaculture businesses have been successful (with very little work conducted outside of experimental work) and at present there is limited focus on the production of varieties of microalgae or zooplankton [42]. Markets exist for live feeds in aquaculture, and therefore research into their exploitation would require some work, but there are many opportunities here that could be exploited, remembering that live feeds effectively underpin hatchery operations, and are capable of providing a significant income stream. It is hoped that by supplying technical information that can be reproduced by aqua culturists, the development of the base of the aquatic food pyramid can take place in inland saline aquaculture situations, thus providing a considerable potential boost to the industry.

ACKNOWLEDGEMENT

We are deeply indebted to Mr. Ben Hokinand Fuji Fuels, Thomastown, for providing some of the materials that were relied on in this investigation.

REFERENCES

1. Conceic L, Yufera M, Makridis P, Morais S, Dinis MT (2010) Live feeds for early stages of fish rearing. Aquaculture Research 41(5): 613- 640.

2. Cho SH, Ji SC, Hur SB, Bae J, Park IS, et al. (2007) Optimum temperature and salinity conditions for growth of green algae Chlorella ellipsoidea and Nannochloropsis oculata. Fisheries Science 73(5): 1050-1056.
3. Nakawagawa H, Sato M, Gatlin DMIII (2007) Dietary supplements for the health and quality of cultured fish, CAB Internatioanl, UK.
4. Partridge GJ, Jenkins GI, Frankish KR (2003) Hatchery Manual for the production of Snapper (Pagrus auratus) and Black Bream (Acanthopagrus butcheri), South Metropolitan College of TAFE (W.A.). Aquaculture Development Unit 152.
5. Bell M (2001) What threat is salt to my farming operation? Australian Grain 1-3.
6. Ghassemi F, Jakerman AJ, Nix HA (1995) Salinisation of land water resources, CAB International, Wallingford, UK, pp. 526.
7. Prime Minister's Science, Engineering and Innovation Council (PMSEIC) (1998) Dryland salinity and its impacts on rural industries and the landscape. A paper prepared for the second meeting of the Prime Minister's Science Engineering and Innovation Council, held in Canberra, Australia, p. 3-26.
8. NLWRA (2001) National Land and Water Resources Audit. Australian Dryland Salinity Assessment 2000, Canberra, Australia.
9. Boland AM, Toulmin M (2010) Integrated water management in the Port Phillip and Western Port (PPW) region, RM Consulting Group, Camberwell, Victoria, Australia.
10. Ingram BA, McKinnon LJ, Gooley GJ (2002) Growth and survival of selected aquatic animals in two saline groundwater evaporation basins: an Australian case study, Aquaculture Research 33(6): 425- 436.
11. Gavin JP, Greg IJ (2002) The effect of salinity on growth and survival of juvenile black bream (Acanthopagrus butcheri). Aquaculture 210(1-4): 219-230.
12. Partridge GJ Creeper J (2004) Skeletal myopathy in juvenile barrumundi Lates calcarifer; cultured in potassium-deficient saline groundwater. J Fish Dis 27(9): 523-530.
13. Robert C (2007) Seaweed agronomy: cropping inland saline groundwater evaporation basins: a report for the Rural Industries Research and Development Corporation. Rural Industries Research and Development Corporation, Canberra, Australia, pp. 1440-6845.
14. Gavine F, Bretherton M (2007) Aquaculture in saline groundwater evaporation basins, RIRDC research paper, No 07/114, Rural Industries Research and Development Corporation, Canberra, Australia, p. 40.
15. Fotedar RK, Harries S, Savage S (2008) Survival growth and osmolality of greenlip Haliotis laevigata (Donovan 1808) when exposed to different ionic profiles of inland saline water. Aquaculture Research 39(5): 441-448.
16. Partridge GJ, Lymbery AJ (2008) The effect of salinity on the requirement for potassium by barramundi (Lates calcarifer) in saline groundwater. Journal of World Aquaculture Society 278(1-4): 164-170.
17. Partridge GJ, Lymbery AJ (2009) Effects of manganese on juvenile mulloway (Argyrosomus japonicas) cultured in water with varying salinity-implications for inland mariculture. Aquaculture 290(3-4): 311-316.
18. Allan G (2005) A fish farming salinity solution. ECOS Magazine 125.
19. Pillay TVR, Kutty MN (2005) Aquaculture: principles and practices. (2nd edn), Blackwell, Oxford, UK, pp. 640.
20. Smith B, Barlow C (1999) Inland saline aquaculture, Proceedings of a workshop held on 6-7 August 1997. Western Australia, ACIAR Proceedings, Canberra, Australia, p. 5-60.
21. APHA-AWWA-WEF (2012) Standard methods for the examination of water and wastewater. (22nd edn), American Public Health Association, American Water Works Association and Water Environment Federation Washington, DC, USA.
22. Daintith M (1996) Rotifers and Artemia for marine aquaculture: a training guide. Turtle Press, Tasmania, Australia, p. 62.
23. Lee CS, O'Brien PJ, Marcus NH (2005) Copepods in Aquaculture, Blackwell, Oxford, UK, pp. 255.
24. Rhodes A, Boyd L (2005) Formulated feeds for harpacticoid copepods: implications for population growth and fatty acid composition. In: Lee et al. (Eds.), Copepods in Aquaculture, Blackwell publishing, Hoboken, New Jersey, USA, pp. 61-73.
25. Lavens P, Sorgeloos P (1996) Manual on the production and use of live food for aquaculture, FAO Fisheries, Technical Paper Rome, Italy, pp. 361.
26. Sahandi J, Jafaryan H (2011) Rotifer (Brachionus plicatilis) culture in batch system with suspension of algae (Nannochloropsis oculata) and bakery yeast (Saccharomyces cerevisiae). Natural Resources 4(4): 526-529.
27. Dabrowski K, Bardega R (1980) Mouth size and predicted food size preferences of larvae of three cyprinid fish species. Aquaculture 40(1): 41-46.
28. Battaglene S, Fielder S (1997) The status of marine fish larval rearing technology in Australia. Hydrobiologia 358(1): 1-5.
29. Faulk CK, Holt GJ (2003) Advances in rearing cobia Rachycentron canadum larvae in recirculating aquaculture systems: Live prey enrichment and green water culture. Aquaculture 249(1-4): 231- 243.
30. Lee CS (2003) Biotechnological advances in finfish hatchery production: a review. Aquaculture 227(1-4): 439-458.
31. Lee CS, Ostrowski AC (2001) Current status of marine finfish larviculture in the United States. Aquaculture 200(1-2): 89-109.
32. Payne MF, Cleary JJ, Rippingale RJ (2001) Cultured copepods as food for West Australian dhufish (Glaucosoma hebraicum) and pink snapper (Pagrus auratus) larvae. Aquaculture 194(1-2): 137-150.
33. Grageda MVC, Kotani T, Sakakura Y, Hagiwara A (2008) Effects of feeding copepod and Artemia on early

growth and behavior of the self-fertilizing fish, Rivulus marmoratus, under laboratory conditions. Aquaculture 281(1-4): 100-105.

34. Rimmer MA, Reed AW, Levitt MS, Lisl AT (1994) Effects of nutritional enhancement of live food organisms on growth and survival of barramundi, Lates calcarifer (Bloch), larvae. Aquaculture and Fisheries Management 25(2): 143-156.

35. Curnow J, King J, Partridge G, Kolkovski S (2006) Effects of two commercial microdiets on growth and survival of barramundi (Lates calcarifer Bloch) larvae within various early weaning protocols. Aquaculture Nutrition 12: 247-255.

36. Yahaya S, Lima LS, Shaleh SRM, Mukaim Y, Anrakuc K, et al. (2011) Ontogenetic eye development and related behavioural changes in larvae and juveniles of barramundi Lates calcarifer (Bloch). Marine and Freshwater Behaviour and Physiology 44(6): 339-348.

37. Speldewinde PC, Cook A, Davies P, Weinstein P (2011) The hidden health burden of environmental degradation: disease comorbidities and dryland salinity. EcoHealth 8(1): 82-92.

38. King AD, Klingaman NP, Alexander LV, Donat MG, Jourdain NC, et al. (2013) Extreme rainfall variability in Australia: patterns, drivers, and predictability. Journal of Climate 27(15): 6035-6050.

39. Hulata G, Simon Y (2011) An overview on desert aquaculture in Israel. In: Crespie, V & Lovatelli A (Eds.), Aquaculture in desert and arid lands: development constraints and opportunities. FAO Fisheries and Aquaculture Proceedings 20: 85-112.

40. Partridge GJ, Lymbery AJ, George RJ (2008) Finfish mariculture in inland Australia: a review of potential water sources, species and production systems. Journal of World Aquaculture Society 39: 291- 307.

41. Milstein A (2010) Integrated Agri-Aquaculture in Israel. Presentation provided at the Integrated Agri-Aquaculture workshop. Growing fish in your dam. Mildura, Victoria, Australia.

42. Kolkovski S (2011) An overview on desert aquaculture in Australia. In: Crespie & Lovatelli (Eds.), Aquaculture in desert and arid lands: development constraints and opportunities. Fisheries and Aquaculture Proceedings, Rome, Italy, FAO, pp. 39-60.

CHAPTER 13 THE ORIGIN OF ALGACULTURE IN DAMSELFISH

Duur K Aanen

COMMENTARY

The transition from a hunter-gatherer to an agricultural lifestyle has had dramatic consequences for human evolution [1]. It paved the way for the advanced division of labor, which characterizes modern societies and which was a precondition for their rise. With that, agriculture ultimately is at the basis of the exponential increase in human population size during the last millennium. What is usually less well appreciated is that this transition has meant an exponential increase in the population size of domesticated organisms compared to their free-living ancestors. For example, in the Netherlands there is about one domesticated pig for each human, so that the domesticated animal outnumbers its wild cousin by some three orders of magnitude. Agriculture therefore fully satisfies the criterion for a mutualism: a mutually beneficial interaction between different species.

More traditional forms of agriculture may shed light on the first steps to agriculture. Amerindian farmers propagate cassava (*Manihot esculenta*) clonally by cuttings. However, recombined wild seedlings also appear spontaneously in fields, and farmers allow them to grow. Later, they recruit some large ones for new cuttings, and it has been demonstrated that the selected plants are more heterozygous. Therefore, the farmers (unconsciously) select for increased heterozygosity [2]. This shows that domestication is an ongoing process and that domesticated organisms are not immediately reproductively isolated from their wild relatives. This has also been shown for domestic cattle where introgression of wild Y chromosomes -but not of mitochondria - has occurred in northern Europe, suggesting continued mating with wild bulls after domestication [3]. And on the human side, although a transition back to a hunter-gatherer lifestyle from an advanced agricultural society now seems hard to imagine, there is evidence for a reversal from a primitive agricultural to a hunter-gatherer lifestyle [4].

A different approach to studying the first stages of agriculture is to look at non-human agricultural mutualisms. Well-known examples are found in two groups of social insects - termites and ants - which have independently domesticated fungi for food [5]. A recent paper by Hata *et al.* [6] in *BMC Evolutionary Biology* describes details of another fascinating agricultural mutualism, this time below the sea surface, between damselfish and filamentous algae. On coral

Citation: Duur K Aanen, the origin of algaculture in damselfish, DOI: 10.1186/1741-7007-8-81

reefs, damselfish individually defend territories from invading grazers and maintain algal turfs, from which they obtain all their food (Figure 1). These turfs are a rich food source and highly productive. Most species grow their algal crops as mixed cultures of mostly Rhodophyta (red algae). The damselfish lack effective enzyme systems or masticatory organs to break down cell-wall material, and hence depend on easily digestible algae. The algae they use typically are fast growers found in the early stages of a succession.

Figure 1 The researcher and the fish. *Stegastes nigricans* chasing Hiroki Hata off its territory in Okinawa, Japan. Photograph kindly provided by Hiroki Hata.

Earlier work by the same group on the damselfish species *Stegastes nigricans* off the coast of Okinawa prefecture in Japan found that it specialized on the filamentous red alga *Polysiphonia* sp. 1, which it cultivated in a near monoculture (Figure 2a). The fish weed out most of the other, indigestible, algae, chase away other herbivores and harvest the alga as their staple food. As this species of *Polysiphonia* is highly susceptible to competition from other algal species, its habitat is limited to *S. nigricans* territories. In this way, the fish and the alga have become highly interdependent and approach the state of an obligate mutualism.

Figure 2 Variations in algaculture by damselfish. (a) A near monoculture of *Polysiphonia* sp. 1 from a territory of *S. nigricans*in Okinawa. (b) *S. nigricans* and its more diverse garden off the coast of Kenya. Photographs kindly provided by Hiroki Hata.

Algal analysis

Hata *et al.* [6] have now carried out a comprehensive DNA-based analysis of the algal composition of 320 territories from 18 species of damselfish, collected throughout the Indian Ocean and the West Pacific. The authors sketch a more complex picture of the mutualism between *S. nigricans* and *Polysiphonia* sp. 1 than initially believed. While *S. nigricans* populations in Japan do indeed maintain an almost exclusive monoculture of this species, in other parts of the fish's distribution most territories contain more diverse mixed cultures (Figure 2b). This diversity is manifest in two ways. First, in other areas *Polysiphonia* species constitute much less of the total algal biomass per territory; second, in most territories, more than one *Polysiphonia* species is found. Consistent with earlier studies, *S. nigricans* still maintains the most specific association of all damselfish studied: in all 117 territories of *S. nigricans*studied, close relatives of *Polysiphonia* sp. 1 belonging to a single clade were found. Almost all the algae from this clade were exclusively found in territories of *S. nigricans* and never outside damselfish territories, suggesting that this mutualism is obligate for both partners. The 17 other damselfish species studied generally cultivate mixed crops of algae, do this in a more extensive fashion than *S. nigricans*, and have less specific associations with*Polysiphonia* algae.

The monoculture found in the earlier studies was initially interpreted as support for a more derived state of farming in *S. nigricans* than in other damselfish. Selective weeding is unique for *S. nigricans*: other species also chase away competing grazers but they do not weed out unpalatable algae [7]. However, the new study shows that this intensive agriculture does not lead to a monoculture throughout the distribution area of this fish. Variation in the intensity of weeding among regions may explain part of the variation, but the composition of the local algal community might also influence algal composition inside the fish territory.

Monoculture versus mixed crop

Theoretical studies have indicated that symbiont relatedness may be an important stabilizing factor for mutualisms (for example, see [8]). The basic idea is as follows. A symbiont's fitness can be partitioned into two components: competitive success relative to other symbionts associated with the host, and overall success of the group of symbionts. As relatedness declines within hosts, a symbiont's success depends more on its competitive ability and relatively less on group success. In many cases, competitive traits will have negative side effects for the host, as symbionts will be selected for fast exploitation of the resources provided by the host. Therefore, theory suggests that by selecting for competitive traits symbiont mixing destabilizes mutualisms.

However, we have to consider the biological idiosyncrasies of a particular mutualism to fully appreciate the significance of symbiont relatedness. In the damselfish-alga mutualism, the algae preferred are those that are fast growing and do not invest in persistent cell walls. In the non-mutualistic situation, these algae are found in the early stages of succession. By their continual weeding, *S. nigricans* continuously recreates these early-succession conditions. The main determinant of competitive ability in these early-succession algae is fast growth. And the fast-growing algae are the ones preferred by the fish. Therefore, this form of competition increases rather than decreases the benefit for the host.

The origin of agriculture

For both humans and non-humans, the transition to agriculture was not a conscious decision. Therefore, we have to identify the consequences of changes occurring in the coevolving organisms that lead to their domestication [1]. Some plausible scenarios have been sketched for domestications by humans. For example, the ears of domesticated wheat and barley do not shatter and drop their seeds when ripe, whereas those of their wild ancestors do, and the seeds are wind dispersed. The non-shattering trait is caused by a single recessive mutation. It seems plausible that humans who collected seeds selected the convenient mutant that concentrated the seeds, and thus selected (unconsciously) for the non-shattering mutation. If some of these seeds were spilled close to human settlements, these mutants gained a fitness benefit as their seeds could be more effectively dispersed, but this depended on intervention by humans [1].

In general, the challenge is to understand the counterintuitive combination of being consumed and simultaneously gaining fitness. A plausible scenario for the origin of termite fungiculture is the initial indiscriminate consumption of unspecialized fungi that inhabited their nest structure. In all extant fungus-cultivating termites, the fungus is propagated by the consumption of nodules, which contain gut-resistant asexual spores, which are mixed with collected plant substrate [9]. It seems plausible that the formation of asexual spores that could survive gut passage was a key innovation in the fungus in response to insect consumption. This change turned the disadvantage for the fungus of being consumed into a benefit [9]. For other insect-fungus mutualisms it has been argued that the interaction started with insects acting as a vector for fungal dispersal [5].

The ancestors of algacultural fish were omnivorous and multiple independent transitions to algaculture have occurred (H Hata, personal communication). This is consistent with the rather unspecialized nature of this domestication. The fish do not plant or inoculate the algae, as has been found in older mutualisms between fungi and insects [5]. Instead, they remove most of the algae from their territories, thereby continuously recreating the early stages of algal succession. The algae thus selected are the algae characterized by fast growth and high palatability. It seems plausible that this mutualism started by weeding or consumption of all the algae in a territory, and that the fast-growing algae remained, and started to specialize on this new niche. With their behavior, the fish show some parallels with cattle, which by grazing create and improve their own niche - grassland. The difference is that the fish seem to remove unpalatable algae without consuming these, and that they are territorial.

Changes in the 'farmer'

According to the Wikipedia definition of domestication ('the process whereby a population of animals or plants, through a process of selection, becomes accustomed to human provision and control'), the farmer is the active partner and the crop or livestock the passive. However, it is increasingly becoming clear that humans themselves have also undergone extensive genetic adaptations after the adoption of an agricultural lifestyle. For example, the persistence of the ability to digest lactose after the weaning phase has arisen independently multiple times in human populations characterized by a history of cattle herding. Likewise, fungus-growing ants have changed from carnivorous to vegetarian after the adaption of fungiculture. To fulfill the definition of coevolution it remains to be shown in which ways the damselfish have genetically changed in response to changes in their cultivated algae. The recent finding that the gut microbiota of Japanese people contains bacteria that aid in digesting the seaweed present in sushi [10] nicely illustrates the fact that adaptations can occur upon the adoption of seaweed in a diet.

REFERENCES

1. Diamond J: Evolution, consequences and future of plant and animal domestication. Nature. 2002, 418: 700-707. 10.1038/nature01019.
2. Pujol B, David P, McKey D: Microevolution in agricultural environments: how a traditional Amerindian farming practice favours heterozygosity in cassava (*Manihot esculenta* Crantz, Euphorbiaceae). Ecol Lett. 2005, 8: 138-147. 10.1111/j.1461-0248.2004.00708.x.
3. Gotherstrom A, Anderung C, Hellborg L, Elburg R, Smith C, Bradley DG, Ellegren H: Cattle domestication in the Near East was followed by hybridization with aurochs bulls in Europe. Proc Biol Sci. 2005, 272: 2345-2350. 10.1098/rspb.2005.3243.
4. Oota H, Pakendorf B, Weiss G, von Haeseler A, Pookajorn S, Settheetham-Ishida W, Tiwawech D, Ishida T, Stoneking M: Recent origin and cultural reversion of a hunter-gatherer group. PLoS Biol. 2005, 3: e71-10.1371/journal.pbio.0030071.
5. Mueller UG, Gerardo NM, Aanen DK, Six DL, Schultz TR: The evolution of agriculture in insects. Annu Rev Ecol Evol Syst. 2005, 36: 563-595. 10.1146/annurev.ecolsys.36.102003.152626.
6. Hata K, Watanabe K, Kato M: Geographic variation in the damselfish-red alga cultivation mutualism in the Indo-West Pacific. BMC Evol Biol. 2010, 10: 185-
7. Hata H, Kato M: Monoculture and mixed-species algal farms on a coral reef are maintained through intensive and extensive management by damselfishes. J Exp Mar Biol Ecol. 2004, 313: 285-296. 10.1016/j.jembe.2004.08.009.
8. Frank SA: Host-symbiont conflict over the mixing of symbiotic lineages. Proc Biol Sci. 1996, 263: 339-344. 10.1098/rspb.1996.0052.
9. Aanen DK: As you reap, so shall you sow: coupling of inoculating and harvesting stabilizes the mutualism between termites and fungi. Biol Lett. 2006, 2: 209-212. 10.1098/rsbl.2005.0424.
10. Hehemann JH, Correc G, Barbeyron T, Helbert W, Czjzek M, Michel G: Transfer of carbohydrate-active enzymes from marine bacteria to Japanese gut microbiota. Nature. 2010, 464: 908-912. 10.1038/nature08937.

CHAPTER 14 SUSTAINABLE TECHNIQUE FOR SELECTED LIVE FEED CULTURE

Zaleha Kassim, Akbar John, Lim Keng Chin, Nur Farahiyah Zakaria and Nur Hidayah Asgnari

INTRODUCTION

Sustainability in the aquaculture industry depends on several factors including the minimum production cost in comparison to the yield, unexpected environmental conditions which affect the farm and practices in the farm management itself. These factors are inter-connected and always incur a synergistic effect on the issue of sustainability. Live feeds as the fundamental needs for larval rearing and fry production have to be prioritised for sustainable farming activity. Dependency on imported sources of live feeds or inert feed will increase the production cost. Thus, the continued activity of screening, stocking and maintaining some local species as an option for live feed production is economically necessary.

Live feeds are an important basic diet for newly-hatched fish and shrimp larvae as they still have an incomplete digestive system and are lacking in enzymes. They are still at a very young stage to generate their own required nutrients or convert them from any pre-cursor obtained from a diet. They need a ready-made diet with readily available nutrient to be absorbed through their digestive system. There have been many species suggested or tested for their potential as live feed. All test animals were mostly zooplankton in nature and must meet the requirement as live feed. They must be in a compatible size with the mouth size or gape of the larvae predator, or they cannot be swallowed. Since larvae are still weak to track down the food, the wave created by the prey will be a great help, thus 'active' swimming prey is preferred. The most important role of a prey is the ability to supply energy and other nutrients which are essential for the larval survival and growth. Live feeds, as the starter diet in larval rearing and fry production must be continuous in supply. Good, nutritious and compatible-size prey must be able to reproduce fast to meet the requirement and adaptable to a simple mass-production technique.

Copepods As Live Feeds

The conventional live feed, brine shrimp and rotifers, are considered unsuitable as live feed if compared to copepods in term of nutritional value. *Artemia* sp. is deficient in polyunsaturated fatty acids (PUFAs), thus it needs to be enriched before feeding [1, 2]. Similarly, rotifer have poor

Citation: Zaleha Kassim, Akbar John, Lim Keng Chin, Nur Farahiyah Zakaria and Nur Hidayah Asgnari (2014). Sustainable Technique for Selected Live Feed Culture, Sustainable Aquaculture Techniques, PhD. Martha Hernandez-Vergara (Ed.), InTech, DOI: 10.5772/57212.

nutritional value and are small in size [3]. On the other hand, copepod diets were proven to increase the growth of larval marine fish compared to diet of rotifers *Brachionus plicatilis*, [4,5] or *Artemia*[4]. The potential use of copepods as live feed due to their excellent fatty acid content has been highlighted by using an example, a paracalanid [6]. They improved the quality of the cultured organism, particularly the larval stages. The superiority of copepods over other live feed such as brine shrimp and rotifers was further confirmed [7]. They have the appropriate ratio of docosahexaenoic acid (DHA) to eicosapentaenoic acid (EPA) which will improve the growth and survival of fish larvae if compared to the conventional live feeds [8]. Nonetheless, it is reminded that copepods could be better used as supplement rather than a sole diet in larval rearing, unless they are used for some high-valued commercial reef fish due to the high operation cost [9]. The possibility of using copepods, particularly the harpacticoids as alternative live feed in fish larval rearing has been stressed [10] and it is confirmed that the nutritional quality of harpacticoid copepods as live feed is extremely high [11,12]. *Macrosetella gracilis*, a planktonic harpacticoid copepod, is also reported to have better diet quality when compared to *Artemia* [13].

Despite these positive findings, rotifers and *Artemia* continue to be the live feeds of choice in commercial hatcheries, because copepods are not currently cultured at sufficient densities to be economically efficient on a commercial scale [14, 15, 16]. In term of culture condition, it was found that the optimum condition for the high production of a tropical harpacticoid copepod,*Pararobertsonia* sp., was at salinity 35psu and temperature of 25°C [17]. The fluctuation in salinity, pH and temperature in the culture vessel would definitely influence the reproduction and population growth of copepods such as the harpacticoids if kept in small containers [18]. Nonetheless, a strategy to produce harpacticoid in large quantities for hatchery use by using a tray-culture method has been suggested [19].

Another copepod group, a Cyclopoida, *Apocyclops dengizicus* was found to increase the survival and growth of *Panaeus monodon* postlarvae when used as live feed [20]. Cyclopoids are omnivorous, and can be fed a mixture of feeds, mainly phytoplankton or a combination of phytoplankton, yeast or other feeds [7]. As for *Apocyclops panamensis*, there is a report on a successful technique for outdoor ponds [15]. Information on the use of copepods in aquaculture, particularly from the tropical *Apocyclops* sp., is still scarce. The species reported in abundance and potentially exploited as live feeds for shrimp post-larvae in Malaysia for example is *A. dengizicus*. A new species, *Apocyclops ramkhamhaengi*, has been described [21] and added to the present report of 3 species of *Apocyclops* recognised from Asia:*A. dengizicus* (Lepeshkin), *A. royi* (Lindberg), and *A. borneonensis* (Lindberg). This new species is found in abundance in eastern Thailand water and has yet to be determined for its potential in aquaculture. Planktonic copepods such as cyclopoids feed on other plankton including planktonic microalgae. To maintain planktonic copepods in the hatchery or aquaculture ponds, a continuous supply of their diet, particularly the microalgae, will definitely be required.

Microalgae

Microalgae are a diverse group of unicellular autotrophs inhabiting almost all aquatic water bodies. Microalgae are rich in many specific and attractive compounds [22] and their nutritional values for aquaculture had been highlighted [23]. Production of microalgae is mandatory in the hatchery as it is a basic and nutritious diet for live feed, specifically the zooplankton. However, its mass production is generally costly due to huge manpower, space requirements and operation which usually related to the cost of the energy used. A good strategy in manipulating the culture environment, particularly during the production process of microalgae would scale down the operational cost.

Light plays a fundamental role in the development of microalgae through photosynthesis. It is one of the major environmental factors which control the performance of microalgae phototrophic growth and productivity [24, 25, 26]. Light may either be natural or supplied by fluorescent tubes giving the maximum effective radiation which can be absorbed by the pigments of the microalgae. Light intensity plays a vital role, but the requirements vary with the culture depth or volume as well as the density of the algae in the culture. At a higher volume, light intensity must be increased to enable it to penetrate through the culture. However, an extreme light intensity may result in photo-inhibition which reduces the photosynthetic rates and growth [27,28]. Furthermore, overheating due to artificial or natural illumination should be avoided in microalgal culture. The most often employed light intensity is 1000 lux which is suitable for Erlenmeyer flasks but 5000-10000 lux is needed for a greater volume of microalgal culture [29, 30]. The duration of illumination can be varied where photosynthesis of microalgae can be enhanced or increased in the light/dark (LD) cycle (usually 12:12 or 14:10 LD, maximum 16:8 LD). For some microalgae, photosynthesis rate could also be increased exponentially with increasing light/dark frequencies where a longer period of dark in relation to the light period can further increase photosynthetic efficiencies but not vice versa [31]. The illuminations also affect the nutrient utilisation in the culture vessel [32].

A cost-effective and nutritionally-adequate alternative to costly maintenance of live microalgae is the production of moist-microalgae concentrates. It is seen to simplify hatchery procedures and has shown promising potential in the aquaculture industry [33,34].The storing of microalgae concentrates in moist form under low temperature can preserve their high nutrient composition and excellent cell viability [35,33]. Juvenile pacific oyster (*Crassostrea gigas*) fed with different algal pastes had shown significant improvement in growth rate than oyster fed with other diets [36]. Concentrates of*Chaetoceros muelleri* and *Tetraselmis pseudonana* refrigerated for 6 weeks at 4°C were found to promote similar survival rates of the tiger prawn *Penaeus monodon* larvae fed with live microalgae [37]. Supplementation of microalgae concentrates to bivalves, oysters and scallops have also recorded the same extent of growth rates as live microalgae [38,34]. It has been documented that most of the demand for mariculture feed in Japan is supplied with live and fresh microalgae which is thickly concentrated and readily stored at 2-4°C for 1-8 weeks with good shelf life [39].

Preparation of concentrated condition of microalgae usually involves centrifugation technique. Nonetheless, although this technique has been successfully applied and utilised for preparing microalgae concentrates, it poses some limitations. First, the process involves exposure of cells to high gravitational and shear forces deteriorating the cell structure with the leaking of nutritional contents. Second, centrifuging large volumes of cultures is time-consuming and requires expensive equipments. Several alternative procedures, less damaging to the cells, which can be applied are filtration [40], foam fractionation [41] and flocculation [33, 34, 38]. Previous studies have observed the excellence of ultrafiltration technique in preserving and retaining the cellular structure and properties of fragile algal cells with little loss of material [42, 43].

The level of natural resources exploitation for aquaculture purposes is commonly high. Coastal land and mangroves forests always become the target area for brackish-water aquaculture ponds. The water source of this area, which is always from the nearby river estuary and lagoon, is also used as the live feeds (zooplankton and microalgae) source. Nonetheless, the supply is always seasonal and could become unavailable unexpectedly due to many factors and natural phenomena. This chapter aims to discuss the possible ways to produce local live feeds, a marine microalgae species and a planktonic copepod, sustainably using a simple technique for larval-rearing purposes. Maintaining local species is hypothesised to be more economical and practical. The usage of the microalgae as an enrichment element for live feed copepods will be proved.

METHODOLOGY

Experiment 1: Production of *Chlorella vulgaris* Concentrate Isolated from Bidong Island and Assessment as Copepod Diet

Seawater samples were obtained from Bidong Island, Terengganu. The collection was made by lowering a Niskin water sampler to a required depth, following the light-penetration depth. Concentrated water samples were then transferred into chilled, white-plastic containers and brought back to the laboratory for microalgae isolation process. Successive plating out on agar plates was performed in order to select the desired marine *Chlorella* colonies. Monospecific colonies were then transferred into trial culture tubes before scaling up into larger volumes of Erlenmeyer flasks.

The microalgae was then cultured for the preparation of moist concentrates in the temperature controlled room (20±2°C) using the standard batch culture method. Triplicate of actively-growing starter cultures were inoculated into 30 litres acrylic tanks enriched with Conway medium under constant illumination (cool-white type; 110 watts). All cultures were started with an initial inoculum of $2x10^6$ cells mL^{-1}. Cultures were aerated continuously using humidified filtered air. Evaporation in the culturing tanks was kept to a minimum by covering the top of the tanks. Cellular density of microalgae cultures was monitored daily using a Neubauer haemocytometer [29]. Scanning electron microscopic observation was also done to determine the ultra structure of the cell. Measurement of radius and height of the target microalgae cells was done under the advanced research microscope (Model Nikon Eclipse 80-i, Japan) and twenty individual cells were measured for the calculation of cell biovolume to avoid biasing results. Cell biovolume was calculated as assumed round-shape volume with the following formula proposed by Sun and Liu [44]:

$$\text{Cell}\, volume = 4/3\pi R^3 \tag{1}$$

Where, π= 3.142, R= radius of cell

Specific growth rate was calculated from the expression as proposed [45] which is shown below:

$$\text{Specific}\, growthrate\ (\mu) = \ln\left(F_1/F_0\right)/t_1\text{-}t_0 \tag{2}$$

Where, μ = specific growth rate, F_1= biomass at time harvest, t_1 and F_0= biomass at time zero, t_0.

Doubling time was computed based on the formula as proposed [45] which is shown below:

$$\text{Doubling}\, time\ (T) = \log(2)/\mu \tag{3}$$

Where, T = doubling time, μ = specific growth rate.

All microalgae cultures were grown to late-logarithmic phase for the preparation of concentrates via ultrafiltration technique. The concentrated aqueous suspensions of microalgae were filtered through a membrane filter (0.1µm pore size) to remove access water from the suspension without rupturing the microalgae, thereby obtaining the microalgae concentrate or paste. Cell viabilities of microalgae concentrates were assessed using Eosin dye as a viability assay on the basis of its penetration into non viable-cells based on the expression as proposed [46]:

$$\text{Cell}\, viability\ (\%) = \frac{Viablecellsx100}{Totalcells} \tag{4}$$

Where, C_B = total number of cells before filtration, C_A = total number of cells after filtration

The harvesting efficiency or percentage recovery (%) was evaluated by comparing the remaining total number of cells in the concentrate with the total number of cells before filtration with the following expression:

$$\text{Harvesting } efficiency/\text{ Percentage } recovery\,(\%) = C_B/C_A \text{x } 100 \quad (5)$$

Where, C_B = total number of cells before filtration, C_A = total number of cells after filtration

Microalgal concentrates were compared to live cultures of the same algae as food for marine copepods. Copepods were obtained from existing culture in UMT's laboratory. Two different sets of cultures were done using a Petri dish where each of them was fed with live and microalgae concentrate respectively. Individual copepods were counted daily under the Leica stereo microscope before being fed (1 drop). The maximum specific growth rate (K) was calculated [47] as shown below:

$$K = \ln\,(X_1/X_0)\,/t_1 - t_0 \quad (6)$$

Where, K = specific growth rate, X_1= the number of copepods at harvest time, t_1 and X_2= the number of copepods at time zero, t_0

The doubling time was computed as:

$$\text{Doubling } time\,(\tau) = \log(2)\,/K \quad (7)$$

Where, τ = doubling time, K = specific growth rate.

Experiment 2: Effects of Photoperiod and Culture Size on *Chlorella vulgaris* Stock Growth

Pure strains of *Chaetoceros* sp. and *C.vulgaris* were obtained from the microalgae maintenance laboratory at Institute of Tropical Aquaculture, Universiti Malaysia Terengganu, Malaysia, while the pure strain of *Nannochloropsis* sp. was obtained from the Fisheries Research Centre, Pulau Sayak, Kedah, and was maintained in autotrophic conditions in liquid and semisolid agar with Conway media [48]. This axenic culture was stored at 25±2°C for 3 days which served as an inoculum for further experiments.

Microalgae were grown in autotrophic conditions as a monospecific axenic culture in different volumes (250mL, 500mL and 2000mL) containing Conway media. 25mL of pure strain with the cell density of ~2 x 10^6 cells mL^{-1} were transferred to each Erlenmeyer culture flask and kept at complete illumination provided by luminescent tubes (1000 Lux). Carbon source was provided by bubbling sterile 2% (v/v) CO_2 in air through the cultures. Culture flasks were maintained at a constant temperature (22°°C± 1°C) with the pH range of 7-8 and salinity of ~35ppt in an air-conditioned laboratory over 2-3 weeks. Daily cell count was calculated using a haemocytometer. To determine the effect of different photoperiods, microalgae cultured in a 2 litre flask containing Conway media was treated at different photoperiods (light/dark) (24:0, 12:12 and 8:16) in replicates and cell count was achieved as mentioned above. Growth curve for each species of algae was constructed and One-way ANOVA with Dunnett's post-test was performed using Graph-Pad Prism.

Mean cell count and specific growth rate were calculated using the formula $X = \frac{\Sigma \chi_i}{n}$ in and $SGR = Ln\,\frac{W_2/W_1}{t_2 - t_1}$ respectively (where, X = mean cell count; χi = total number of cells; 'n'= number of cell counts; SGR = specific growth rate; W_1= Initial cell density, $cell^{-1}$; W_2= Cell density at late exponential phase, $cell^{-1}$; t_1 = Time at initial cell density, $cell^{-1}$; t_2 = Time at late exponential phase, $cell^{-1}$).

Experiment 3: Low-Cost Commercial Fertiliser for Mass Culture of Marine *Chlorella vulgaris*: Manipulation of N:P:K Ratio

An investigation was made to see the adaptability of the local marine *C. vulgaris* to the natural conditions in an aquaculture farm. This means that they need to adapt to different fertilisers other than Conway media, different salinity regimes and uncontrolled temperatures. Preparation of NPK-based fertiliser was made by manipulating the ratio of nitrogen, phosphorus and potassium source as summarises in Table 1. Each of the different N:P:K ratio treatments was prepared in triplicate. Source of nitrogen was obtained by using urea fertiliser.

Culture containers were well-cleaned with bleach and rinsed thoroughly before filling up with 1L of the farm water (salinity of between 20-25ppt). The marine *C.vulgaris* concentrate was prepared and 1mL of it was inoculated into the container and 1mL of the fertilizer was added. The containers were vigorously aerated to provide required quantity of oxygen and to keep cells and media in suspension. The containers were kept in the open under 100% outdoor light exposure.

TABLE 1. Type and fertiliser used in N:P:K ratio for mass culture of marine *Chlorella vulgaris*

N:P:K ratios			Type and fertiliser used		
N	P	K	Urea (g)	P^+(mL)	$Potash^+$(g)
1	1	1	0.98	0.98	0.98
15	15	15	14.7	14.7	14.7
8	8	2	7.84	7.84	1.96
16	8	6	15.68	7.84	5.88
12	6	4	11.76	5.88	3.92
12	8	4	11.76	7.84	3.92

Sampling of microalgae cells was done daily and counting was carried out using a Neubauer Hemocytometer covered with glass slide under a compound microscope.

The growth rate, divisions per day, and generation time or doubling time was calculated following [49]

$$\text{Growth rate; } K' = \text{Ln } (N_t / N_o) / (t_2 - t_1) \quad (8)$$

$$\text{Divisions } per day; \text{ Div. } day^{-1} = K'/\text{Ln}2 \quad (9)$$

$$\text{Generation } time \text{ (days) ; Gen'} t = 1/ \text{Div. } day^{-1} \quad (10)$$

$$\text{Generation } time \text{ (hours) ; Gen'} t = 24 \left(1/ \text{Div. } day^{-1}\right) \quad \mathbf{(11)}$$

Where, No and Nt = final and initial populations at time t1 and time t2, respectively.

Since sample was collected daily, therefore, t2 – t1 = 1.

Experiment 4: Egg Production, Growth and Development of *Apocyclops ramkhamhaengi* Fed on Marine *Chlorella vulgaris*

Detailed observation on the reproduction performance of a zooplankton depending solely on a*C.vulgaris* diet was planned to prove the important role played by this local microalgae in live-feed production. Samples of copepods were collected from Sungai Semerak (N 05° 51.737, E 102° 30.809'), Tok Bali, Kelantan using a zooplankton net. This area receives sea water from the South China Sea, which is near to the Thailand coast where the copepod species was first identified and reported. Live copepods were maintained and adapted to the laboratory environment. Sand-filtered sea water from the Marine Hatchery, Universiti Malaysia Terengganu was diluted with deionised water to be at salinity of 25ppt and was further filtered through a GFC membrane filter and then autoclaved at 121°C for 15 minutes [11]. Salinity was measured using a portable hand-refractometer (ATAGO, Japan) Microalgae diet for the copepod was prepared from the marine algae *C. vulgaris* stocked at the Marine Hatchery. The microalgae were cultured in 29-31ppt Conway medium with 24h-light, room temperature of 25-27°C and continuous aeration for 7days. The cell concentration in each 500ml conical flask was determined by using Neuber haemocytometer (0.25mm^2 x 0.1 mm) under a compound microscope. The algal production was done weekly and supplied to *A. ramkhamhaengi*culture.

The investigation on the reproduction performance started with fifteen gravid females of *A.ramkhamhaengi* placed into two sets of triplicate of 250mL beakers. The diet constituted,1mL of Baker's yeast (0.02g/L) and 1mL of *C. vulgaris* at density 1x10^{6}cells/mL which were introduced into both sets of the beakers and covered with parafilm layer to avoid contamination. Three subsamples (approximately 1mL) from each beaker of the cultures were observed daily. The number of the copepods at all stages, including nauplii, copepodite, adult and gravid female, were counted under a dissecting microscope (Leica ZOOM 2000) and then returned to the culture. Changing of approximately 80% of the culture medium was done every alternate day by passing the copepods culture through 100 and 40 microns nylon net which would retain all stages of copepods (the smallest size of 60 microns) but remove most of the waste.

The population growth of *A. ramkhamhaengi* was studied for 30 days. The specific growth rates (K) of all stages of the copepods in both diets given were calculated by using the formula [50]:

$$K = \frac{\ln Nt - \ln No}{t} \quad (12)$$

Where, t is the culture days, No and Nt is the number of copepods at the initial and final selected time interval. The doubling time (Dt) was calculated by dividing $\log_e 2$ by the population growth rate (K) of all stages of *A. ramkhamhaengi* in both diets given:

$$Dt = \frac{\log_e 2}{K} \quad (13)$$

Although cyclopoid copepods are known to suspend in water column, *A. ramkhamhaengi* showed its adaptability to swim on the near bottom of its culture vessel. The culture for this experiment was started by introducing a gravid female on the experimental petri dish. The adult was removed after the eggs hatched and the nauplii were monitored until they reached copepodite-v stage and were ready to mate. Adult females and males from the culture were prepared for the experiment. A pair of male and female was put into each set of glass Petri dish filled with 15mL sea water. The use of a Petri dish instead of a beaker eased the daily observation of different stages of the copepod in the population. The cultures were maintained at room temperature of 25-27°C without additional oxygen supply or aeration. Observation was done twice per day under a dissecting microscope (Leica ZOOM 2000) before feeding to avoid the disturbance of the diet materials during individual or population counting. The culture medium was changed approximately 80% daily, and culture containers were subsequently changed every 4days. Daily feeding was done in the morning and evening by dropping 1mL of 1 x 10^{6}cells/ml *C.vulgaris* into the culture. The time taken for the females to become gravid was based on the observation recorded twice per day (morning/evening). Once the females become gravid, the male broodstocks were removed, and the female were left alone inside the Petri dish in order to determine the number of eggs per female from its first copulation. Observation on the development time from nauplii to adult, maturation time and generation time of *A. ramkhamhaengi* were recorded coupling with the numbers of offspring produced and percentage of hatching.

RESULT AND DISCUSSIONS

Experiment 1: Production of *Chlorella vulgaris* Concentrate Isolated from Bidong Island and Assessment as Copepod Diet

The ultra-structure of the *C.vulgaris* isolated and cultivated in this study is shown in Figure 1. The scanning electron micrographs displayed the characteristic features of green single cells with spherical shape and possession of rigid cell wall. There are some differences found in the present specimen if compared to some other established species. The outer shell is rough if compared to the latest SEM of*C.sorokiana* [51]. The feature is almost the same as found in SEM of *C. vulgaris* [52]. In terms of size, the specimen was found to be in between the size of marine *C.vulgaris* (2.1µm) and estuarine*C. vulgaris* (2.3µm) from Korean waters [53].

Kingdom PLANTAE
Phylum CHLOROPHYTA
Class CHLOROPHYCEAE
Order CHLORELLALES
Family CHLORELLACEAE
Genus *CHLORELLA*

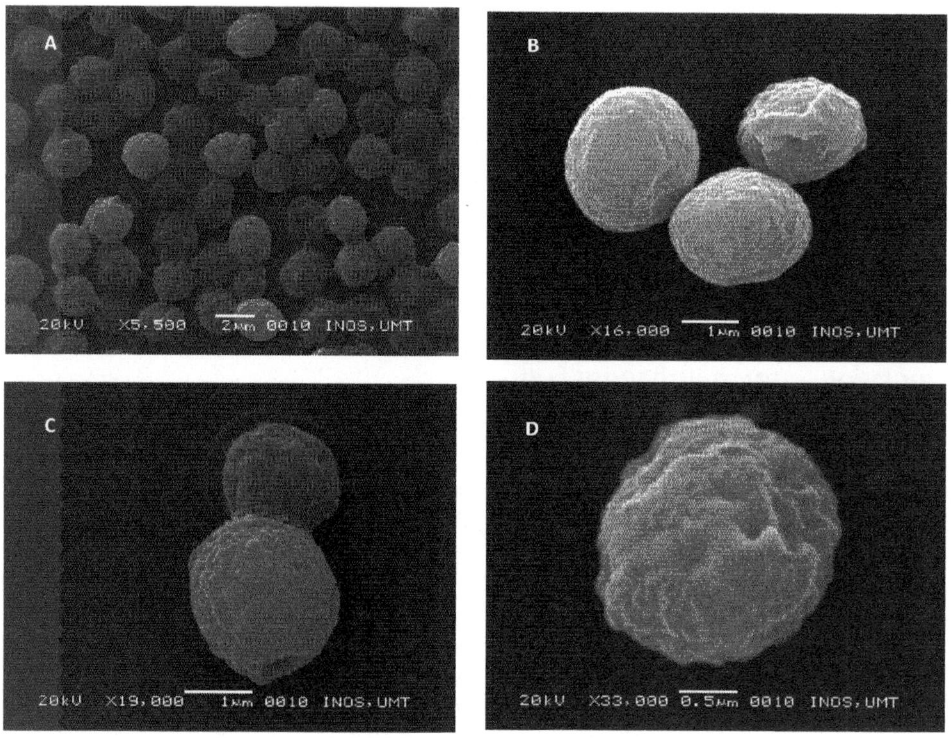

Figure 1. The scanning electron micrographs of *C.vulgaris* under different views and various magnifications. A, Cell aggregation. B, *C.vulgaris* cells under different magnification C. Cell divison in *C.vulgaris*. D, Single cell and cell wall structures.

The cell has an average cell biovolume of 5.26±0.87 µm^3. The cell densities changed following the culture period in both culture of concentrates (paste) and live condition (although they both started at the same density). Nonetheless, they followed more or less the same growth patten. The variation in cell densities during the experimental period is shown in Figure 2. Cell density of *C.vulgaris* increased rapidly to 227.22±0.87 x10^6 cells mL^{-1} prior to stationary phase. After that, the cell densities maintained at this point for ten days before decreasing significantly thereafter (Figure 2). The average specific growth rate (SGR, µ) achieved during the exponential phase was 0.660±0.001 day^{-1} with the doubling time (T) of 0.580±0.004 hour which then decreased drastically to 0.126±0.001 day^{-1} during the retardation phase with the doubling time of 2.420±0.019 hour before the death phase. Based on cell density and growth rate observed, the following growth phase is described:

- Exponential (log) phase (days 0-6),
- Declining of relative growth rate phase (days 6-12),
- Stationary phase (days 12-22),
- Death phase (days 22-26).

C.vulgaris paste was successfully concentrated from the pure culture isolated from Bidong Island. This concentrate contains cell density of approximately 58.46±2.44 x10^9 cells mL^{-1} - 227.22±0.82 x10^{6}cells mL^{-1}. The present result also showed that this *C.vulgaris* concentrate can still be inoculated after refrigeration for a duration of six weeks and exhibited similar growth characteristics as the live culture (Figure 2). The cells had very high viability even after 6 weeks of storage in chilling conditions (4ºC) as shown in Figure 3. It is interesting to note that the paste had recorded the highest cell viabilities of 99.51±0.57% and continued to display slow and steady decrement of cell viabilities to 83.28±0.58% on the sixth week of storage. Microscopic examination also indicated that the cells were in single forms without any aggregation occurring and can be readily dispersed in seawater medium as single suspension of cells upon inoculation (Figure 1A). The harvesting efficiency of the ultrafiltration technique using membrane filter had recorded a very high percentage recovery of 93.14±1.35% showing the effectiveness of this technique for harvesting and concentrating the microalgae biomass.

Copepod species, *Apocyclops* sp., showed a higher population density when fed with *C.vulgaris* paste (60±4.36 individual mL^{-1}) than with the live culture (14.33±0.58 individual mL^{-1}) (Figure 4). In addition, the copepod populations fed with this concentrate exhibited a higher instantaneous growth rate, K=0.455±0.008 day^{-1} and faster doubling time (0.662±0.012 hour) than live culture which recorded an instantaneous growth rate of 0.296±0.005 day^{-1} and doubling time of 1.108±0.016 hour. Results from one-way ANOVA test has shown that there is a significant difference ($P<0.005$) between the copepod population densities fed with the *C.vulgaris* concentrate and live culture.

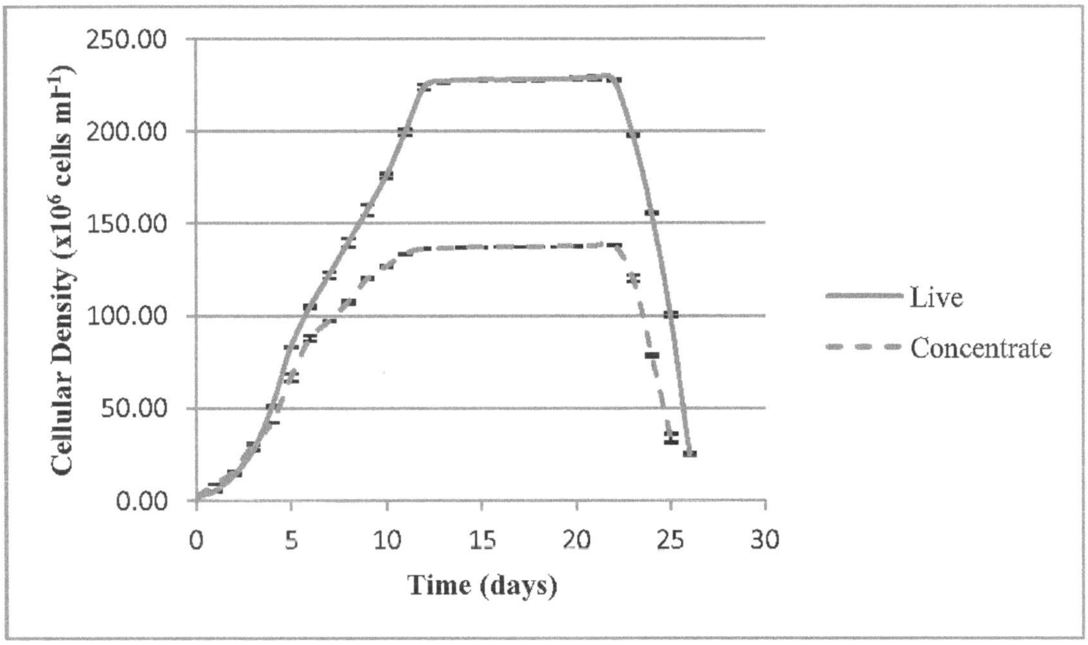

Figure 2. Cellular densities of live *C.vulgaris* (30L) and concentrate/paste (after reinoculation in 5L) cultured under laboratory conditions with Conway medium. Data are mean value and standard deviation of 3 repetitions.

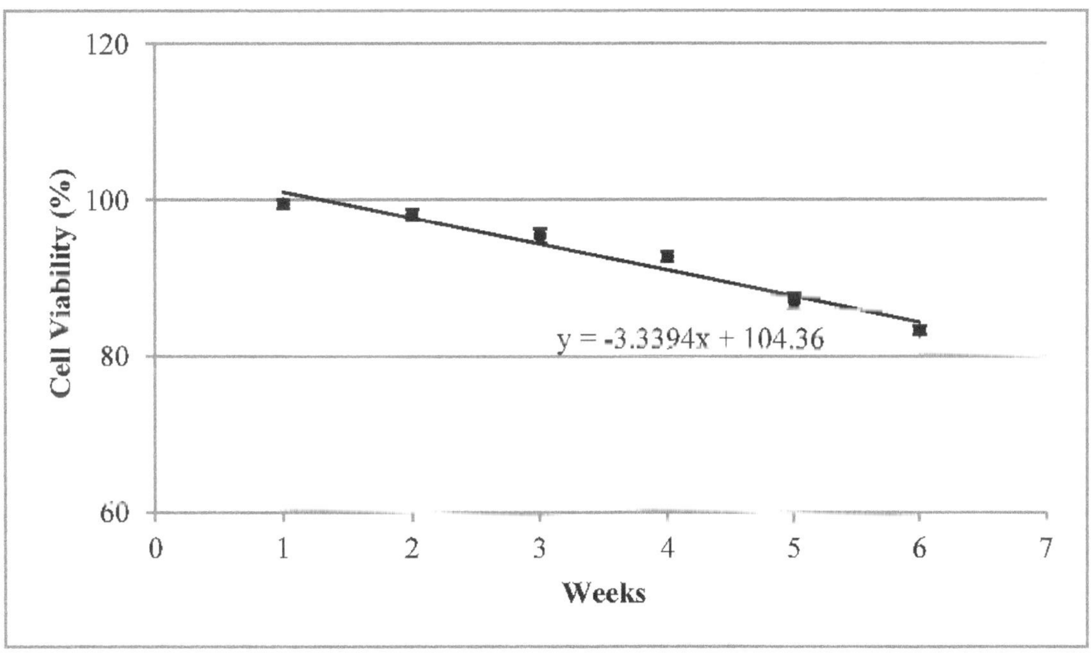

Figure 3. Variations in the cell viabilities of *C.vulgaris* microalgae concentrate over 6 weeks of storage in chilling condition at 4° C. Data are mean value and standard deviation of 10 repetitions.

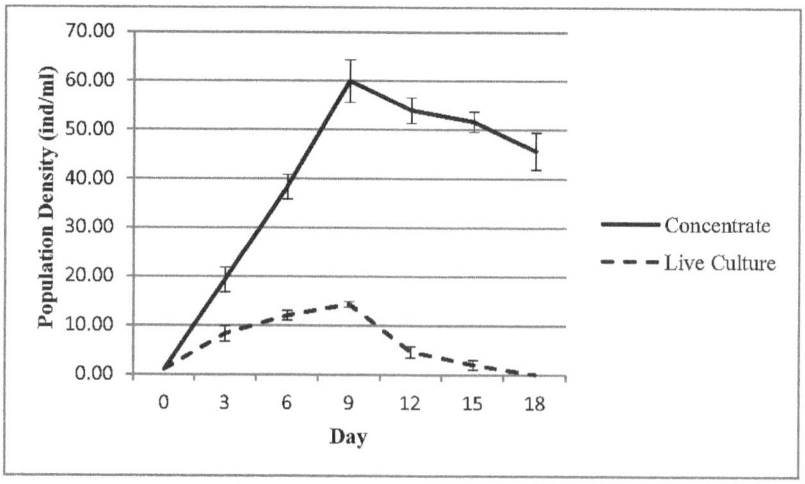

Data are mean value and standard deviation of 3 repetitions.

Figure 4. Population densities of *Apocyclops* sp. fed with live culture and microalgae concentrate of *C.vulgaris*.

C.vulgaris isolated from Bidong Island had exhibited an extremely rapid growth rate. This might be attributed to its relatively small cell volume with an average of 5.26±0.87 μm^3. This is indeed much smaller than the cell sizes recorded for some other species; *Chaetoceros calcitrans* (8μm diameter, volume 276.95 μm^3) and *Isochrysis galbana* (4μm diameter, volume 33.49 μm^3) [54]. Small size species of microalgae grow faster with a rapid growth rate. This is due to the reason that the greater surface-to-volume ratio of smaller size cells facilitates assimilation of nutrients at a relatively faster rate. In addition, the smaller size cells may achieve high density because they occupy less space. Apart from that, it was also cultured with optimal values for all environmental factors in the laboratory, thus promoting favourable environmental conditions for the cells to grow to extremely high density. In this high-density culture, the possibilities of contamination were excluded. The sudden collapse in the growth rate after day six could be mainly the result of the depletion of the nutrient in the culture. Growth rate declines and growth of microalgae ceases when the nutrient in shortest supply relative to the metabolic needs of algal population [27,55]. The populations of *C.vulgaris* cells then entered the stationary phase of the growth cycle and collapsed after day twenty two. The long stationary phase of this culture might indicate that contamination was absent during the culture period. It has been reported that this stationary phase can last for several weeks if there is no contamination in the culture [30].

The ultrafiltration technique which was used to concentrate the *C.vulgaris* cells in this study can be applied to concentrate a range of other microalgae species used as aquaculture feeds. Concentrating and storing the microalgae concentrate in moist form preserves its high nutritional value through maintaining excellent cell viability [33, 35]. The cells were readily re-suspended upon dilution in sea water with high cell viability which was proven by their ability to be inoculated even after storage for a duration of 6 weeks. The efficiency of ultrafiltration through this study was ≥90% which is very comparable to the reported efficiency of ≥80% for flocculation technique by Knuckey *et al.* [34]. There has been no comparative assessment of concentrates prepared by ultrafiltration with those prepared by centrifugation. However, from a practical and theoretical point of view, it is proven that the centrifugation method possesses some disadvantages due to its exertion of shear gravitational forces rupturing the microalgae cell structure during harvesting procedure. This reduces their nutritional values due to leaking of nutritional contents. On the other hand, microalgae concentrates prepared by ultrafiltration are not subjected to the same gravitational forces during harvesting. As reported earlier [38], the major production cost of centrifuged concentrates may exceed US$10,000 (RM32,620) which is unaffordable for small-scale hatcheries and is likely to be limited to larger hatcheries with specialised equipments or facilities specifically set up to produce microalgae concentrates to hatcheries. Advantages of the ultrafiltration technique used in this study is that it is a relatively simple, inexpensive and volume-independent process which can be readily adopted by small-scale hatcheries to prepare their own microalgae concentrates on site.

The use of *C.vulgaris* concentrate as diet for cyclopoid copepods increases population density, instantaneous growth rate as well as doubling time and it was proven as a better diet than the *C.vulgaris* live cultures. This might be possibly due to the significantly higher cellular density of the microalgae concentrates. The rates of ingestion and egg production in copepods are dependent on the quantity of the provided microalgae [56, 57] implying that quantity of food is the most important factor regulating the productivity of copepod culture. Other studies have also demonstrated that the rate of egg production of calanoid copepod, *Acartia tonsa,* increases with increasing food concentrations [58, 59]. Essential substances such as cholesterol, HUFA and PUFA are present or exist abundantly in microalgae, and, copepod production is positively related to the lipid levels or DHA: EPA ratio in the diet [60]. Thus, microalgae concentrate could be useful as a replacement for live or fresh microalgae. This is extremely important as a stable and continuous supply of live feed for aquaculture hatcheries must always be provided.

Experiment 2: Effects of Photoperiod and Culture Size on *C.vulgaris* Stock Growth

It is very important for hatcheries to be able to maintain the stock for microalgae for their sustainable live-feeds supply. Batch cultures need to be maintained under optimal environmental conditions and in a suitable culture vessel which will not affect the cell density and quality. Comparison on the effect of photoperiod and culture sizes between *C. vulgaris* and other microalgae was made to investigate the adaptability of the species to simple stock handling in the laboratory or hatchery conditions. No significant difference in the cell density was noted in *Nannochloropsis* sp. (Figure 5A) and *C.vulgaris*(Figure 5C) cultures grown in different volume flasks while *Chaetoceros* sp. Figure 5B showed significant variation in cell-density level at similar culture conditions ($P < 0.001$). However, stationary phases of all cultured species were achieved earlier in 250mL flask compared to the cultures in 500mL and 2L flasks. All cultures showed greater response towards daylight variations whereby higher cell density was noted in culture flasks exposed to continued illumination (24:0 L/D), and it was followed by 12:12 L/D and 6:18 L/D condition. *Nannochloropsis* sp. (Figure 5D) responded less towards the treatment compared to the other 2 species (Figure 5E & F) which clearly showed a specific response towards culture conditions.

It is well-documented that, in natural conditions, microalgae growth is not curtailed by ambient environmental conditions because the growth rate is just enough for species survival. However, their multiplication rate is highly influenced by various environmental parameters. In an *In vitro* setup, the proper maintenance of optimum culture condition triggers the metabolic pathway of target species in a unidirectional fashion to achieve high cell density. In this study, a higher cell density of microalgae in low volume flask culture together with early stationary phase was observed could be used to obtain continuous harvest of selected microalgae.

Highest cell density and specific growth rate were recorded in selected species cultured in 250mL culture flask compared to the cultures in 500mL and 2000mL flasks (Table 2). The highest cell density was achieved during the end of the log phase. Cell density of early stationary phase, which is the end of the log phase for *Nannochloropsis* sp., *Chaetoceros* sp. and *C.vulgaris* was achieved on the 10th day of culture in 250mL. After the 10th day, density decreased and the lowest level was different for different species and different culture volumes. Similar results were also noted for the specific growth rate values. Significant variation in cell density and specific growth rate was observed between the cultures in different sizes of culture flask ($P < 0.05$ or $P < 0.001$). This observation might probably be due to the light-penetration efficiency in the culture flask. Similar observation was noted for the culture of *Nannochloropsis* sp. in 2000mL flasks which produced greater cell density compared to the culture in 20L carboys [61]. The effect of light saturation could decrease in the denser culture and the average irradiance in the culture reduced due to absorption from other cells [62]. The large volume culture needed higher light intensity to allow light penetration while the smaller volumes were less affected by the light penetration.

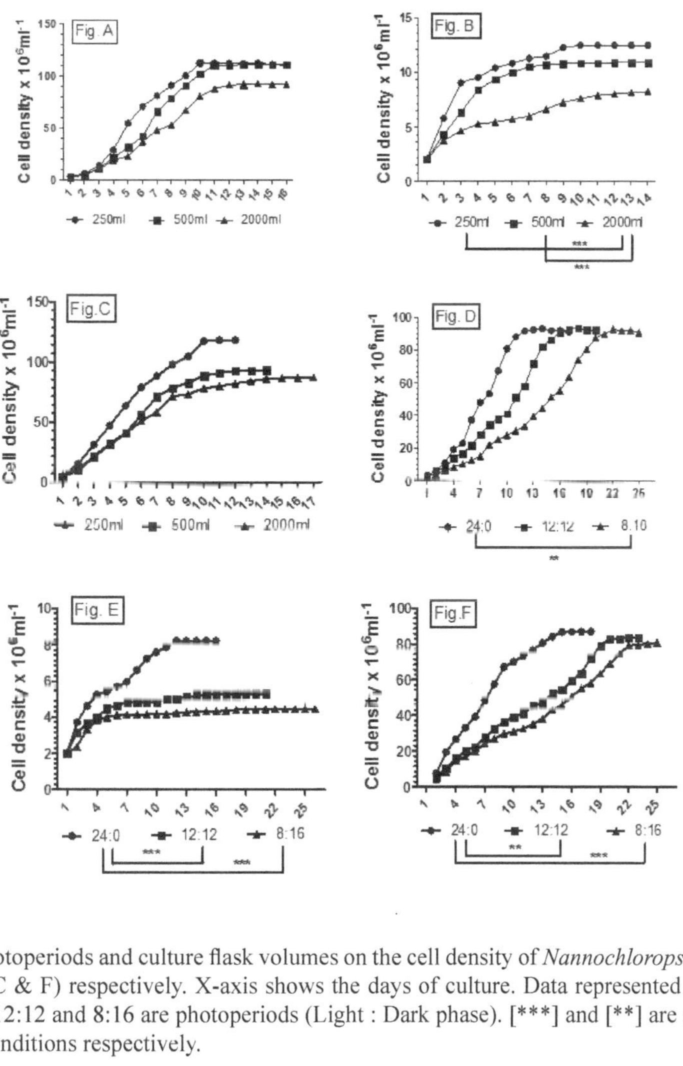

Figure 5. Influence of different photoperiods and culture flask volumes on the cell density of *Nannochloropsis*sp. (Fig A & D), *Chaetoceros* sp. (Fig B & E) and *C.vulgaris* (Fig C & F) respectively. X-axis shows the days of culture. Data represented as 250ml, 500ml and 2000ml are culture flask volumes while 24:0, 12:12 and 8:16 are photoperiods (Light : Dark phase). [***] and [**] are significantly different at $P < 0.001$ and $P < 0.05$ level in the culture conditions respectively.

Table 2. Cell density and specific growth rate of selected microalgae cultured at different flask volumes.

Microalgae	Flask volume (mL)	Early stationary phase	Cell density ($x10^6$cells mL^{-1})	Specific growth rate (K')
Nannochloropsissp.	250	Day 10	112.5 ± 2.3^{6}a	0.52 ± 0.0^{1}a
	500	Day 11	110.17 ± 1.7^{7}a	0.46 ± 0.0^{4}b
	2000	Day 13	92.2 ± 0.8^{7}a	$0.34 \pm 0.0^{1c*\,(ac)}$
Chaetoceros sp.	250	Day 10	12.460 ± 0.01^{8}a	0.203 ± 0.00^{2}a
	500	Day 12	$10.889 \pm 0.01^{3b**\,(ab)}$	$0.145 \pm 0.00^{1b*\,(ab)}$
	2000	Day 14	$8.225 \pm 0.00^{1c**\,(ac\,\&\,bc)}$	$0.037 \pm 0.00^{3c**\,(ac),\,*\,(bc)}$
C.vulgaris	250	Day 10	117.53 ± 0.8^{4}a	0.4749 ± 0.000^{7}a
	500	Day 11	91.0 ± 0.5^{5}a	0.4081 ± 0.000^{2}b
	2000	Day 14	86.13 ± 0.8^{1}a	$0.3166 \pm 0.000^{7c*\,(ac)}$

[i] - *Note*: Data represented in Mean ± SD. [*] and [**] indicates significant difference at $P < 0.05$ and $P < 0.001$(respectively) level between different superscripts depicted for each species.

Table 3. Cell density and specific growth rate of selected microalgae cultured at different photo periods.

Microalgae	Photo period (Light : Dark phase) in hours	Early stationary phase	Cell density ($x10^6$cells ml^{-1})		Specific growth rate (K')
Nannochloropsissp.	24:0	Day 13	112.5 ± 2.3^{6}a		0.34 ± 0.0^{1}a
	12:12	Day 17	110.17 ± 1.7^{7}a		0.25 ± 0.0^{1}b
	8:16	Day 23	92.2 ± 0.8^{7}a		$0.19 \pm 0.0^{2c*\,(ac)}$
Chaetoceros sp.	24:0	Day 12	8.225 ± 0.00^{1}a		0.129 ± 0.00^{3}a
	12:12	Day 17	$5.293 \pm 0.00^{9b**\,(ab)}$		$0.061 \pm 0.00^{2b*\,(ab)}$
	8:16	Day 22	$4.453 \pm \pm 0.00^{3c**\,(ac)}$		$0.037 \pm 0.00^{3c**\,(ac)}$
C.vulgaris	24:0	Day 14	86.60 ± 0.1^{7}a		0.3170 ± 0.000^{1}a
	12:12	Day 19	$83.04 \pm 0.1^{9b**\,(ab)}$		0.2313 ± 0.000^{1}b
	8:16	Day 21	$79.23 \pm 0.2^{1c**(ac)}$		$0.2010 \pm 0.000^{1c*\,(ac)}$

Note: Data represented in Mean ± SD. [*] and [**] indicates significant difference at $P < 0.05$ and $P < 0.001$ (respectively) level between different superscripts depicted for each species [i] -

Highest cell density and specific growth rate were recorded in all cultured species that were exposed to continued illumination (24:0. L/D) followed by 12:12 and 6:18 L/D respectively. Early stationary phases differed for *Nannochloropsis* sp. (day 13), *Chaetoceros* sp. (day 12) and *C.vulgaris*. (day 14) respectively while the corresponding specific growth rate was also highest under 24hours illumination. Significant variation in both growth parameters was observed between the cultures exposed to different photoperiods ($P < 0.05$ or $P < 0.001$) (Table 3). Photosynthetic efficiency of microalgae can be enhanced by sudden alteration between light and dark phase [63]. During this process, the fast reduction of e-acceptors, Qa and Qb, associated to photosystem II (PSII) followed by their oxidation in the dark period will take place that will ultimately maximise the proton-accepting capacity of PSII during sudden irradiant of light [64].

C.vulgaris proved its adaptability to different culture volumes and lighting periods with good growth performance comparable to *Nannochloropsis* sp. and better than *Chaetoceros* sp. The cells responded positively towards continuous illumination of light by producing higher cell density and specific growth rate in the culture media. It was also noted that the culture in the low-volume flask produced an early stationary phase due to high penetration of light and continuous sharing of available nutrients in the media for faster growth and survival. On the other hand, *C. vulgaris* consistently grew at significant cell densities even in larger volume containers and shorter period of illumination than dark condition (comparable to *Nannochloropsis* sp. and better than *Chaetoceros* sp.). In another study to analyse the effect of photoperiod to the cellular essential fatty acid in these species, the photoperiod of 12:12h L/D regime is recommended for the fast and economical technique for batch culture production [65]. A better ratio of essential fatty acid accumulated in *C. vulgaris* exposed in the 12:12h if compared to 24:0 or 8:16 L/D photoperiod.

Experiment 3: Low-cost Commercial Fertiliser for Mass Culture of Marine *Chlorella vulgaris*: Manipulation of N:P:K Ratio

C.vulgaris showed its adaptability to grow well when fertilised with a low-cost commercial N:P:K plant fertiliser (Figure 6). Duration of the log phase for *C.vulgaris* varied among treatments. The 12:6:4 and 12:8:4 ratios had a result of 3 days while the longest period was in the 15:15:15 treatment. Combined applications of urea, P+ and K+ (N:P:K; 12:6:4) produced the highest cell number (4.0×10^6cells mL^{-1}) during log period at 5 days while N:P:K; 15:15:15 (control) produced highest cell number (4.16×10^6 cells mL^{-1}) at 7 days of log period. Different ratios of N:P:K, 12:8:4, 8:8:2 and 16:4:6 resulted in decrease of cell density, 3.3×10^6, 3.0×10^6 and 2.7×10^6 cells mL^{-1}, respectively.

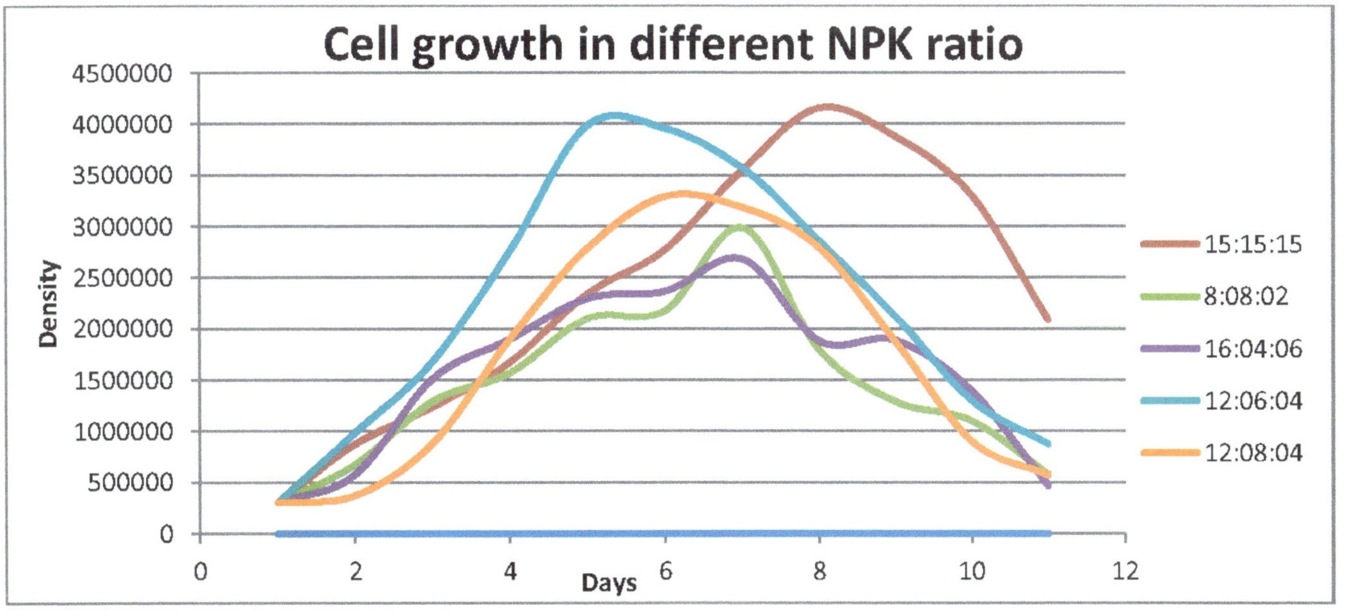

Figure 6. Density (cells mL^{-1}) of *C.vulgaris* cultured with commercial fertiliser of different N:P:K ratios

Table 4. Growth rate of *C.vulgaris* cultured with commercial fertiliser of different N:P:K ratios

		Days in Log Phase					
NPK ratio	2nd	3rd	4th	5th	6th	7th	
		Growth rate; K'					Average
15:15:15	1.07	0.35	0.30	0.34	0.17	0.25	0.41
8:08:02	0.81	0.65	0.20	0.29			0.49
16:04:06	0.65	0.96	0.23	0.19			0.51
12:06:04	1.19	0.53	0.50				0.74
12:08:04	0.20	0.87	0.77				0.61

Table 5. Division per day of *C.vulgaris* cultured with commercial fertiliser of different N:P:K ratios

		Days in Log Phase					
NPK ratio	2nd	3rd	4th	5th	6th	7th	
		Division/ Day; Div. day-1					Average
15:15:15	1.54	0.50	0.44	0.48	0.25	0.36	0.59
8:08:02	1.16	0.94	0.28	0.42			0.70
16:04:06	0.94	1.39	0.33	0.28			0.73
12:06:04	1.72	0.77	0.72				1.07
12:08:04	0.29	1.25	1.11				0.89

The 12:6:4 ratios showed the best average (74%) growth rates of natural increase at log phase. The second was 12:8:4 with 61% average growth rate.15:15:15 NPK ratio showed the lowest average growth rate of 41% (Table 4). The *C.vulgaris* cell in 12:6:4 NPK ratio recorded an average division per day by 107% which was the best compared to others. In 12:8:4 ratios the average cell division was 89% and decreasingly followed by 16:4:6 and 8:8:2 for 73% and 70% respectively. The control ratio which was 15:15:15 showed the lowest average division which was 59% (Table 5).

Table 6. Generation time (days) for *C.vulgaris* cultured with commercial fertiliser of different N:P:K ratios

		Days in Log Phase					
NPK ratio	**2nd**	3rd	4th	5th	6th	7th	
		Generation time (days); Gen't					**Average**
15:15:15	0.65	2.00	2.28	2.06	4.07	2.80	2.31
8:08:02	0.86	1.06	3.55	2.38			1.96
16:04:06	1.06	0.72	3.04	3.62			2.11
12:06:04	0.58	1.31	1.39				1.09
12:08:04	3.41	0.80	0.90				1.70

Table 7. Generation time (hour) for *C.vulgaris* cultured with commercial fertiliser of different N:P:K ratios

		Days in Log Phase					
NPK ratio	**2nd**	3rd	4th	5th	6th	7th	
		Generation time (Hour); Gen't					**Average**
15:15:15	15.56	47.99	54.84	49.55	97.66	67.32	55.49
8:08:02	20.60	25.55	85.22	57.05			47.11
16:04:06	25.44	17.28	73.01	86.79			50.63
12:06:04	13.97	31.34	33.37				26.22
12:08:04	81.76	19.14	21.54				40.82

Measurement of generation time for *C.vulgaris* is summarised in Table 6 and Table 7. *C.vulgaris*cultured with fertiliser of the ratio 12:6:4 only took 1.09 days (26.22 hour) to complete one generation of replication, the shortest time compared to other treatments. The longest generation time was when using the15:15:15 ratio which was completed in 2.31 days (55.49 hour). The other three intermediate treatments recorded 1.70 (40.82 hour), 1.96 (47.11 hour) and 2.1 days (50.63 hour) for 12:8:4, 8:8:2 and 16:4:6 respectively. When comparing the performance by using all of the growth parameters, N:P:K; 12:6:4 ratio gave the best result with average growth rate per day (74%), maximum growth rate day^{-1} (107%), maximum cell density (4.0×10^{6} cell/mL), division's day^{-1} (107%) and generation time (1.09 day; 26.22 hour). *C.vulgaris* in control treatment (15:15:15) exhibited the poorest growth performance. Nonetheless, it is interesting to note that they experienced longer log period which could give more time for reproduction activity, thus the density did not decrease drastically as when cultured using other ratios. The fluctuation of temperature and different salinities could be the reason why cell densities were not as high as the first and second experiment.

Numerous nutrient media have been use for the culture of pure *Chorella* sp. Most of those were for laboratory use and/or for low-grade production of algae. Majority of these media are composed of pure nutrients (N-8). Commercial fertilisers are least considered for *Chlorella* culture because of the conception that they do not provide required nutrients for algal growth and are mostly suitable for crop (land) agriculture. Nevertheless, it has been proved that the commercial plant fertiliser could support a freshwater *Chlorella* [66]. The use of N:P:K fertiliser could be a better choice if compared to the organic fertiliser. Organic matter has its own limitations and depends on the microbial activity to release the inorganic nutrient and it cannot be compared to the performance of pure nutrients. Despite good growth performance, the short period of the log phase when *C.vulgaris* is cultured using N:P:K; 12:6:4 need specific and efficient up-scaling or harvesting method, indicating that other ratios such as 15:15:15 could be a better choice.

Experiment 4: Egg Production, Growth and Development of *Apocyclops* ramkhamhaengi Fed on*C.vulgaris*

Different diets gives significantly ($P<0.05$) different densities of *A. ramkhamhaengi*. The mean gravid production of *A. ramkhamhaengi* fed on *C.vulgaris* and Baker's yeast was highest on 23rd day with 1.11ind./ml and 0.67ind./ml respectively. The production peaked on 11th, 23rd, and 26th day and on 17th, 20th and 23rd day when fed with *C.vulgaris*. and Baker's yeast respectively (Figure 7). In this 30 days culture, the highest mean population density of *A. ramkhamhaengi* fed with *C.vulgaris* was recorded on the 9th day with 3.31ind./ml and when fed on Baker's yeast was on day 19th with 1.83ind./ml (Figure 8). *A. ramkhamhaengi* fed on *C.vulgaris* showed the higher instantaneous growth rate (K) than when fed on Baker's yeast (Table 8). The period taken to double their population (Dt) was shorter in *A. ramkhamhaengi* fed *C.vulgaris* (8days) than Baker's yeast (11days).

Table 8. The instantaneous growth rate and doubling time of *A. ramkhamhaengi* fed on different diets.

Diets	Instantaneous growth rate (K)	Doubling Time(Dt)(day)
C.vulgaris	0.1150	8
Baker's yeast	0.0756	11

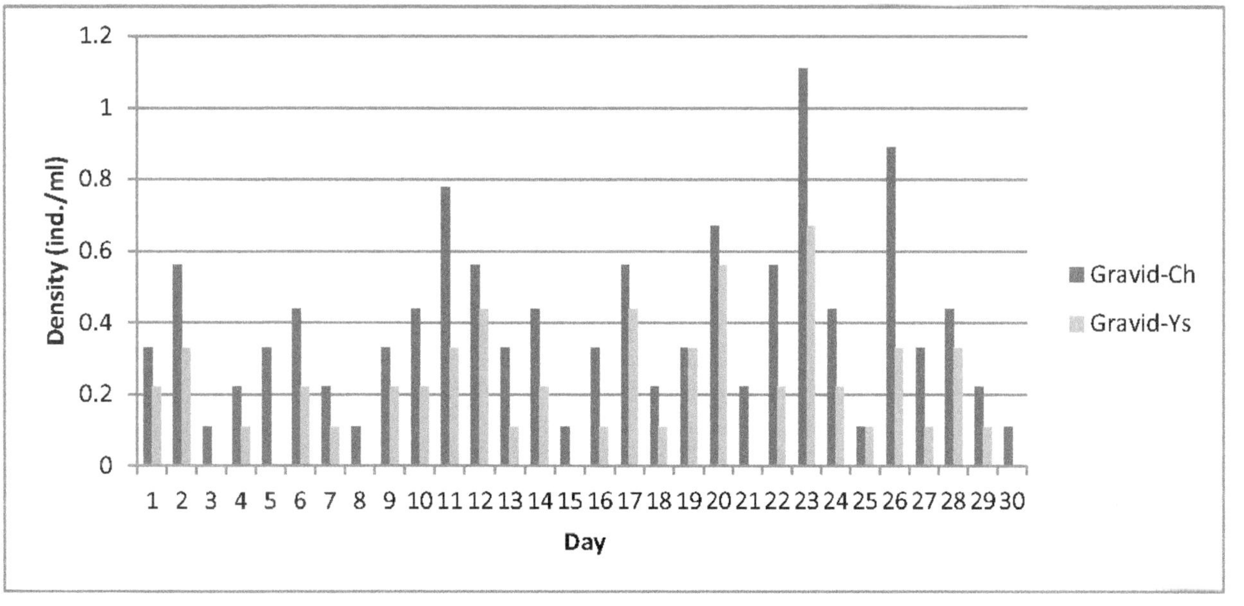

Figure 7. Mean density of gravid female of *A. ramkhamhaengi* fed with *C.vulgaris* (Gravid-Ch) and Yeast (Gravid-Ys) in 30 days.

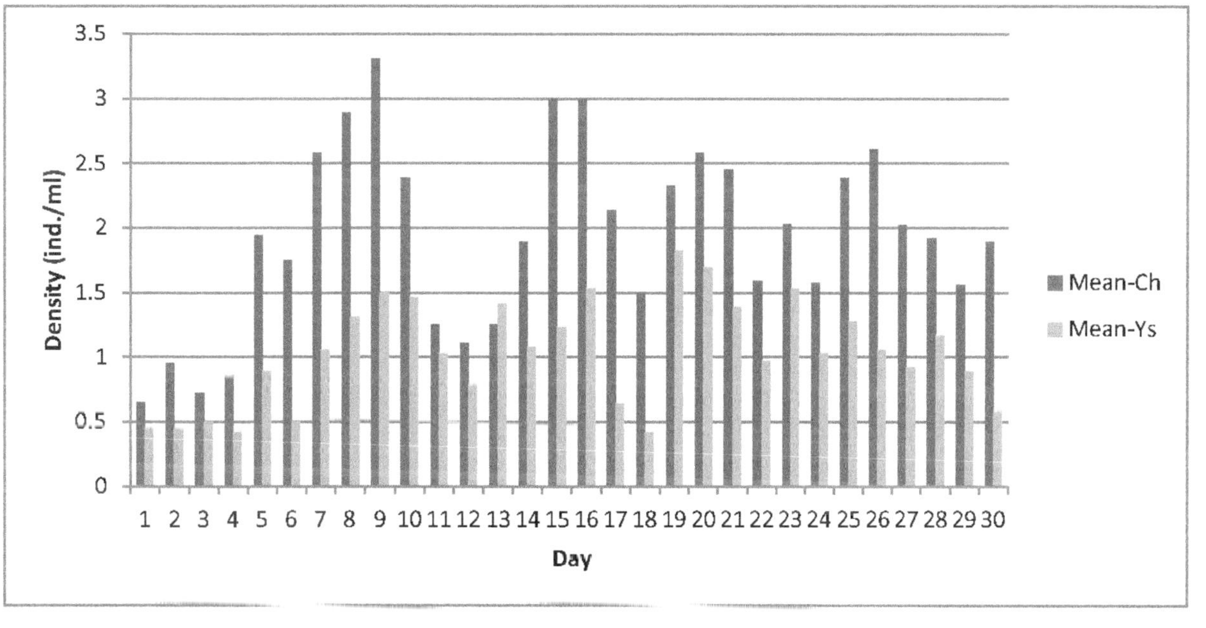

Figure 8. Mean total density of *A. ramkhamhaengi* fed with *C.vulgaris* (Mean-Ch) and Yeast (Mean-Ys) in 30 days.

The development times for nauplii, copepodite, adult and gravid female were observed separately using the copepod culture fed on *C.vulgaris*. The longest period was at copepodite stage (7.33 ± 2.08days) and the shortest period was the naupliar stage which needed only 1.33 ± 0.58 days (Figure 9). The mean number of eggs produced was 21.33 ± 1.53. Hatching percentage of the three individuals of *A. ramkhamhaengi* was 96.82 ± 2.77 % (Table 9). Maturation time which is the time between the appearance of eggs and their hatching time was 1.33 ± 0.58 days. The time taken to become gravid female from the produced nauplii was about 20.67 ± 3.51days and it is known as generation time.

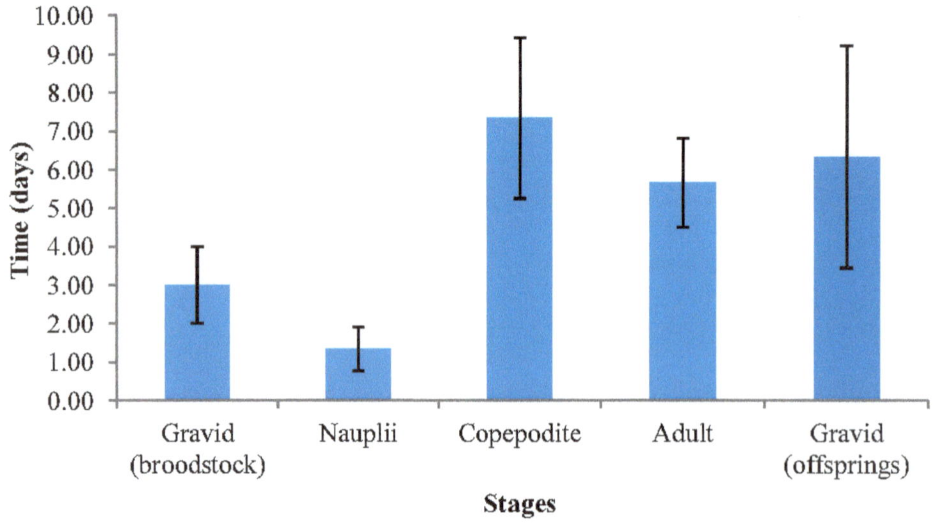

Figure 9. Mean development time in each stage in *A. ramkhamhaengi* life cycle

Table 9. The number of observation (N), mean and standard deviation (SD), minimum and maximum of total number of eggs, percentage of hatching of *A. ramkhamhaengi*, maturation time and generation reared in laboratory-controlled condition.

Parameter	N	Mean ± SD	Mini-mum	Maxi-mum
Number of eggs	3	21.33 ± 1.15	20	22
% hatching	3	96.82 ± 2.77	95	100
Maturation time(days)	3	1.33 ± 0.58	1	2
Generation time(days)	3	20.67 ± 3.51	17	24

Production of the gravid females and the population density obviously increased when *A.ramkhamhaengi* was fed on *C.vulgaris* as compared to Baker's yeast. This finding is in agreement with the previous study [20] on the population growth and production of *A. dengizicus* fed on different diets. Population reached its peak in term of total density for several periods in the 30day culture condition indicates the existence of different populations. These populations reached their peak density in accordance with the diet taken where *Chlorella*-fed population were found to grow faster than those fed on Baker's yeast. It seemed that *A. ramkhamhaengi* has the potential to become more nutritional when enriched, thus growing faster when fed on a good-quality diet such as microalgae if compared to Baker's yeast. The nutritional value has been shown to increase when cyclopoid nauplii stage such as in*A. panamensis* was offered an enriched diet [15]. Although other microalgal diets such as *Tetraselmis*sp. and *Isochrysis* sp. could be a better choice for *Chlorella* sp. [67], at least the present finding is able to prove the potential of the species to reproduce and grow when fed on the marine *C.vulgaris* as used in this study.

A female of *A. ramkhamhaengi* fed on *C.vulgaris* could produce between 20 and 22 eggs with about 97% hatching success. This is more than what has been reported before for *A. panamensis* [68], and it could be related to many environmental factors and culture procedure. Environmental parameters such as temperature, food availability and predation were reported to influence the life-history strategy in copepods [69]. Binary diet of *Nannochloropsis* sp. and T-ISO improved the hatching rate by 88.1 ±2.1 % in a calanoid copepod, *Acartia sinjiensis* [70]. The brackish water cyclopoid, *A. ramkhamhaengi* has shown its potential to be cultured and reproduced under controlled conditions. The population adapted very well to the introduced diet, a marine *C.vulgaris* and a common Baker's yeast. *Chlorella*-fed population of *A. ramkhamhaengi* grow faster and need fewer number of days to double its population than those fed on Baker's yeast.The number of eggs produced was 21.33 ± 1.15 eggs at the maturation time of 1.33 ± 0.58 days and generation time was 20.67 ± 3.51 days. The species show great potential to be cultured together with *C.vulgaris* for hatchery and farm use. A more comprehensive study is essential to investigate the reproductive biology of this species, particularly in a large-scale production system to verify its suitability in aquaculture.

CONCLUSION

C.vulgaris isolated from Bidong Island exhibited a rapid growth rate under optimum environmental conditions in the laboratory culture and was able to achieve an extremely high density when cultured in bigger containers. The photoperiod of 24:0 proved to be the best condition for cells growth but 12:12 L/D photoperiod could be the more economical. The high-density culture could be harvested using a relatively cheap, inexpensive and simple ultrafiltration technique for other use or reinoculation. This will save the space and long period of maintaining live algae for unexpected use. The cells collected using a ultrafiltration technique showed high viability and long shelf life when kept in 4°C refrigerator. The product is called as *C.vulgaris* paste or concentrate which could be used to enrich or maintain the zooplankton live feeds for aquaculture purposes. The *C.vulgaris* also showed its best growth performance when cultured using a common commercial plant fertiliser with certain ratio of N:P:K. This was shown by their ability to perform cell division and grow and easily adapted to certain ratio such as 15:15:15 and 12:6:4. Nonetheless, the cells density is very much lower than those cultured with the specific chemical fertilizer, Conway media. This problem could be overcome by further investigation on their ion requirement when cultured openly in hatchery or ponds. The suitability of C.vulgaris as enriched diet for a zooplankton potentially used as live feed, A. ramkhamhaengi was proved by the population increase and reached high individual density with good reproduction performance. Maintaining local species of microalgae and zooplanktons in hatchery and ponds will definitely support the continued supply of live feeds for larval rearing and the aquaculture industry.

ACKNOWLEDGEMENTS

The authors are grateful to the Ministry of Science, Technology and Innovation (MOSTI) for the funding given under the ABI-MOSTI grant (2009-2011), "Mass fry production technology for grouper (Epinephelus sp.)", the Ministry of Education Malaysia for the Knowledge Transfer Programme grant (2012-2013) Using Microalgae and Copepod Live Feeds for Brackish Water Aquaculture Farm, UMT-PPKJBS.

REFERENCES

1. Sorgeloos, P.The use of brine shrimp Artemia in Aquaculture. The brine shrimp, Artemia. 1980. Proceedings the International Symposium on the Brine shrimp, *Artemia salina*. Universa Press, Wetteren, Belgium. Pp. 25- 46.
2. Leger, P., Bengtson, D. A., Simpson, K. L. & Sorgeloos, P. The use and nutritional value of Artemia as a food source. Marine biology and oceanography: an annual review 1986; 24, 521–623
3. Chu, K. H. & C. K. Shing. Feeding behaviour of the shrimp, *Metapenaus ensis*, on Artemia nauplii. Aquaculture 1986; 58, 175-184.
4. Kuhlmann, D. G., Quantz and Witt, U. Rearing of turbot larvae (*scopthalmus maximus*) on cultured food organism and postmetamorphosis growth on natural and artificial food. Aquaculture 1981; 23. 183-196
5. Watanabe, T., and Kiron, V. 1994. Prospects in larval fish dietetics. Aquaculture 1994; 124: 223-251.
6. McKinnon, A.D., Duggan,S., Nichols, P.D., Rimmer, M.A., Semmens, G. and Robino, B.The potential of tropical paracalanid copepods as live feeds in aquaculture. Aquaculture 2003; 223(1-4), 89-106.
7. Stottrup, J.G. Review on status and progress in rearing copepods for marine larviculture and advantages and disadvantages among calanoid, harpacticoid and cyclopoids copepods. 2006. VIII Symposium Internacional Acuicola 15 17 Noviembre, Universidad Autonoma de Nuevo Leon, Mexico.
8. Rajkumar & Kumaraguru vasagam. Suitability of the copepod, *Acartia clausi* as a live feed for Seabass larvae (*Lates calcarifer* Bloch): Compared to traditional live-food organisms with special emphasis on the nutritional value. Aquaculture 2006; 261(2), 649-658.
9. Schipp, G. 2006. The use of Calanoid Copepods in semi-intensive, tropical marine fish larviculture. VII Simposium Internacional de Nutricion Acuicola. 15-17 Noviembre. Universidad Autonoma de Nuevo Leon, Monterrey, Nuevo Leon Mexico.
10. Zaleha,K., Ibrahim, B., Akhbar John, B., Kamaruzzaman, B.Y. Generation times of some marine harpacticoid species in laboratory condition. Journal of Biological Sciences 2012; 12(8), 426-432. DOI: 10.3923/jbs.2012.433.437
11. Carli, A., Mariottini, G. L. & Pane, L. Influence of nutrition on fecundity and survival in *Tigriopus fulvus* Fischer (Copepoda, Harpacticoida). Aquaculture 1995; 134,113-119.
12. Cutts, C.J. Culture of harpacticoid copepods: potential as live feed for rearing marine fish. Advances in Marine Biology 2003; 44:295-316.
13. Ananth, S. and Santhanam, P. Laboratory culture and biochemical profile of marine copepod *Macrosetella gracilis* (Dana). Aquaculture 2011; 12(1): 49-55.
14. Rhodes. A. 2003. Methods for high density batch culture of *Nitokra lacustris*, a marine harpacticoid copepod.The Big Fish Bang. In: Howard I. Browman, A., Skiftesvik, B. (eds.) Proceedings of the 26th Annual Larval Fish Conference. Institute of Marine Research, Bergen, Norway.
15. Sumiarsa, G.S. Cyclopoid copepod nauplii Apocyclops panamensis: mass production method in outdoor ponds and their fatty acid profiles.VDM Verlag Dr.muller GmhH & Co. KG. Germany. 2011.
16. Drillet,G., Frouël,S., Sichlau, M.H., Jepsen, P.M., Jonas K., Højgaard, J.K., Joarder,A.K., Hansen, B.W. Status and recommendations on marine copepod cultivation for use as live feed. Aquaculture 2011; 315:155-166
17. Zaleha, K., Farahiyah, I. J. Culture and growth of a marine harpacticoid, *Pararobertsonia* sp in different salinity and temperature. Sains Malaysiana 2010; 39(1): 135–140
18. Kassim, Z.,Ibrahim, B. Culture of Harpacticoid Copepods: Understanding the Reproduction and Effect of Environmental Factors. In: Muchlisin, Z. (ed.) Aquaculture. Rijeka: Intech; 2012.p343-360.
19. Rhodes. A. Methods for high density batch culture of *Nitokra lacustris*, a marine harpacticoid copepod.The Big Fish Bang. In: Browman, H.I., Skiftesvik, A.B. (eds.). Proceedings of the 26th Annual Larval Fish Conference. Institute of Marine Research, Postboks 1870 Nordnes, N-5817, Bergen, Norway. ISBN 82-7461-059-8. 2003
20. Farhadian, O., Yusoff, M.F., Arshad, A. Population growth and production of *Apocyclops dengizicus*(Copepoda: Cyclopoida) Fed on Different Diets. Journal

of the World Aquaculture Society, 2008; 39(2), 384-396. DOI: 10.1111/j.1749-7345.2008.00172.x

21. Chullasorn, S., Kangtia, P., Pinkaew, K.,Ferrari, F.D. *Apocyclops ramkhamhaengi* sp. nov. (Copepoda: Cyclopoida) in a Culture Originating from Brackish Waters of Chang Island, Trat Province,Thailand. Zoological Studies 2008; 47(3),326-337.

22. Apt, K. E., Behrens, P. W. 1999. Commercial developments in microalgal biotechnology. *Journal of Phycology, 35*(2), 215-226. doi: 10.1046/j.1529-8817.1999.3520215.x

23. Geudes, A.C., Malcata, F.X. Nutritional values and uses of microalgae in aquaculture. In: Muchlisin, ZA (Ed); InTech Open Acess Publisher

24. Muller-Feuga, A., Moal, J. and Kaas, R. 2003. The microalgae of aquaculture. In: Stottrup, J.G and Mcevoy, L.A (eds.) Live Feeds in Marine Aquaculture. Blackwell Publishing Ltd., USA, p206-252.

25. Mata, T.M., Martins, A.A., Caetano, N.S. 2010. Microalgae for biodiesel production and other applications: a review. Renewal Sustainable Energy Review 2010; 14 : 217-232.

26. Chen, C.Y., Yeh, K.L., Aisyah, R., Lee, D.J. and Chang, J.S. 2011. Cultivation, photobioreactor design and harvesting of microalgae for biodiesel production: a critical review. *Bioresource Technology*, 102, pp. 71-81.

27. Lee, Y.K. and Shen, H. 2004. Basic culturing techniques for microalgae. In: *Handbook of Microalgal Culture, Biotechnology and Applied Phycology* (ed. A. Richmond). Blackwell Publishing Ltd., USA, pp. 40-56.

28. Chen, C.Y., Yeh, K.L., Aisyah, R., Lee, D.J. and Chang, J.S. 2011. Cultivation, photobioreactor design and harvesting of microalgae for biodiesel production: a critical review. *Bioresource Technology*, 102, pp. 71-81.

29. Becker, E.W., 1995. Microalgae biotechnology and microbiology. *Cambridge University Press*, Great Britain, pp.56- 62.

30. Borgne, Y.L. 1996. Culture of microalgae. In: *Aquaculture* (ed. G. Barnabe). *Ellis Horwood Ltd*., England, pp. 197-206.

31. Grobbelaar, J.U., Nedbal, L. and Tichy, V. 1996. Influence of high frequency light/dark fluctuations on photosynthetic characteristics of microalgae photoacclimated to different light intensities and implications for mass algal cultivation. *Journal of Applied Phycology*, 8, pp. 335-343.

32. Meseck S.,L., Alix J.,H., Gary, H., Wikfors G.,H., 2005. Photoperiod and light intensity effects on growth and utilization of nutrients by the aquaculture feed microalga Tetraselmis chui (PLY429). *Aquaculture* 246: 393–404.

33. Heasman, M., Diemar, J., O'Connor, W., Sushames, T. and Foulkes, L. 2000. Development of extended shelf-life microalgae concentrate diets harvested by centrifugation for bivalve mollusks- a summary. *Aquacult. Res*., 31, pp. 637-659.

34. Knuckey, R., Brown, M., Robert, R. and Frampton D. 2006. Production of microalgal concentrates by flocculation and their assessment as aquaculture feeds. *Aquacult. Eng*., 35, pp. 300-313.

35. Montaini, E., Zitelli, G.C., Tredici, M.R., Molina Grima, E.M., Fernandez Sevilla, J.M. and Sanchez Perez, J.A. 1995. Long-term preservation of*Tetraselmis suecica*: influence of storage on viability and fatty acid profile. *Aquaculture*, 134, pp. 81-90.

36. McCausland, M.A., Brown, M.R., Barrett, S.M., Diemar, J.A. and Heasman, M.P. 1999. Evaluation of live and pasted microalgae as supplementary food for juvenile Pacific oyster (*Crassostrea gigas*). *Aquacult. Res*., 174, pp. 323-342.

37. D'Souza, F.M.L., Knuckey, R.M., Hohmann, S. and Pendrey, R.C. 2002. Flocculated microalgae concentrates as diets for larvae of the tiger prawn*Penaeus monodon* Fabricius. *Aquaculture Nutrition*, 8, pp. 113-120.

38. Brown, M. and Robert, R. 2002. Preparation and assessment of microalgal concentrates as feeds for larval and juvenile Pacific oyster (*Crassostrea gigas*). *Aquaculture*, 207, pp. 289-309

39. Iwamoto, H. 2004. Industrial production of microalgal cell-mass and secondary products-Major industrial species: *Chlorella*. In: *Handbook of Microalgae culture, Biotechnology and Applied Phycology* (ed. Richmond, A.). Blackwell Science Ltd., Oxford, UK, pp. 255-263.

40. Rossingol, N., Vandanjon, L., Jaouen, P. and Quemeneur, F. 1999. Membrane technology for the continuous separation microalgae culture medium: compared performances of cross-flow microfiltration and ultrafiltration. *Aquacult. Eng*., 20, pp. 191-208.

41. Csordas, A. and Wang, J.K. 2004. An integrated photobioreactor and foam fractionation unit for the growth and harvest of *Chaetoceros* sp. in open systems. *Aquacult. Eng*., 30, pp. 15-30.

42. Petrusevski, B., Bolier, G., Van Bremen, A.N. and Alaerts, G. J. 1995. Tangential flow filtration: a method to concentrate freshwater algae. *Water Res*., 29, pp. 1419-1424.

43. Rossingol, N., Vandanjon, L., Jaouen, P. and Quemeneur, F. 1999. Membrane technology for the continuous separation microalgae culture medium: compared performances of cross-flow microfiltration and ultrafiltration. *Aquacult. Eng*., 20, pp. 191-208.

44. Sun, J. and Liu, D.Y. 2003. Geometric models for calculating cell biovolume and surface area for phytoplankton. *J. Plankton Research*., 25, pp. 1331-1346.

45. Muller-Feuga, A. 2000. The role of microalgae in aquaculture: situation and trends. *J. Appl. Phycol*., 12, pp. 527-534.

46. Baraniak, B., Niezabitowskab, M., Pieleckic, J. and Wojcika, W. 2004. Evaluation of usefulness of Magnafloc M-22S flocculant in the process of obtaining protein concentrates from peas. *Food Chem*., 85, pp. 251-257.

47. Omori, M. and Ikeda, T. 1984. Methods in Marine Zooplankton Ecology. *Wiley*, New York, 322 pp.

48. Usha, T.L., Sarada, R., and Ravishankar, G.A. 2002. Effect of culture conditions on growth of green alga *Haematococcus pluvialis* and astaxanthin

production. Acta Physiologiae Plantarum 24 (3): 323-329

49. Fogg and Thake 1987. Algal Cultures and Phytoplankton Ecology. University of Wisconsin Press.

50. Abu-Rezq, T. S., Yule, A. B. & Teng, S. K. 1997. Ingestion, fecundity, growth rates and culture of the harpacticoid copepod, *Tisbe furcata*, in the laboratory. *Hydrobiologia, 247*, 109-118.

51. Imase, M., Ohko, Y., Takeuchi, M., Hanada, S. 2013. Estimating the viability of Chlorella exposed to oxidative stresses based around photocatalysis. International Biodeterioration and Biodegradation. 78: 1-6

52. Dwaish, A.S., Mohammed, D.Y., Jawad, A.M., Al-kubaicy, A.A. Determine the Uptake of Lead in Chlorella vulgaris Isolated from Tigris River in Baghdad City. International Journal of Scientific and Engineering Research, 2011, 2(9): 1-4

53. Bae, J.H. and Hur, S.B. 2011. Selection of suitable species of Chlorella, Nannochloris and Nannochloropsis in high- and low temperature seasons for mass culture of the rotifer *Brachionus plicatilis*. Fisheries and Aquatic Sciences, 14(4):323-332.

54. Phatarpekar, P.V., Sreepada, R.A., Pednekar, C. and Achuthankutty, C.T. 2000. A comparative study on the growth performance and biochemical composition of mixed culture of *Isochrysis galbana* and *Chaetoceros calcitrans* with monocultures. *Aquaculture*, 181, pp. 141-155.

55. Barsanti, L. and Gualtieri, P. 2006. Algae: anatomy, biochemistry and biotechnology. *CRC Press*, New York, 301 pp.

56. Stottrup, J.G. and Jensen, J. 1990. Influence of algal diet on feeding and egg-production of the calanoid copepod *Acartia tonsa* Dana. *J. Exp. Mar. Biol. Ecol.*, 141, pp. 87-105.

57. Stottrup, J.G. 2003. Production and nutritional value of copepods. In: *Live Feeds in Marine Aquaculture* (eds. Stottrup, J.G and Mcevoy, L.A.).*Blackwell Publishing Ltd.*, USA, pp. 145-205.

58. Kiorboe, T. 1997. Population regulation and role of mesozooplankton in shaping marine pelagic food webs. *Hydrobiologia*, 363, pp. 13-27.

59. Dam, H.G., Peterson, W.T. and Bellantoni, D.C. 1994. Seasonal feeding and fecundity of the calanoid copepod *Acartia tonsa* in Long Island Sound: is omnivory important to egg production? *Hydrobiologia*, 292/293, pp. 191-199.

60. Payne, M.F. and Rippingale, R.J. 2001. Intensive cultivation of the calanoid copepod *Gladioferens imparipes*. *Aquaculture*, 201, pp. 329-342.

61. Hernandez, D.E. and Martinez. S.F. 2001. The effect of the culture conditions on the growth and lipid contents of two strains of Nannochloropsis sp. to be used in aquaculture. Hidrobiologia 11(2):163-168.

62. Sandnes, J.M., K°allqvist, T., Wenner, D. and Gislerød. H.R. 2005. Combined influence of light and temperature on the growth rates of*Nannochloropsis oceanica*: linking cellular responses to large-scale biomass production. Journal of Applied Phycology 17: 515–52.

63. Nedbal, L., Tichyacute, V., Xiong, F. and Grobbelaar. J. U. 1996. Microscopic green algae and cyanobacteria in high frequency intermittent light. Journal of Applied Phycology, 8(4/5), 325-333.

64. Matthijs, H.C.P., Balke, H., van Hes, U.M., Kroon, B.M.A., Mur, L.R. and Binot. R.A. 1996. Application of light emitting diodes in bioreactors: Flashing light effects and energy economy in algal culture (Chlorella pyrenoidosa). Biotechnology and Bioengineering, 50(1), 98-107.

65. Lim, K.C and Zaleha, K. 2013. Effect of Photoperiod on the Fatty Acid Composition of Three Tropical Species of Marine Microalgae. *Malaysian Applied Biology Journal,* 42(1): 41-49

66. Ashraf, M., Javaid, M. Rashid, T., Ayub, M., Zafar, A., Ali, S., Naeem,M. 2011. Replacement of expensive pure nutritive media with low cost commercial fertilizers for mass culture of freshwater algae, *Chlorella vulgaris. International Journal of Agriculture and Biology*, 13: 484–490

67. Lee, K.W., Park, H.G., Lee, S.& Kang, H. 2006. Effects of diets on the growth of the brackish water cyclopoid copepod *Paracyclopina nana*Smirnov. *Aquaculture, 256,* 346-353. http://dx.doi.org/10.1016/j.aquaculture.2006.01.015

68. Phelps, R.P., Sumiarsa, G.S., Lipman, E.E., Lan, H.-P., Moss, K.K., & Davis, A.D. 2005. Intensive and extensive production techniques to provide copepod nauplii for feeding larval red snapper *Lutjanus camperchanus*. In: Copepods in Aquaculture (Ed. By C.-S. Lee, P.J. O'Bryen, N.H. Marcus), pp. 151-168. Blackwell Publishing Ltd. Oxford, UK.

69. Da Grac, M., Melao, A.G., & Rocha, O. 2004. Life history, biomass and production of two planktonic cyclopoid copepods in a shallow subtropical reservoir. *Journal of Plankton Research*, *26(8),* 909-923.

70. Milione, M., & Zeng, C. 2008. The effects of temperature and salinity on population growth and egg hatching success of the tropical calanoid copepod, *Acartia sinjiensis. Aquaculture,275,*116-123.

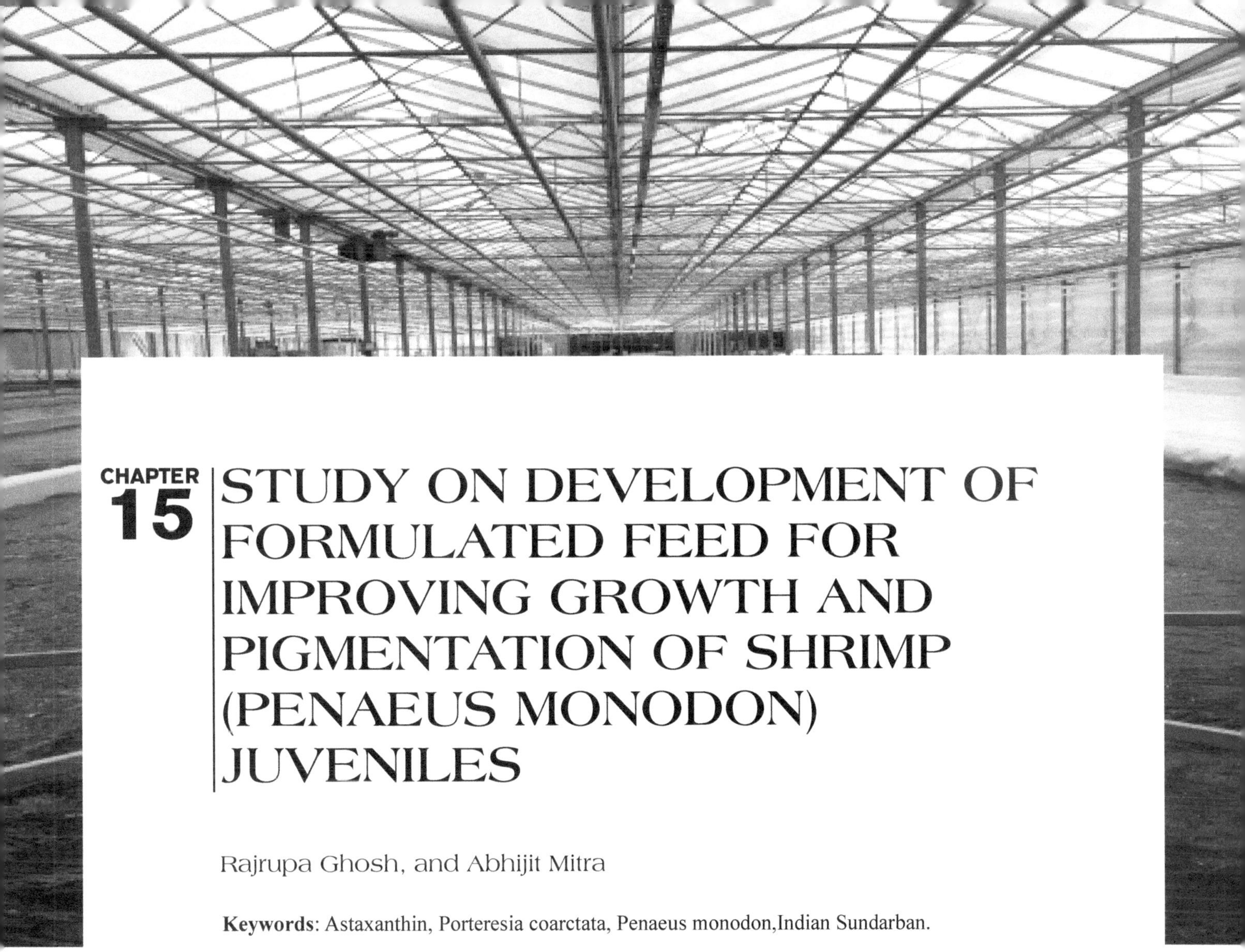

CHAPTER 15

STUDY ON DEVELOPMENT OF FORMULATED FEED FOR IMPROVING GROWTH AND PIGMENTATION OF SHRIMP (PENAEUS MONODON) JUVENILES

Rajrupa Ghosh, and Abhijit Mitra

Keywords: Astaxanthin, Porteresia coarctata, Penaeus monodon,Indian Sundarban.

INTRODUCTION

The culture of black tiger shrimp Penaeus monodon, Fabricius 1798 (Crustacea: Decapoda) is practiced in the brackish water system of the Indian subcontinent with the aim to increase the fish production basically for human consumption. Feed is a significant factor in increasing the productivity and profitability in the aquaculture sector (Jamu and Ayinla, 2003). From economic point of view, feed cost and feed management accounts for at least 60% of the production cost and appears to be one of the major constraints against the greater expansion of aquaculture (Kaushik, 1990). Today, fish meal replacement in the diets by another protein sources for fish and shrimp is becoming necessary and the main trend for sustainable aquaculture. In which, SM (Soybean meal) and PBM (poultry by-product meal) are basically considered as the appropriate alternatives of FM (Fish meal) in the diets for many cultured aquatic species (Hasan et al., 1997; Gatlin and Hardy, 2002; Emre et al., 2002; James and Sampath, 2003; Weeks and Garling, 2010; Milamena, 2004; Stankovic et al., 2011). Considering the marine source of astaxanthin, the present paper is aimed at formulation of shrimp feed by using astaxanthin derived from coastal floral sources (Avicennia marina, Avicennia alba, Avicennia officinalis, Sonneratia apetala, Porteracia coarctata, Enteromorpha intestinalis, Ulva lactuca, Catenella repens and Sueda sp.) This is the first hand approach to study the effect of floral based formulated feed on water quality, shrimp quality and production, survival rate and FCR in a shrimp culture unit of Indian Sundarbans.

MATERIALS AND METHODS

Screening of mangroves for astaxanthin

The entire network of the present programme encompassed the sampling of the leaves of ten dominant mangrove species during the low tide period from the Jharkhali Island during May, 2012 to April, 2013. Leaves of the selected species were collected from two different portions (submerged lower zone and exposed upper zone) of the same plant. The lower region of the tree gets inundated during the high tide condition and upper region of the same plant remains unexposed to tidal water. In addition to true mangrove species, the astaxanthin

Citation: Rajrupa Ghosh,* and Abhijit Mitra, Study on Development of Formulated Feed for Improving Growth and Pigmentation of Shrimp (Penaeus Monodon) Juveniles, Online ISSN:2345-4385

level of few associates like Porterasia coarctata, Enteromorpha intestinalis, Ulva lactuca, Catenella repens and Sueda sp. were also monitored. Salinity, pH, temperature, dissolved oxygen and nutrient load of the ambient water were analysed simultaneously to pinpoint the hydrological parameters to which the plant species are exposed in natural condition. The collected leaves were thoroughly washed with ambient water followed with deionized water and oven dried at 1100C overnight. The extraction of astaxanthin was done in organic solvent as per the standard method and analysed spectrophotometrically. The mean results of all the analyses of 12 months; May, 2012 to April, 2013) are shown in Table 1.

Table 1: Mean astaxanthin content in mangrove and associated species collected from Jharkhali island of Indian Sundarbans during May, 2012 to April, 2013

Mangrove species	Astaxanthin content (mg/kg)		Percentage increase
	Submerged	Exposed	
Avicennia officinalis	396.56±8.9	297. 75±7.8	33.19 %
Avicennia alba	441.33±9.6	328.02±8.6	34.54%
Avicennia marina	435.59±7.9	319.10±7.9	36.51%
Sonneratia apetala	162.80±3.2	103.49±3.7	57.31%
Aegiceros corniculatum	120.86±2.1	98.15±1.2	23.14%
Aegialitis rotundifolia	105.92±1.8	82.41±0.99	28.53%
Ceriops decandra	91.09±0.99	67.44±0.87	35.07%
Heritiera fomes	761.00±10.6	398.54±6.7	90.95%
Rhizophora apiculata	84.90±0.98	56.22±0.77	51.01%
Bruguiera gymnorrhiza	461.38±3.4	397.11±5.8	16.18%
**Porterasia coarctata*		607.56±4.6	
**Enteromorpha intestinalis*		120.78±0.99	
**Ulva lactuca*		56.43±0.67	
**Catenella repens*		129.05±1.4	
**Sueda* sp.		87.55±0.69	

Values are the mean ± SD

*Remark: Mangrove associates (denoted with * sign) were not monitored during 2 tidal phases. They were collected during low tide condition*

SHRIMP NUTRITION

Experimental Design and Layout

The study area for culturing shrimp (P. monodon) was selected in the central part of Indian Sundarbans in Jhharkhali located in South 24 Parganas district of the state of West Bengal (22o16/40.6// N latitude & 88o38/18.4// E longitude) during January to April, 2014. The culture site is located in Sundarban Biosphere Reserve (SBR) on the bank of River Matla adjacent to Herobhanga Island (River Salinity – 22.50 ±0.68 psu). Two ponds were selected in the study site out of which one was treated as control (C) which is 390m2 and the other was treated as experimental (E) which is 780m2 . On the basis of astaxanthin concentration, Porteresia coarctata was selected for feed preparation because of its high astaxanthin content. The cultured species (P. monodon) in the experimental pond was provided with the Porteresia coarctata based formulated feed and the control pond were provided with traditional feed. Good quality shrimp seeds obtained from a local shrimp farm were stocked after proper acclimatization with ambient environmental conditions. The stocking density was 5 PL20/ m2 in both the control and experimental ponds. The shrimps were fed initially at 15% of the biomass in each pond and the ration was then adjusted to actual consumption every day, thus reducing uneaten feed to a minimum. Water quality parameters were monitored continuously 90 days culture period (15th January to 15th April, 2014) which were in acceptable ranges. Dissolved oxygen (D.O.), pH, transparency and nutrients were analyzed following the standard spectrophotometric method (Strickland and Parsons, 1968, 1972). Phytopigment concentration (Chl a) was analysed as per the method of Jeffrey and Humphrey, 1975. Organic carbon content of pond bottom soil was estimated by the standard titration method (Walkey and Black, 1934).

Diet Preparation

The experimental pond was provided with Porterasia coarctata dust to replace totally the fish and shrimp meal component (Table 2) of the traditional fish feed. Commercial diet from local market was given to the control pond. The ingredients used in the feed were accurately weighed and the feeds were prepared following the method by Jayaram and Shetty (1981).

Table 2: Ingredients (gm) and proximate composition (% DW) of diets of experimental pond and control

Component	Control (gm)	Experimental (gm)
Rice bran	250	250
Wheat bran	250	250
Fish meal	150	Salt marsh grass dust = 300
Shrimp meal	150	
Sun flower oil cake	50	50
Mustard oil cake	50	50
Vit. + Min. Mixture	50	50
Binder	30	30
Shark oil	20	20
Proximate Analysis (%)		
Crude protein	34.37±1.2	36.78±1.5
Soluble carbohydrate	32.67±2.6	38.10±0.67
Crude Fat	7.1±0.80	6.5±0.22
Ash	9.1±0.12	13.1±0.34
Moisture	8.9±0.10	10.7±0.98

Zoo-Technical Parameters and Statistical Analysis

Individual weights and lengths of shrimps were taken at fortnightly interval for 90 days culture period and the relevant response variables were determined for each control and experimental ponds. Condition Index (C.I.) was analysed at fortnightly interval during the culture period as per the expression; C.I.= W/L3×100, where W= weight of the cultured species (in gm) and L = length of the cultured species (in cm). Percentage weight gain was calculated as the difference in weight from the average final weight with respect to the initial weight; weight gain=[(average individual final weight– average individual initial weight)/average individual initial weight]×100. Feed consumption reported was the total of the consumption estimated for 90 days period. The survival rate was determined as percentage of the difference of stocking number and number recovered at the end of the experimental trial. Feed Conversion Ratio (FCR) was analysed after the harvesting of shrimps as per the expression: FCR = $\Delta f/\Delta b$, where, Δf = Change in feed biomass and Δb = Change in body biomass of the cultured species. Astaxanthin in the shrimp tissue was estimated as per the standard spectrophotometric method (Schuep and Schierle, 1995) and body pigmentation was assessed for each treatment on shrimp cooked for 5 min in boiling water and comparing the orange colouration with Roche SalmoFanTM colour score. All data were expressed in terms of mean and standard deviations / range (±SD/ range). Analysis of variance (ANOVA) was computed between all the selected parameters (physico- chemical and zoo technical) considering both control and experimental ponds to evaluate the differences caused by inclusion of Porteresia coarctata dust in the feed.

RESULTS AND DISCUSSION

The astaxanthin level in the selected mangrove species (collected from Jharkhali region) exhibited significant variations. It is of the order Heritiera fomes > Bruguiera gymnorrhiza > Avicennia alba > Avicennia marina > Avicennia officinalis > Sonneratia apetala > Aegiceros corniculatum > Aegialitis rotundifolia > Ceriops decandra > Rhizophora apiculata (Table 1). The relatively greater astaxanthin content in the submerged leaves of mangroves confirms the synthesis of astaxanthin content under stressful condition. Our results agree with several reports of decrease content of carotenoids by salinity as reported in a number of glycophytes (Gadallah., 1999; Agastian et al., 2000, Mitra and Banerjee, 2010). However, more studies are needed to confirm the influence of tidal influx and subsequent salinity fluctuation on astaxanthin level in the mangrove floral parts. The present data may serve as baseline information on the regulatory role of tidal submergence on astaxanthin level in the estuarine and coastal vegetation as mentioned by Mitra et al., 2006. The enhancement of astaxanthin production under stressed condition of organisms is a matter of interest and several researches are still being undertaken to pinpoint the reaction pathway of astaxanthin production by inducing stress of varied nature (Mitra et al., 2006). Many types of yeast have been described with an increase ability to produce carotenoids when they grow under unfavourable environment (Certik et al., 2005). Several workers have reported both in the dark and light the enhancement of the accumulation of astaxanthin in cysts of Haematococcus pluvialis under salt stress conditions (Harker et al., 1996; Lee, 1999; Boussiba, 2000; Park and Lee, 2000, 2001; Steinbrenner and Linden, 2003 and Chio et al., 2003). The present study points to higher astaxanthin level in those leaves of the mangroves that are inundated for 10 to 12 hours by tidal waters of Jharkhali station having typical estuarine water characteristics (salinity = 10 - 25.85 psu, pH = 7.98 – 8.28, temperature = 29.8 - 31.5°C, dissolved oxygen = 5.93 - 6.10mg/l, NO_3 = 15.09 - 21.04 µgat/l, PO_4 = 1.12 -1.39 µgat/l and SiO_3 = 64.44 - 83.16 µgat/l). The enhancement of astaxanthin level in the inundated Sundari leaves (Heritiera fomes) clearly reflects the highest degree of stress posed by water salinity on this species. Heritiera fomes, being fresh water loving mangrove species cannot tolerate high salinity (Mitra and Pal, 2002) and thus acceleration of astaxanthin production may probably be a part of its adaptation to cope with the stenohaline condition of coastal and estuarine environment that becomes acute during high tide. The astaxanthin level of mangrove flora is thus a function of its physiological system, which is extremely species specific. Highest astaxanthin was recorded in mangrove associate Porteresia coarctata (commonly known as salt marsh grass) in comparison to other species and therefore considered for feed preparation.

Shrimp Nutrition and Growth

Shrimps fed with Porteresia diet exhibited higher final weights and better weight gain (Table 3) at the end of the experiment (27.2 gm final weight) in comparison to control pond (20.5 gm final weight). Condition Index values of shrimp were also higher in experimental ponds (3.87 ± 0.67) than control pond (2.98 ± 0.55) (Table 3). The FCR value for control pond was 1.67 and for experimental pond was 1.27. The survival rate was found to be 58.4% in the control pond and 69.2% in experimental pond. The present study speaks in favour of healthy pond environment, better growth, higher survival rate and low FCR values through use of Porteresia based feed. An important factor governing the consumer acceptance and market value of many cultivated fish and shrimp species is the pink or red colouration of their flesh or boiled exoskeleton (Brun and Vidal, 2006). In the wild, this colouration is achieved through the ingestion of carotenoid pigments particularly astaxanthin contained within invertebrate food organisms (Johnson et al., 1977; Ibrahim et al., 1984). The Porteresia based feed in the present study resulted in higher astaxanthin values in shrimps of experimental pond (15.32 ± 1.22 ppm; n = 10) as reflected through darker orange-red colouration of shrimp exoskeleton in comparison to control pond (8.66 ± 0.78 ppm). The Roche SalmoFanTM colour score showed the color value of 23 and 29 for shrimps from control pond and experimental ponds respectively (Table 3), which confirms the variation of astaxanthin level due to different feed ingredients.

Table 3: Growth parameters of Penaeus monodon fed Porteresia based feed

Parameters	Control pond	Experimental pond
Condition Index of shrimp	2.98 ± 0.55	3.87 ± 0.67
Wt. at harvest (g)	20.2	27.5
Astaxanthin in shrimp (ppm)	8.66 ± 0.78	15.32 ± 1.22
Roche SalmoFan^{TM} colour score	23	29
Survival rate (%)	58.4	69.2
FCR	1.67	1.27

Water Quality Analysis

Hydrological parameters of the shrimp culture ponds are a reflection of the quality of feed provided and its response in morphometric parameters. (Maceina and Murphy, 1998) (Table 4).Surface water temperature in both the culture ponds showed more or less parallel trend of variation throughout the study period. The uniformity in temperature profile is due to the location of both the ponds in the same site that experienced similar weather conditions. Water temperature plays a major role in shrimp enzyme kinetics which may have a regulatory influence on their growth (Mitra et al., 2006). It also affects the process of molting during the post larval stage of shrimps (WWF-India, 2006). The salinity of the Hugli-Matla estuarine complex is known to exhibit intensive variations (Saha et al., 1995). The difference in salinity between ponds may be attributed to the different soil salinity (as substratum of the pond) that leaches soluble salt to the pond water. The relatively higher C.I. values in the experimental pond with less salinity proves the efficiency of formulated feed in combating the stress posed by salinity. Shrimp culture directly affects the pH of the pond bottom through deposition of excess feed, shrimp excreta, dead shrimps, etc. This shifts the soil and overlying aquatic pH towards acidic condition. In the present study, such condition was not observed owing to the traditional practice of liming at a regular interval of time. Dissolved oxygen (D.O.) is a vital parameter regulating the aquatic life. The shrimp health is a direct function of dissolved oxygen and its diurnal variation. Excessive organic load in the system results in lowering the D.O. value during night/dawn posing threat to the survival of aquatic life. In the present study the D.O. level in the control pond showed lower value owing to deposition of organic carbon at the bottom of the pond. The significant variation of D.O. between ponds may be attributed to different growth rate of the culture species and also the use of different types of feed. Traditional feed contains dry fish dust and trash shrimp dust which lowers the D.O. due to their utilization for oxidizing the residual matter (BOD value Rajrupa Ghosh and Abhijit Mitra, Global Journal of Animal Scientific Resear increases under this situation). Floral based feed, on the other hand generates very limited residue due to which D.O. remains almost unaltered. Transparency controls the phytoplankton standing stock in shrimp culture ponds due to their dependency on the solar radiation for photosynthesis. The experimental pond provided with formulated feed showed increased transparency due to its unique binding property. The ready acceptance of the Porteresia based feed by the cultured species in the experimental pond may be the basis of reduced suspended particulate matter in the aquatic phase of the experimental pond. Nutrients (comprising of nitrate, phosphate and silicate) budget in the aquatic phase of the culture ponds are regulated through quantum of excretory products of the cultured species, left over feed and also by the churning of the pond bed (due to run-off from the adjacent land masses). High concentration of nitrate in the control pond may be due to leaching of the feed ingredients (particularly from animal component in traditional feed) in pond water and also the faecal matter that generates ammonia (Mitra and Choudhury, 1995). The phosphate concentration during the study period showed no significant variation between the ponds owing to ban imposed on washing utensils, clothes and other daily house-hold activities during the culture period. The silicate level of the ponds may be attributed to substratum or pond bottom composition. In both the control and experimental ponds no significant variation in silicate was observed. Soil organic carbon was greater in control pond due to more generation of residual feed and excreta in the absence of any feed management. On contrary, the lower value of organic carbon in the experimental pond is an indication of better acceptability of Porteresia based feed by shrimp due to which wastage was minimum.

Table 4: Variation in the physico-chemical parameters of the culture ponds

Parameters	Control pond	Experimental pond
Surface water temperature (°C)	29.4 ± 0.22	29.4 ± 0.22
Surface water salinity (psu)	18.56 ± 0.20	15.08 ± 0.13
pH	7.88 ± 0.35	8.08 ± 0.03
Transparency (cm)	17.02 ± 2.02	24.3 ± 2.58
Dissolved oxygen (mgl^{-1})	4.66 ± 0.77	5.47 ± 0.12
Nitrate ($\mu gatl^{-1}$)	19.5 ± 1.21	16.6 ± 1.01
Phosphate ($\mu gatl^{-1}$)	2.22 ± 0.05	2.19 ± 0.29
Silicate ($\mu gatl^{-1}$)	64.01 ± 2.23	64.32 ± 2.67
Chlorophyll *a* (mgm^{-3})	1.74 ± 0.51	1.92 ± 0.16
Soil organic carbon (%)	1.18 ± 0.16	0.97 ± 0.10

Values are the mean ± SD

The replacement of dietary fish meal in the diets of fish and shrimp by another protein sources derived from animal, plant and biology has been considered as a practical solution for sustainable aquaculture development. Thus, the finding of alternatives of fish meal is absolutely in compliance with the trend of using another protein sources to replace fish meal in the diets of aquatic animal. Recent studies proved that several plant and animal protein sources can be used for partial and totally fish meal replacement in the diets for many aquatic species. In which, PBM and SM have been commonly used as the suitable sources to replace FM and other animal protein in the diets for fish and shrimp. Yesilayer et al., (2011) found that it had no adverse effects on growth performance of koi juveniles fed the diets containing 0% FM and 16% FM, correspondingly with 45% SM and 22.5% SM in the diet. Ismail et al., 2013 reported that 100% fish meal in the diet of Malaysian mahseer (Tor tambroides) can be replaced by poultry offal meal without adversely effecting survival and growth and body composition. Tri and Davis (2009) asserted that dietary FM can be totally replaced in commercial diets for tilapia juveniles by de-hulled soybean meal (solvent extracted) and expelled soybean meal (pressed) without adversely effecting growth and feed utilization.

Yang et al. (2004) recommended that PBM could replace up to 500 g·kg-1 of FM protein in diets for gibel carp (Carassius auratus gibelio) without negative effects on growth. Likewise, Saadiah and Abol-Munafi (2011) found 100% dietary FM can be replaced by PBM in the diets for cobia, Rachycentron canadum without adversely affecting growth performance, but an optimal replacement level at approximately 60% was recommended for better growth and efficient feed utilization. Whereas, the replacement of FM with PBM in the diets for mirror carp fingerling (Cyprinus carpio) with the range of PBM from 12-36% showed significantly differences and as increasing the PBM content in the diets resulted in lower WG and higher feed utilization (Emre et al., 2003). Not only SBM (Soybean meal) and PBM (poultry by- product meal), but also the other protein sources derived from plant and biology have been used as potential protein sources to replace dietary fish meal for reducing feed cost as well as exhausted exploitation of wild fish. Korkmaz and Cakirogullari, (2011) found that Dried Baker's Yeast can be used to replace up to 30% dietary fish meal in the diets for koi carp fingerlings that did not show significant differences in growth compared to those at the control diet without FM replacement. Similarly, Mazurkiewicz (2009) reported that replacement of dietary FM by plant-derived protein with the ration of plant protein (legume seeds and extracted rapeseed meal) up to 26% in the diets (35% protein, 22% FM) for carp fingerlings and early juveniles without adversely affecting the rearing results. Kim et al., (2013) indicated that spirulina can replace dietary fish meal up to 26% in the diet for parrot fish (Oplegnathus fasciatus) without negative effects in weight gain and feed efficiency

CONCLUSIONS

Aquaculture has become a peak industry in the present millennium, which involves seafood farming with shrimp, cuttle fish, squid, lobster and other such culinary delights actually "cultivated" in aquatic enclosures under scientifically controlled conditions (Rajkhowa, 2005). The use of nutrient-rich feed continues to gain wide acceptance in the aquaculture industry in order to boost up the quality of the aquacultural products. The use of quality feed results in substantial reduction in the overall variable cost of an operation through improved animal performance, better FCR and improved water quality due to a reduction in the amount of nutrients and solids (i.e., feces and uneaten food) in the waste water effluent. P. coarctata based formulated feed showed better growth performance of the cultured species with respect to condition index values and survival rate. Body pigmentation improved in the cultured species of experimental pond and showed significantly higher astaxanthin level than the controlled pond. A series of experiments are still needed for time testing the results and make the programme sustainable for the poor island dwellers of lower Gangetic delta.

REFERENCES

1. Agastian, P., S. J. Kingsley, and M. Vivekanandan. 2000. Effect of salinity on photosynthesis and biochemical characteristics in mulberry genotypes. Photosynthetica. 38: 287–290.
2. Boussiba, S. 2000. Carotenogenesis in the green alga Haematococcus pluvialis: cellular physiology and stress response. Physiol. Plant. 108: 111–117.
3. Brun, H.L., and F. Vidal. 2006. Shrimp pigmentation with natural carotenoids. Feed Technology, Aquaculture Asia-Pacific Magazine. pp: 34-35.
4. Certik, M., S. Masrnova, V.M. Sitkey, and B. Emilia. 2005. Biotechnological production of astaxanthin. Chem. Listy. 99: 237-240. Choi, S.L., I.S. Suh, and C.G. Lee. 2003.
5. Lumostatic operation of bubble column photobioreactors for Haematococcus pluvialis cultures using a specific light uptake rate as a control parameter. Enzyme Microbiol. Technol. 33: 403–409.
6. Emre, Y., H. Sevgili, and I. Diler. 2003. Replacing Fish Meal with Poultry By-Product Meal in Practical Diets for Mirror Carp (Cyprinus carpio) Fingerlings. Turkish Journal of Fisheries and Aquatic Sciences. 3: 81-85.
7. Gadallah, M. A.A. 1999. Effects of proline and glycinebetaine on Vicia faba in response to salt stress. Biological Plant. 42: 249–257.
8. Gatlin, D.M.I., and R.W. Hardy. 2000. Manipulations of diets and feeding to reduce losses of nutrients in intensive aquaculture, in: J.R. Tomasso (Ed.), Aquaculture and the Environment in the United States U.S. Aquaculture Society, A Chapter of the World Aquaculture Society, Baton Rouge, LA, pp: 155-165.
9. Harker, M., A.J. Tsavalos, and A. Young. 1996a. Factors responsible for astaxanthin formation in the chlorophyte Haematococcus pluvialis. Bioresource Technol. 55: 207-214.
10. Harker, M., A.J. Tsavalos, and A. Young. 1996b. Autotrophic growth and carotenoid production of Haematococcus pluvialis in a 30 liter airlift photobioreactor. J. Ferment. Bioeng. 82:113-118.
11. Hasan, M.R. 2001. Nutrition and feeding for sustainable aquaculture development in the third millennium. In: Aquaculture in the third millennium. Technical proceedings of the conference on aquaculture in the third millennium, Bangkok, Thailand, 20-25 February 2000, pp: 193-219. NACA, Bangkok and FAO, Rome.
12. Hasan, M.R., D.J. Macintosh, and K. Jauncey. 1997. Evaluation of some plant ingredients as dietary protein sources for common carp (Cyprinus carpio L.) fry. Aquaculture. 151: 55-70.
13. Ibrahim, A., C. Shimizu, and M. Kono. 1984. Pigmentation of culture red sea bream, Chrysophrys major, using astaxanthin from Antarctic krill, Euphausia superba, and a mysid, Neomysis sp. Aquaculture. 38: 45-57.
14. James, R., and K. Sampath. 2003. Effect of animal and plant protein diets on growth and fecundity in ornamental fish, betta splendens (regan). The Israeli Journal of Aquaculture Bamidgeh. 55 (1): 39-52.
15. Jamu, D.M., and O.A. Ayinla. 2003. Potential for the development of aquaculture in Africa. NAGA. 26: 9- 13.
16. Jeffrey, S.W., and G.R. Humphrey. 1975. New spectrophotometric equations for determining chlorophyll a, b, c1 and c2 in higher plants, algae and natural phytoplankton. Biochemical Physiology Pflanzen Bd. 167: 191-194.
17. Johnson, E.A., D.E. Conklin, and M.J. Lewis. 1977. The yeast Phaffia rhodoyma as a dietary pigment source for salmonids and crustaceans.
18. Journal of Fisheries Research Board of Canada. 34: 2417-2421.
19. Kaushik, S. 1990. Use of alternative protein resources for the intensive rearing of carnivorous fish. In: Mediterranean aquaculture (Eds: R. Flos, L. Tort and P. Torres), Hellis Horwood Ltd., Chichester, pp: 125-138.
20. Kim, S.S., S. Rahimnejad, K.W. Kim, and K.J. Lee. 2013. Partial replacement of fish meal with Spirulina pacifica in diets for parrot fish (Oplegnathus fasciatus). Turkish Journal of Fisheries and Aquatic Sciences. 13: 197-204.
21. Lee, Y.K., and D.H. Zhang. 1999. Production of astaxanthin by Haematococcus. In: Cohen Z, ed. Chemicals from Microalgae. Taylor and Francis, London. pp: 175-195.
22. Maciena, M.J., and B.R. Murphey. 1988. Variation in the weight to length relationship among Florida and Northern large mough bass and their interspecific F1 hybrid. Transitions American Fisheries Society. 117: 232-237.
23. Mazurkiewicz, J. 2009. Utilization of domestic plant components in diets for common carp, Cyprinus carpio L., Arch. Pol. Fish. 17: 5-39.
24. Milamena, O.M. 2004. Replacement of fish meal by animal by – product meals in a practical diet for grow-out culture of grouper (E. coioides), in: M.A. Rimmer, S. McBride, K.C. Williams (Eds.), Advances in Grouper Aquaculture, ACIAR, Canberra, Australia, pp: 110-112.
25. Mitra, A., and A. Choudhury. 1995. Causes of water pollution in prawn culture farms. Journal of Indian Ocean Studies. 2 (3): 230- 235.
26. Mitra, A., R. Chakraborty, K. Banerjee, A. Banerjee, N. Mehta, and H. Berg. 2006. Study on the water quality of the shrimp culture ponds in Indian Sundarbans. Indian Science Cruiser. 20(1): 34-43.
27. Mitra, A., and S. Pal. 2002. In: The oscillating mangrove ecosystem and the Indian Sundarbans (editors: Shakti Banerjee & Farida Tampal), Published by WWF-India, WBSO.
28. Mitra, A., S. Basu, K. Banerjee, and A. Banerjee. 2006. Impact of tidal submergence on astaxanthin content

of mangroves. Ultra Scientist of Physical Sciences. 18(2):117-122

29. Mitra, A., and K. Banerjee. 2010. Pigments of Heritiera fomes seedlings under different salinity conditions: perspective sea level rise, Mesopot. J. Mar. Sci. 25 (1): 1-10

30. Park, K.H., and C.G. Lee. 2000. Optimization of algal photobioreactors using flashing lights. Biotechnol Bioprocess Eng. 5:186–190.

31. Park, K.H., and C.G. Lee. 2001. Effectiveness of flashing light for increasing photosynthetic efficiency of microalgal cultures over a critical cell density. Biotechnol Bioprocess Eng. 6:189–193.

32. Rajkhowa, I. 2005. Action in aquaculture – opportunities in a quaint specialization. Business Today (May 22 Issue) .131.

33. Saadiah, I., and C.U. Abol-Munafi. 2011. Replacement of fishmeal in cobia (Rachycentron canadum) diets using poultry by-product meal. Aquaculture International. 19 (4): 637-648.

34. Saha, S.B., B.B. Ghosh, and V. Gopalakrishnan. 1995. Plankton of Hooghly estuary with special reference to salinity and temperature.b Journal of Marine Biological Association of India. 17(1): 107-120.

35. Saufinas, I., K. Mohd Salleh, and R.F. Ehsan. 2013. Performance of commercial poultry offal meal as fishmeal replacement in the diet of juvenile malaysian mahseer, tor tambroides, Asian Journal of Animal & Veterinary Advances. 8 (2): 284-292.

36. Schuep, W., and J. Schierle. 1995. Astaxanthin determination of stabilized, added astaxanthin in fish feeds and pre-mixes. In: Carotenoids Isolation and Analysis, 1: 273-276. Birkhauser Verlag Basel. Stankovic, M.B., Z.P. Dulić, and Z.Z. Marković. 2011. Protein sources and their significance in carp (Cyprinus Carpio) nutrition. Journal of Agricultural Sciences. 56 (1): 75-86.

37. Steinbrenner, J., and H. Linden. 2003. Light induction of carotenoid biosynthesis genes in the green alga Haematococcus pluvialis: regulation by photosynthetic redox control. Plant Mol. Biol. 52: 343–356.

38. Strickland, J.D.H., and T.R. Parsons. 1968. A manual for seawater analysis. Bulletin of Fisheries Research Board Canada. 167.

39. Strickland, J.D.H., and T.R. Parsons. 1972. A practical handbook of seawater analysis. Journal of Fisheries Research Board of Canada, Ottawa. 167: 311p.

40. Tri, N.N., and D.A. Davis. 2009. Evaluation of alternative protein sources to replace fish meal in practical diets for juvenile Tilapia, Oreochromis spp., Journal of the World Aquaculture Society. 40 (1): 113-120.

41. Walkey, A., and I.A. Black. 1934. An examination of the effect of the digestive method for determining soil organic matter and a proposed modification of the chronic and titration method. Soil Science. 37: 29-38.

42. Weeks, C., and D. Garling. 2010. The Effect of feeding varying levels of soybean meal in high-nutrient-density diets on growth performance and body composition of juvenile atlantic salmon. North American Journal of Aquaculture. 72: 279-289.

43. WWF-India. 2006. Training manual onecofriendly and sustainable fishery, New Delhi, India, 130 p.

44. Yang, Y., S. Xie, Y. Cui, W. Lei, X. Zhu, Y. Yang, and Y. Yu. 2004. Effect of replacement of dietary fish meal by meat and bone meal and poultry by-product meal on growth and feed utilization of gibel carp, Carassius auratus gibelio. Aquaculture Nutrition. 10 (5): 289-294.

45. Yesilayer, N., O.Z. Meryem, Z. Karsli, O. Aral, A. Karacuha, and O.Z. Unal. 2011. Growth performance and feed utilization of koi carp (Cyprinus carpio L., 1758) fed partial or total replacement of fish meal with hazelnut meal and soybean meal. Journal of Animal and Veterinary Advances. 10 (15): 1956-1961.

INDEX

L

M

N

P

R

S

T

W